Aircraft Hardware Standards

AF893362

Manual
&
Engineering Reference

by

Stanley J. Dzik

COPYRIGHT 1971 BY THE HECTOR CERVANTES INC.

All rights reserved. No Part of this manual may be used or reproduced in any manner whatsoever without the prior written permission of the publisher or his authorized agent or assigns.

PUBLISHED IN THE UNITED STATES OF AMERICA
BY

**AVIATION
PUBLICATIONS**

P. O. BOX 357
217 E. WASHINGTON ST.
APPLETON
WISCONSIN 54911
U.S.A.

WORLD'S FAMOUS AIRCRAFT SERIES

Curtiss JN-4D Jenny
Ford Trimotor
Lockheed P-38 Lightning
Bell P-39 Airacobra
Curtiss P-40 Warhawk
Republic P-47 Thunderbolt
North American F-51 Mustang
Northrop P-61 Black Widow
Chance Vought F4U Corsair
Northrop YB-49 Flying Wing
Messerschmitt 262 Sturmvogel
de Havilland Mosquito
Supermarine Spitfire
Hawker Hurricane
Boeing B-17 Flying Fortress
Bell P-63 King Cobra
X-15 Research Airplane
F-82 Twin Mustang
Piper J-3 Cub
AT-6 Texan
Lockheed F-80 Shooting Star
Grumman F6F Hellcat
Consolidated B-24 Liberator

AVIATION PUBLICATIONS

P. O. BOX 357
217 E. WASHINGTON ST.
APPLETON
WISCONSIN 54911
U.S.A.

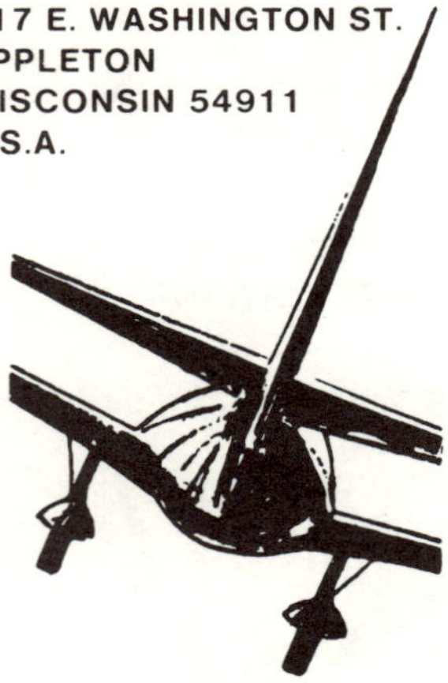

Litho in U.S.A. at Graphic Communications Center
with production by Badger Printing Division
Appleton, Wisconsin 54911

TABLE OF CONTENTS

Introduction to National AEROSPACE STANDARDS

The user of this manual can further obtain additional information on the following AN, AMS, AND, MS and NAS standards which can be purchased in volume form or single specification sheets. For further information and schedule of prices, write to:

NATIONAL STANDARDS ASSOCIATION
1321 FOURTEENTH ST. N.W.
WASHINGTON, D.C. 20005

AMS
Developed by :
Aeronautical Material Specifications Div.
of SAE Aeronautics Committee

Distributed by:
Society of Automotive Engineers, Inc.
485 Lexington Avenue
New York, New York 10017
Aeronautical Material Specifications are material and process specifications for aircraft components conforming to established engineering and metallurgical practices in the aircraft components. All specifications are prefixed by "AMS".

AIR FORCE-NAVY
AERONAUTICAL STANDARD DRAWINGS (AN STANDARDS)
Distributed by:
Commanding Officer
Naval Aviation Supply Depot
700 Robbins Avenue
Philadelphia, Pennsylvania 19111
Attention: CDS
Dimensional standards for aircraft fasteners developed by the Aeronautical Standards Group. All drawings are prefixed by "AN".
MIL
Distributed by:
Commanding Officer
Naval Aviation Supply Depot
700 Robbins Avenue
Philadelphia, Pennsylvania 19111
Attention: CDS All drawings are prefixed by "MIL"

AN, MS, MIL, NAS and AMS drawings and specifications the scope of such a subject involves several thousand drawings in over thirty volumes and keep in mind these are only for so called standard fasteners and in modern aircraft design standard fasteners only comprise about thirty percent of the total fasteners used.

At the beginning of World War Two the standard fastener was the backbone of airframe propeller and engine design and in most cases proved adequate for the design and operational limit loads placed on such fasteners. They were primarily divided into three categories, tensile, shear and corrosion and heat resistant fasteners and seldom exceeded 200,000 P.S.I. tensile and were manufactured from well known low alloy steels such as 4130, 4140, 4340, 300 and 400 series stainless grades and processed by rather simple methods of heat treating and surface treatment.

The fact that early day work horses of the Air Force such as the Boeing B-17 and the Douglas C-47 experienced such tremendous records of dependability and achievement can largely be attributed to stress design overfactor, that is, if design factor were based on 100% then a safety factor of two or three hundred percent was often incorporated in the end product with the results that such aircraft flew to infinity under some of the most adverse conditions imaginable.

As the tactical and logistic requirements for aircraft advanced so did the state of the art for all the components until today's concept encompasses fasteners in the 300,000 plus tensile range and operational temperatures at load of 1200° F and higher. The raw materials available for these requirements run the gamut from the so called exotics thru the various maraging and tool steel lists and are supplemented by titanium, molybdenum, boron, zirconium and cobalt fasteners all of which are used in various areas of the airframe or engine to meet severe demands of tensile, shear, fatigue, corrosion, torsional, temperature and weight factors to which they are exposed.

The aircraft and engine manufacturing stalwarts included names synonymous with the Industry such as Douglas, Northrup, Convair, Grumman, Curtis Wright, Chance Vought, Pratt & Whitney, General Electric and many others and they were paralleled by aircraft fastener manufacturers such as Standard Pressed Steel, Voi Shan Manufacturing, Cooper Precision Products, Briles Manufacturing, H.M. Harper, Dumont Aviation, Russell Bolt, V B Products and dozens of others and the hardware produced by these companies and the airframes and engines they hold together will span the space of man's wildest imagination in the future.

A complete "Assembly Engineering Master Catalog" with specifications and Engineering Referance may be purchased for $15.00 per copy from:

HITCHCOCK PUBLISHING CO.
POST OFFICE BOX 3001
HITCHCOCK BUILDING
WHEATON, ILLINOIS 60187

I would like to express my gratitude and appreciation to the following who have been of great assistance in the preparation of this manual.

National Aeronautics and Space Administration
Van Dusen Aircraft Supplies
VB Products Inc.
Russell Burdsall & Ward, Bolt and Nut Co.
Hitchcock Publishing Co.
Amerace Esna Corp.
Aerospace Industries Association of America In.

Stanley J. Dzik

Nut & Bolt Talk

People talk about "getting down to the nuts and bolts" when describing something in minute detail. An apt phrase. It recognizes that standard fasteners are the basic and fundamental component of assembled metal products.

Despite their uncommon importance to all industry, these commonplace fasteners are too seldom viewed as a highly engineered item. Frequently, they are applied by "rule of thumb" rather than by slide-rule. The engineering talent behind them is often overlooked Less than the maximum economy and performance inherent in standard fasteners is bound to result.

SELECTING THE RIGHT GRADE OF SCREWS

With few exceptions, the true function of a bolt is to clamp members together, and not to act as an axle or fulcrum. The residual tension set up in the bolt keeps joints tight.

WHAT'S THE RIGHT TORQUE FOR BOLTS?

This is one of the toughest questions we're asked. Too many variable conditions. But the following may help.

The bolt takes two stresses during wrenching: (1) Torsion (2) Tension. Tension is what you want. Torsion is the necessary evil due to friction. Probably 90% of applied torque goes to overcome friction.

With the friction factor changed by lubrication, plating, etc., the torque needed to produce a given tension is hard to predict. However, a useful empirical formula exists for normal friction conditions.

INCH-LBS. TORQUE = 0.2 x BOLT DIAMETER x BOLT TENSION

Many tests show that the 0.2 torque coefficient is approximately constant for the usual friction conditions, for all diameters, and for both coarse and fine thread. Average deviation is about 7%. But when are conditions "normal"? The only sure way to check torque is to set up a pilot assembly and try it out.

In pilot testing for rigid joints, tighten a few bolts with torque wrench to failure, and then set torque at 60% of that load; or even at yield strength, since torsion component vanishes when the tightening operation ceases, leaving bolt under tension only, which is well below ultimate strength.

BOLTS TAKE GREATEST STRESS DURING WRENCHING

If a bolt doesn't fail when being wrenched up tight, it won't fail in service (assuming bolts and joint have been designed adequately for the loads).

That's because two forces put stress on bolts (and hex screws) as they tighten: Tension due to bolt stretch; torsion due to friction. But only tension remains after wrenching. In a rigid joint, if this tension exceeds external forces, bolts will never experience any further strain, and will therefore not loosen or fail.

WHY SOME FAILURES?

Obviously, unusual unforeseen loads cause trouble. The instant they exceed residual tension, they add to the stress placed on the bolt and can cause immediate failure. Or they can cause loosening, leading to stress change, which in turn causes fatigue and failure. That's why you've got to torque bolts tight. . .and the tighter the better.

An exception: A flexible joint. With high cyclic loading, again loosening and fatigue cause trouble. Since you shouldn't tighten a metal gasket too much, sometimes the only remedy is to take out the flexible element and put in a rigid joint. (A metal to metal flange connection instead of a gasketed one, for example.)

THE PROPER LOADING OF BOLTS

The pre-load, or residual tension, in a tightened bolt means more to assembly strength than the actual strength of the bolt itself.

In a joint, a bolt torqued to its proper load level resists a maximum amount of external load without loosening. Designers can take advantage of this fact and assure better results, and at the same time, cut costs.

AVOIDING BOLT FAILURE UNDER DYNAMIC LOADS

Dynamic loading in a tightened bolt may vary from no stress at all to that exceeding the bolt preload.

A classic example of dynamic loading is a connecting rod in a reciprocating engine. But you encounter such cyclic stress wherever you have fastened members that move or vibrate.

It has been shown that when the fluctuating stress approaches or exceeds actual bolt tension early fatigue failure can be expected.

DEMONSTATION

In tests bolts tightened to a 1420 pound tension and stressed cyclically to 9215 pounds, failed after only 5960 cycles. Identical bolts tightened to 8420 pounds, and cyclically stressed to the same 9215 pounds, went 4.65 million cycles before failing.

Had the bolts been tightened to beyond 9215 pounds, it would have been impractical to try to cycle them to failure.

So, if you want to avoid fatigue failure, be sure bolt tension exceeds the maximum dynamic load, known or estimated. This way, the bolt's life under dynamic loading will approach its life under static loading

Take advantage of the high residual tension available to you in today's high strength bolts and screws, and thereby avoid fastener failures.

A WORD ABOUT BOLT LOAD LEVELS

If you tighten a bolt beyond its proportional limit it deforms plastically. Materials engineers call the stress where this permanent set is measurable the bolt's yield strength.

NO HARM IN OVERTIGHTENING

You can tighten up to yield strength — and even beyond. Where permanent connections are made this is recommended.

Naturally, pitch of the external threads increases as a screw stretches. Conversely, because a nut undergoes compression the pitch of its internal threads decreases.

This results in a cumulative off-lead condition. In most instances, it's not practical to re-use a bolt stressed beyond its elastic limit.

TIGHTEN BOLTS TO YIELD STRENGTH?

Actually, it is safer to "overtighten" than to undertighten. The following explains why.

Up to yield point, the strain in a bolt is proportional to stress. Beyond the elastic limit, the bolt goes into its "plastic range". Some permanent stretch takes place. Yet while the bolt will not return to original length, note that residual tension is fully maintained. And remember, it's this force that keeps a bolt tight, determines joint strength.

The permanent set starts at section with highest unit stress--which is at the unengaged threads. Ultimately, this throws thread pitch off and nut locks, subjecting bolt to torsion (rather than further tightening). This force disappears with wrench-removal. Thus, a bolt can even be torqued well into its plastic range if it won't be reused or need adjustment.

FACTOR OF SAFETY -- MAKE IT A REALITY

You can't calculate that with a bolt having yield strength of 4 times the working load you're automatically getting a safety factor of 4. Far from it. Only when the bolt is tightened to four times working load do you get it.

That's because rigidly fastened members can be externally loaded to full value of residual tension in bolts without any separation or extra bolt stress.

Suppose you need a bolt for a 5000 lb. working load. For a X4 safety factor, you use a bolt of 20,000 lb. capacity, and tighten it to 20,000 lb. tension. If you tighten to only 10,000 lbs., any external load larger than this causes loosening, and progressive bolt failure from fatigue. So your safety factor is really only 2.

PRODUCTION MAN FOLLOW-THROUGH IMPORTANT

Factor of safety, then, is not established on the drawing board. It can only be put into the product by the shop man with wrench. In short, a bolt is no better than the supervision of its tightening.

A NOTE ON FLEXIBLE JOINTS

This case is different. You tighten such a joint just to working load. So use a bolt capable of this plus any added stress multiplied by your safety factor.

ROLE OF FRICTION IN FASTENING

Friction is both friend and foe.

In the one case, it complicates the torque/tension relationship; makes it a variable. In the other case, it makes for a better joint.

FASTENER FRICTION

There's no problem until you develop high thread tension. Then friction becomes the major resistance to further rotation. It exists between head and bearing surface; and between mating threads.

Lubrication, either from plating or oil, changes this ratio. That's why you can't use the same torque value for dry, oiled, and plated fasteners. What can you do? Test for the proper torque with pilot assemblies. Don't let friction stop you short; get full usable strength from high strength fasteners.

JOINT FRICTION

In structural engineering, a friction-type joint means one where bolts have been so tightened that load transfers from one connected member to the other by friction, rather than by bearing on the bolts.

There's no slippage, no loosening under such conditions. The full nut section of the structural member becomes available to support the shear load.

BOLT TENSION FROM TURN-OF-NUT METHOD

Product designers think in terms of torque when specifying how much to preload a bolt. Not so with the structural steel people.

More times than not, they go by "Turn-of Nut". Might be useful to you to understand the validity of this method.

First, the fundamental fact: Within its elastic limit, a steel bolt stretched 0.001 in. for each inch of grip length is loaded to 30,000 psi.

As nut is turned it advances, stretching bolt and loading it in tension. Since thread pitch is fixed, and increment of nut turn from the snug tight position produces a fixed increment in tension. Coarse thread nuts obviously need less turning than fine thread nuts since they have fewer threads per inch.

In steel structures, nuts are coarse thread exclusively. They are wrenched 1/2 or 2/3 turn depending on length of bolt. This is sufficient to load the bolt to 100% of its proof load, or beyond, into the bolt's plastic range.

The significance?

It results in a safe, strong joint because turn-of-nut method eliminates the effect of variation in coefficient of friction due to presence or absence of lubrication on threads or bearing surfaces.

THE CHOICE BETWEEN A HEX SCREW AND SOCKET HEAD SCREW

If you have the choice, make it a socket screw when you have a space problem, but choose the hex screw for more holding power per dollar.

No question that you should apply socket screws in counterbored holes; and that you can clear tight spaces. What you can't do is utilize the full strength of the alloy needed for the socket head. And remember, strength of a connection depends on how you preload the fasteners, and not on the strength of the fastener material alone.

THE HANDICAPS OF ONE

Internal wrenching rarely develops as much torque as needed for proper preloading. And if high torque is developed, the smaller bearing area of the socket head tends to crush the surface on which it bears rather than increase tension or preload.

THE ADVANTAGES OF THE OTHER

If you design to take advantage of heat treated hex screws you'll get a stronger connection at lower cost. These standard fasteners have ample bearing and wrenching surfaces, and can be torqued right up to yield strength.

WHY HEX SCREWS INSTEAD OF STUDS?

Many types of products once fastened with studs are today assembled with hex screws. Why the change in preference?

Studs show certain advantages in large diameters and in high temperature applications. But when it comes to smaller sizes used in tapped holes, designers have found more advantages in hex screws.

ASSEMBLY FACTORS

In fastener selection for a given application, you have to consider assembly as well as joint strength.

Obviously, studs require two wrenchings (first stud, then nut). Also, there's a more expensive close tolerance tapping job since a stud takes an interference fit to stay tight and not withdraw when nut is backed off.

Hex screws require only a clearance fit. Used in a tapped casting, they can be repeatedly inserted and unscrewed without damage to threads.

HOW NOT TO USE STUDS

Studs are not meant to serve as dowels to locate and line up for fastening. To line up numerous studs and bring two pieces together raises assembly cost. Use dowel pins for alignment and use hex screws for fastening to achieve greater economy.

PRIMER ON ROLLED THREADS

You roll-thread a bolt blank by squeezing it between rotating or reciprocating dies. Displaced metal then flows into the dies' threads, which equal in diameter the bolt's shank.

Rolled threads date back to 1851. At one time rolled threading used to produce threads with larger diameter than the shank. It was necessary to cut threads to have diameters the same. And so, rolled or cut threads used to identify screws with undersize or full-size body shank, respectively.

But this is no longer true. Today all types of threads are rolled with precision. In most respects, rolled threads are better than cut threads, because, because cold-working the metal results in continous flow lines and improves physical properties— as well as burnishing the surface of the thread.

A high production process, roll threading is used on 95% of today's output of screws. And here's where the nomenclature confusion comes in.

For if you specify a "rolled-thread screw" you'll get one with an undersize body every time.

But if you don't specify, you'll get one having a fullsize body. It will have been roll threaded, with this exception: Screws up to 6-ins. long have rolled (literally) threads. Screws larger and longer usually have cut (machined) threads because rolling them becomes uneconomical.

COARSE THREADS BETTER THAN FINE FOR MANY JOBS

The load and stress concentrations on threads are lower in standard coarse thread fasteners than in fine threaded ones. Flank engagement is also greater because threads are deeper. Except in such cases where fine adjustments are needed, coarse threads are, therefore, preferable to fine threads. They have greater resistance to stripping and, consequently, can be more highly torqued to make a stronger assembly.

PRODUCTION SAVINGS

Coarse thread fasteners tighten with only two-thirds the revolutions needed for fine threads. So your assembly time is faster, too. Coarse thread bolts enter nuts or mating holes with less tendency to cross thread when not truly positioned. In hard-to-reach areas, this ease of starting can often be your deciding factor. Bear in mind, too, that coarse threads need less "babying" in handling since they're less apt to be damaged.

All in all, coarse threaded standard fasteners prove best for an assembly because of their additional clamping strength- and best for the assembler because of their extra economy and production advantages.

NUTS - THEIR USE AND ABUSE

With bolts tightened to high load levels, nut performance becomes critical.

A nut produces bolt tension by rotating and advancing on the bolt threads. To do this, there there must be a mating condition of threads, which is influenced by thread lead. Lead is a matter of tolerance only before bolt is stressed. When tightened, the nut is then under compression and threads tend to contract; the bolt is in tension, and threads tend to stretch. Lead of thread is affected elastically before yield point, permanently beyond it.

This shortening of one lead and lengthening of the other has two effects. (1) The load distributes unequally along the threads. (2) Torsion on bolt increases. Something has to give. For high tensile bolts especially, it is better for the nut to do so. A nut therefore should be soft enough so that it deforms plastically and compensates for off-lead. If it does, it distributes the load and can advance to increase tension.

"SOFT" NUTS DO MOST JOBS

"Soft" nuts do adjust more readily than hard ones under these severe conditions. While such nuts may not be as strong in shear as heat treated ones, the important point is the bolt tension they produce. As long as the nut can pull the bolt well into its plastic range, it is doing more than its share of the job.

THE STRENGTH OF NUTS

With a hard, heat treated nut, ability to plastically adjust and distribute load over many threads diminishes. High loads tend to concentrate on first thread severely enough to cause stripping...or fracture of first thread, which causes locking. Nut cannot increase tension in bolt.

Untreated nuts are strong enough for most needs and don't pose this problem.

A CAUSE OF STRIPPING

Upon tightening, a nut both compresses, and dilates. Dilation can be overcome by wall thickness only, not by added height.

Dilation is important since a reduction in area of bolt under tension accompanies it. Threads pull away from each other, from their stronger base to weaker tips. The shallowness of fine threads can cause progressive shear.

That's why High Nuts (with fine threads) are no answer where "Heavy" Nuts (with coarse threads) are.

WRENCHING STRESS HIGHEST

Rotation of nut produces both tension and torsion in bolt. The force applying this combined stress is about 20% more direct pull than it sustained during the tightening.

HOW THREADS AFFECT STRENGTH OF NUTS

Any nut should withstand stripping loads equal to the minimum ultimate tensile strength requirement of bolt with which it's used.

Theoretically, using stress areas, fine thread bolts should be stronger than coarse thread bolts. But they're not. Because failure by thread stripping is the greater risk. For two reasons: (1) there's 231% of average stress on all threads concentrated on first two fine threads, compared to 179% on coarse threads. (2) Nut dilates under load, shifting stress toward weaker tips. Both invite progressive shear.

In short, it's not stress area, but rather volume of thread metal remaining in actual engagement under load and ability to distribute load over all threads, that determine stripping strength.

ENGAGED THREAD METAL GOVERNS STRENGTH

There are two ways to look at mating threads of a nut and bolt and their ability to take high loads (high thread tension) without stripping.

First, there's thread profile. The thread can be considered as a triangle of metal - with an apex and a root. Obviously, the closer to the apex that the load exerts its force the less metal there is to sustain it.

It's better to have more flank engagement of threads than less (as long as you don't get binding). Especially since nuts tend to dilate under high loads.

Coarse threads give you more latitude and reliability because there's more flank engagement. (It's significant that latest SAE specs. rate the proof load of coarse thread Grade 5 nuts at 120,000 psi for fine thread nuts.)

Now look at threads as a helical inclined plane. The longer the portion of the "inclined plane" in contact with the mating member, the greater the distribution of load and the less the stress concentration on any individual section (or thread).

That's why nuts should be made from metal that enables threads to "give" and thereby engage more thread metal to carry the load.

SHOULD NUT BE ABLE TO BREAK BOLT?

Practically, the nut should be the strongest member of a bolt. Nut proof load is rated at equal to the minimum ultimate tensile strength of the bolt if suitably mated.

But strength of materials isn't the only factor governing which fails first. Threads-per-inch complicate the picture. As this graph shows, stress variation differs for coarse and fine threads.

The nut with higher stress on first thread- and shallower thread- runs more of a risk of failure through progressive shear.

Now let's apply this to a possibility on the production line. Power wrenched to excessively high load, fine thread nut generally strips before breaking bolt. Nut locks; wrench kicks out. Installer can't tell nut is stripped.

With coarse thread at same load, bolt would break. It would be replaced. To provide for this automatic "inspection," use coarse threads. You'll get higher "actual" proof load. And more reliability for critical joints.

STRIPPING STRENGTH vs. TENSILE STRENGTH

A long, threaded anchor or "nut" made from Delrin was mated to a steel screw and tensile tested. The screw broke before the plastic stripped.

Mind you, Grade 2 steel has 7 times the tensile strength of Delrin. Seems incredible. But not if you study thread geometry and behavior under load.

LOCATION OF BREAK

Almost always, bolts break not at shank, but at an unengaged thread root. Understandably. This is the minor diameter of the external thread, therefoe the smallest cross section. It's the weakest link.

When equally loaded, a bolt experiences the same tension along its entire grip length. It's the same at any cross sectional plane. Overloaded, the weakest area is first to strain beyond its ultimate strength. It "necks" down. Finally snaps.

Thread tension, however, causes a plastic adjustment of nut threads. Bolt load distributes itself along the entire height of the nut. Thus, shearing stress on an individual internal thread is far less than on a plane across the minor diameter of external thread.

That's why, paradoxically, a nut of weaker material can match the proof load of a stronger screw. Unless nut dilation is excessive, its threads continue to do the job of developing high thread tension in the fastener.

SELECTING THE RIGHT GRADE OF NUTS

"Workhorse" among nuts is the standard "FINISHED" series. It gives good seating area; sufficient height to sustain high thread tension; enough wall thickness to control elastic nut dilation under load.

"HEAVY" nuts are wider than "Finished" nuts in all sizes by only 1/8 in. across flats. Thus, the value diminishes as size increases. Most effective in 1/2 in. to 1 1/2 in. range, they satisfy applications involving excessive clearance holes, unusual loads, and certain boiler codes.

MATERIALS? The regular carbon nut steel (non heat treated). It makes nuts strong enough to pull bolts (through Grade 5), beyond yield point, lets threads distribute load to avoid stripping.

NUTS WITH "SPECIAL" FUNCTION

JAM NUTS are made for position locking. Use of two together forms a superior and economical locking device. When used to lock a regular nut, jam nut should be adjacent to work surface. Otherwise, the jam nut takes the load- a job it wasn't designed for.

LOCKING POWER OF LOCK NUTS

Self locking depends on: (1) An increase in thread friction; (2) an integral anchoring device; or both.

"FREE SPINNING" LOCK NUTS

This type turns easily until it begins to seat. Further tightening then brings into play the locking action caused by pressure on the bearing surface.

Since these nuts must be tight to lock, they also get extra resistance to back-off from higher thread friction resulting from high thread tension.

"PREVAILING TORQUE" LOCK NUTS

Resistance to back-off starts as soon as locking device reaches thread. Once engaged, this type stays put anywhere along the screw.

Usually, there's some plastic insert, spring action or deformed thread that develops the prevailing torque. Of course, this type of nut can also be used under thread tension, adding thread friction to its back-off resistance.

Prevailing torque lock nuts won't rattle off when not snugly seated. Free spinning types, once loosened, become free turning again; but resistance to initial loosening far surpasses that obtainable from prevailing torque alone.

OFF-TORQUE DISTINCTIONS

Effort to break loose an axially loaded nut is the Break-Loose Torque.

Effort required to start in motion a non-axially loaded nut is the Break-Away Torque.

Effort to overcome frictional resistance of locknut already in rotation is the Prevailing Torque.

FASTENING OF RIGID JOINTS

Theoretically, there's no such thing as a rigid joint. There's always some elasticity of the fastened metals. For practical purposes, you can consider a joint rigid when the bearing areas of the metal-to-metal fastened members will not crush or yield before the full load-carrying capacity of screw or bolt is developed.

PRODUCT BENEFITS

Rigid joints afford a definite product advantage. They can take high strength hex screws or bolts tightened up to fastener yield strength or beyond. Under such tension, fasteners have demonstrated that they'll stay tight despite vibration. They are resistant to fatigue from the constant load reversals. Hight Strength Hex Screws cut costs and speed assembly, too, since you can reduce size of the fasteners or their number, while actually improving joint strength.

INSTALLATION HINTS

When you're connecting steel members of fairly heavy section, you've no problem getting a rigid joint. Just clamp them to the full fastener capacity. Thin sections can be reinforced and similarly fastened. And in joining milder steels or softer metals, use of a plate washer will distribute bolt load, prevent crushing and give the desired effect of rigidity.

FASTENING OF "VIBRATION" JOINTS

With static assemblies, your only fastening worry is enough safe load capacity. But with dynamic loading, you also have to guard against loosening.

HIGH THREAD TENSION

You'll find one of the best locking mechanisms within the fastener itself: high thread tension. Obviously, the tighter the fastener, the higher the tension and the better the "lock."

Using split lock washers give no such guarantee against loosening. It becomes equivalent to a solid washer at a relatively low load level. Once the screw has loosened to the point where a washer becomes a spring, it's too loose for safety.

When you can't fully utilize a high strength fastener, go to a bolt with prevailing torque lock nut; or to a unit that combines high thread tension along with a high off-torque value.

With vibration joints involving thin gauge or sheet metal, it's better to use a thread-cutting screw, than a machine screw in tapped hole. It gives 100% depth of thread contact for more thread friction to resist back-off.

FASTENING OF JOINTS IN CORROSIVE ENVIRONMENT

For a fastener to stand up in a corrosive environment it must either be coated, or made from anti-corrosive metal.

THE ENDURING METALS

While each metal poses problems in certain applications, stainless steel, aluminum, and silican bronze all offer distinct advantages.

Silicon bronze is popular for electrical uses due to its unusual strength, resistance to stress-corrosion resistance, good conductivity.

Widely used for fasteners, steel assures good strength and excellent corrosion resistance in most atmospheres.

WHAT THE DESIGNER WANTS OF A THREADED FASTENER SYSTEM

Threaded fasteners have proved to be the most efficient and effective way to put together shapes that would be impractical as one piece structures. They also simplify, or reduce costs, or make practical the disassembly of such structures for safety inspection, cleaning, repair or rebuilding.

When a vehicle, structure or machine is overhauled, its worn and corroded parts may be replaced as a routine maintenance operation. However, the threaded fasteners are not generally regarded as things that wear out or deteriorate. They are expected to be reusable.

Fasteners must carry loads, of course. They resist various combinations of tension and shear loading, usually without permitting any significant relative movement of the fastened parts. Most threaded fasteners are screwed up tight so that they clamp the fastened parts together. It is desirable to maintain this initial clamping force — or as large a portion of it as possible.

Although their primary function is to permit convenient assembly and disassembly, threaded fasteners are expected to stay in place between those events. Without fail!

Characteristics the designer wants, then, are reliability, strength, tightness and convenience in service. He looks for ways to combine all these ideals economically.

FASTENERS FIRST LOOSEN WITHOUT TURNING!

"Settling down" is the term ESNA laboratory people use to describe a relaxation phenomenon that occurs at room temperature under static conditions. During the first few hours after a nut and bolt are tightened on an arbor, the fastener system has lost some of its pre-stress.

The mechanical fits and finishes involved determine to a great extent how much initial clamping load will be lost. With precise, well-finished parts, this relaxation may be limited to 2 or 3 per cent of pre-stress. With rough surfaces, loose thread tolerances and lack of squareness,

Schematic of all-metal fastener testing histories shows how bolt tension is lost first in the static condition and then, more rapidly, after vibration begins. The nut turns at a critical tension level.

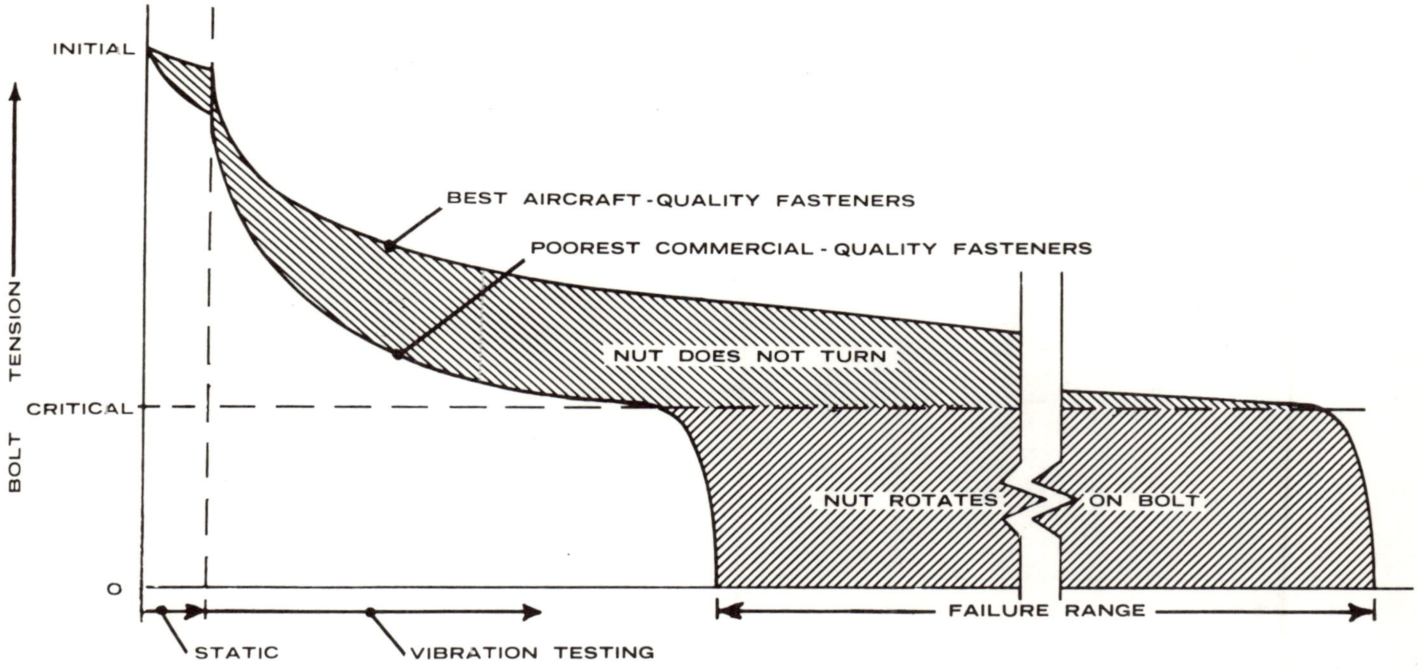

as much as 10 per cent of the original loading may be lost.

This stress relaxation is analogous to high-temperature creep. We suppose that it involves a gradual adjustment as localized zones stressed beyond the yield point transfer load to adjacent metal. Another mode of relaxation was observed by Professor Wright[4] of the University of Illinois in connection with torque-tension studies of bolting for structural steel. He found that part of the pre-stress loss that occurred within a few hours after tightening was accounted for by rotation of one end of the bolt as it recovered from being twisted during tightening.

When a retightened fastener system is subjected to a vibration test, loss of clamping force continues, but at a rate much slower than occurred initially. The nut does not turn on the bolt while that is happening. Nevertheless, bolt tension decreases as the vibration test progresses.

The mechanism of loosening in this phase — before any relative thread turning occurs — is not known accurately. It is probably a combination of local yielding and wear. The vibrations tend to hammer down and wear away minute metal projections on all bearing surfaces.

For any given nut type — plain or self-locking — and a given level of vibration intensity, resistance to loosening is influenced in large measure by bolt tension at the start of the test. At the critical tension level, the nut begins to turn, then comes off very soon.

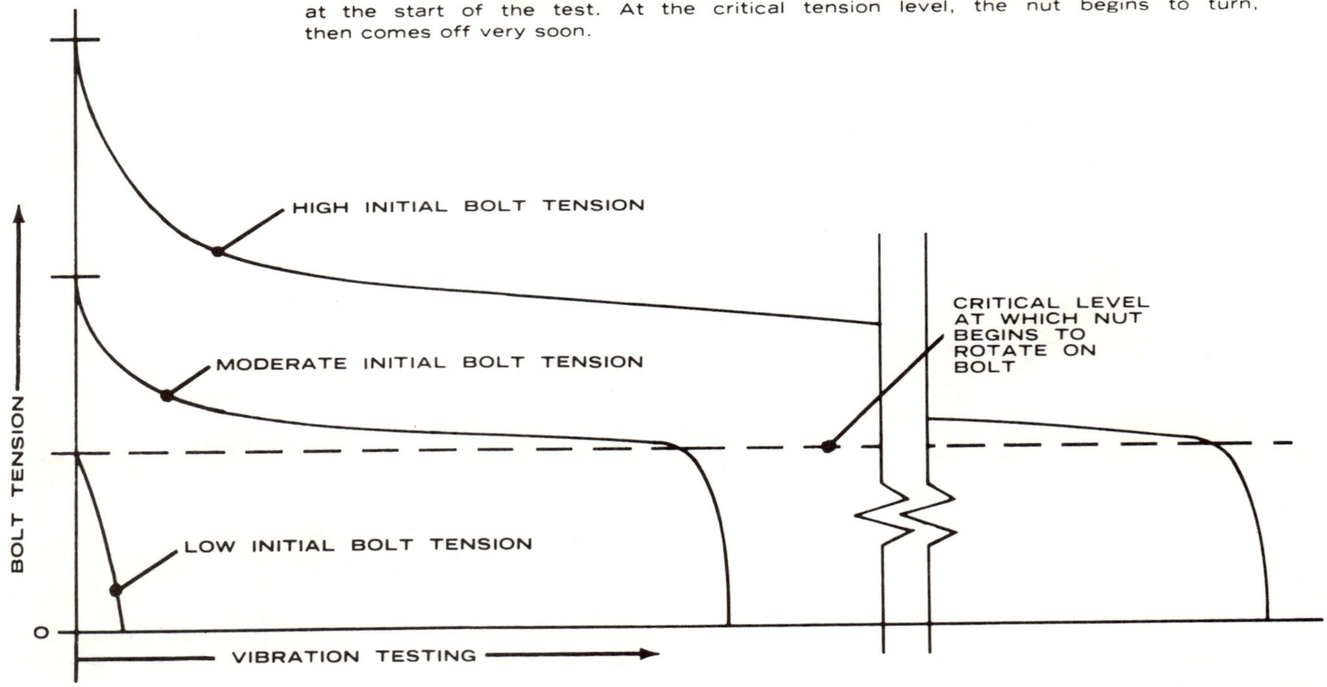

AEROSPACE FASTENERS

N A S

NATIONAL AEROSPACE STANDARD

A N

AIR FORCE - NAVY AERONAUTICAL

M S

MILITARY STANDARD

NUMERICAL LISTING OF PARTS

NAS

NAS 333 thru NAS 340
100° Flush Head Bolt - Phillips Recess Only
SPS Santa Ana, Calif.

NAS 428
Crown Hex Head Bolt
SPS Santa Ana, Calif.

NAS 443
Internal Wrenching Nut
SPS Jenkintown, Pa.

NAS 464
Hex Head Bolt
SPS Santa Ana, Calif.

NAS 501
S/S Hex Head Bolt
SPS Santa Ana, Calif.

NAS 517
100° Flush Head Bolt
SPS Santa Ana, Calif.

NAS 560
Screw - Hi Temp 100° Flush Head
321 S/S, A286 or Inconel "X" as required
SPS Santa Ana, Calif.

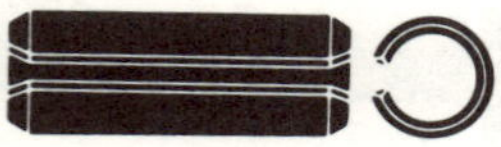

NAS 561
Spring Pin
S/S and Carbon Steel
SPS Jenkintown, Pa.

NAS 563 thru NAS 572
Hex Head Bolt
SPS Santa Ana, Calif.

NAS 583 thru NAS 590
100° Flush Head Hi-Torque Bolt
95,000 psi Min. SS
SPS Santa Ana, Calif.

NAS 607
Unbrako - Dowel Pin
SPS Jenkintown, Pa.

NAS 608 - NAS 609
Unbrako - Std. Socket Head Cap Screw
SPS Jenkintown, Pa.

NAS 624 thru NAS 644
12 Point Bolt
SPS Jenkintown, Pa.

NAS 653 thru NAS 658
Hex Head Bolt, Titanium
SPS Santa Ana, Calif.

NAS 663 thru 668
Flush Head Bolt, Titanium
Hi-Torque and Phillips Recess
SPS Santa Ana, Calif.

NAS 673 thru 678
Hex Head Bolt, Titanium
SPS Santa Ana, Calif.

NAS 679 (96)
Lightweight Sheet Metal Locknut
NAS 679C, 900°F (97)
SPS Santa Ana, Calif.

NAS 680 and NAS 681
Nutt-Shel, Self-locking Plate Nut
Two Lug
SPS Santa Ana, Calif.

NAS 682 and NAS 683
Nutt-Shel Self-locking Plate Nut
One Lug
SPS Santa Ana, Calif.

NAS 684 and NAS 685
Nutt-Shel Self-locking Plate Nut
Corner
SPS Santa Ana, Calif.

NAS 686
Nutt-Shel Self-locking Plate Nut
Floating - Two Lug
SPS Santa Ana, Calif.

NAS 687
Nutt-Shel Self-locking Plate Nut
Floating - One Lug
SPS Santa Ana, Calif.

NAS 696
Nutt-Shel Self-locking Plate Nut
Miniature - One Lug
SPS Santa Ana, Calif.

NAS 697
Nutt-Shel Self-locking Plate Nut
Miniature - Two Lug
SPS Santa Ana, Calif.

NAS 698
Nutt-Shel Self-locking Plate Nut
Miniature - Corner
SPS Santa Ana, Calif.

NAS 1003 thru NAS 1020
Hex Head Bolt
SPS Santa Ana, Calif.

NAS 1021 and NAS 1022
Flexloc Nut - Self-locking
Std. Regular Height and Thin Nuts
SPS Jenkintown, Pa.

NAS continued

NAS 1023 and NAS 1024
Nutt-Shel Self-locking Plate Nut
Two Lug
SPS Santa Ana, Calif.

NAS 1025 and NAS 1026
Nutt-Shel Self-locking Plate Nut
One Lug
SPS Santa Ana, Calif.

NAS 1027 and NAS 1028
Nutt-Shel Self-locking Plate Nut
Corner
SPS Santa Ana, Calif.

NAS 1031
Nutt-Shel Self-locking Plate Nut
Floating - Two Lug
SPS Santa Ana, Calif.

NAS 1032
Nutt-Shel Self-locking Plate Nut
Floating - One Lug
SPS Santa Ana, Calif.

NAS 1033
Nutt-Shel Self-locking Plate Nut
Floating - Right Angle
SPS Santa Ana, Calif.

NAS 1067
Nutt-Shel Self-locking Plate Nut
Miniature - Modified Base
SPS Santa Ana, Calif.

NAS 1068
Nutt-Shel Self-locking Plate Nut
Floating Miniature - Two Lug
SPS Santa Ana, Calif.

NAS 1081
Hex Socket Set Screw - Self-locking
SPS Jenkintown, Pa.

NAS 1083 thru 1088
100° Flush Head Bolt, Titanium
Phillips and Hi-Torque
SPS Santa Ana, Calif.

NAS 1096
Screw, Hex Head, Recessed
Full Thread
SPS Santa Ana, Calif.

NAS 1100
Pan Head Bolt, Torq-Set Recess
Alloy Steel and A286
SPS Santa Ana, Calif.

NAS 1101
Flat Fillister Head Screw,
Torq-Set
Alloy Steel and A286
SPS Santa Ana, Calif.

NAS 1102
100° Flush Head Screw,
Torq-Set
Alloy Steel and A286
SPS Santa Ana, Calif.

NAS 1103 thru NAS 1120
Hex Head Bolt
SPS Santa Ana, Calif.

NAS 1121 thru 1128
Flat Fillister Close Tolerance Bolt
Torq-Set
Alloy Steel and A286
SPS Santa Ana, Calif.

NAS 1131 thru 1138
Pan Head Bolt, Close Tolerance,
Torq-Set
Alloy Steel and A286
SPS Santa Ana, Calif.

NAS 1131 thru 1138 (Titanium)
Pan Head Bolt, Close Tolerance,
Torq-Set
SPS Santa Ana, Calif.

NAS 1141 thru 1148
Brazier Head Bolt, Close Tolerance,
Torq-Set
Alloy Steel and A286
SPS Santa Ana, Calif.

NAS 1151 thru 1158
100° Flush Head Bolt,
Close Tolerance, Torq-Set
Alloy Steel and A286
SPS Santa Ana, Calif.

NAS 1151 thru 1158 (Titanium)
100° Flush Head Bolt,
Close Tolerance, Torq-Set
SPS Santa Ana, Calif.

NAS 1161 thru 1168
100° Flush Head Bolt,
Self-locking, Torq-Set
SPS Santa Ana, Calif.

NAS 1171 thru 1178
Pan Head Bolt, Self-locking,
Torq-Set
SPS Santa Ana, Calif.

NAS continued

NAS 1181 thru 1188
Flat Fillister Head Bolt
Self-locking, Torq-Set
Alloy Steel and A286
SPS Santa Ana, Calif.

NAS 1189
100° Flush Head Bolt,
Self-locking, Torq-Set and Phillips
Alloy Steel and A286
SPS Santa Ana, Calif.

NAS 1190
Pan Head Screw, Self-locking,
Torq-Set and Phillips
Alloy Steel and A286
SPS Santa Ana, Calif.

NAS 1191
Flat Fillister Head Screw,
Self-locking, Torq-Set and Phillips
Alloy Steel and A286
SPS Santa Ana, Calif.

NAS 1202 thru NAS 1210
Flush Head Bolt
SPS Santa Ana, Calif.

NAS 1216
Pan Head Screw, Hi Torque
Alloy Steel, 302 S/S and A286
SPS Santa Ana, Calif.

NAS 1217
Pan Head Screw, Hi-Torque
Alloy Steel, 302 S/S and A286
SPS Santa Ana, Calif.

NAS 1218
Pan Head Screw, Hi-Torque
Alloy Steel, 302 S/S and A286
SPS Santa Ana, Calif.

NAS 1219
100° Flush Head Screw,
Hi-Torque
Alloy Steel, 302 S/S and A286
SPS Santa Ana, Calif.

NAS 1220
100° Flush Head Screw,
Hi-Torque
Alloy Steel, 302 S/S and A286
SPS Santa Ana, Calif.

NAS 1221
100° Flush Head Screw,
Hi-Torque
Alloy Steel, 302 S/S and A286
SPS Santa Ana, Calif.

NAS 1223 thru 1235
Hex Head Bolt, Self-locking
Alloy Steel and A286
SPS Santa Ana, Calif.

NAS 1243 thru 1250
100° Flush Head Bolt,
1/64 Oversize, Hi-Torque
SPS Santa Ana, Calif.

NAS 1253 thru 1260
100° Flush Head Bolt,
1/32 Oversize, Hi-Torque
SPS Santa Ana, Calif.

NAS 1261 thru NAS 1265
Hex Head Close Tolerance
Short Thread Bolt
Titanium Alloy - 6AL-4V
SPS Santa Ana, Calif.

NAS 1266 thru NAS 1270
Hex Head Close Tolerance Bolt
Titanium Alloy - 6AL-4V
SPS Santa Ana, Calif.

NAS 1271 thru NAS 1280
12-Point Bolt
Titanium Alloy - 6AL-4V
SPS Jenkintown, Pa.

NAS 1291
Lightweight, Low Height
Hexagonal Locknut
Alloy Steel and Stainless Steel
550°F and 800°F
SPS Jenkintown, Pa.

NAS 1292 thru NAS 1296
Threaded Shear Pin, 100° Flush He
SPS Jenkintown, Pa.

NAS 1297
Hex Head Shoulder Bolt
SPS Santa Ana, Calif.

NAS 1298
Brazier Head Shoulder Screw
Phillips Recess
SPS Santa Ana, Calif.

NAS 1299
100° Flat Head Shoulder Screw
Phillips Recess
SPS Santa Ana, Calif.

NAS 1303 thru NAS 1320
Hex Head Bolt
SPS Santa Ana, Calif.

NAS continued

NAS 1322 thru NAS 1326
Threaded Shear Pin,
Protruding Head
SPS Jenkintown, Pa.

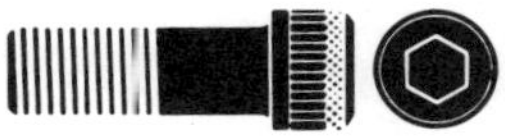

NAS 1351 and NAS 1352
Socket Head Cap Screw
(1960 Series)
SPS Jenkintown, Pa.

NAS 1393
82° Flat Head Bolt
Torq-Set Recess
Alloy Steel, A286
SPS Santa Ana, Calif.

NAS 1402 thru 1406
Pan Head Screw, Phillips
SPS Santa Ana, Calif.

NAS 1473
Nut, Cap, Self-Sealing,
Self-locking, Two Lug Floating,
Plate
SPS Santa Ana, Calif.

NAS 1474
Nut, Cap Reduced Rivet
Spacing, Self-Sealing, Self-
Locking, Two Lug Floating,
Plate
SPS Santa Ana, Calif.

NAS 1503 thru NAS 1510
100° Flush Head Hi-Torque Bolt
95,000 psi Min. SS
SPS Santa Ana, Calif.

NAS 1578
Bolt, Shear, Pan Head
1200°F
SPS Jenkintown, Pa.

NAS 1579
Screw, Pan Head, Full
Threaded, 1200°F
SPS Jenkintown, Pa.

NAS 1580
Bolt, Tension, Flush Head
1200°F
SPS Jenkintown, Pa.

NAS 1581
Bolt, Shear, Reduced 100° Flush Head,
1200°F
SPS Santa Ana, Calif.

NAS 1582
Screw, Flush Head, Full Threaded,
1200°F
SPS Santa Ana, Calif.

NAS 1586
Bolt, Tension, 12 Pt.
Ext. Wrenching, 1200°F
SPS Jenkintown, Pa.

NAS 1587
Washer, Plain and CSK,
1200°F
SPS Jenkintown, Pa.

NAS 1588
Bolt, Shear, Hex Head,
1200°F
SPS Santa Ana, Calif.

NAS 1603 thru NAS 1610
1/32 Oversize Hi-Torque and Phillips
SPS Santa Ana, Calif.

NAS 1620 thru NAS 1628
100° Flat Head Bolt,
Torq-Set Recess
SPS Santa Ana, Calif.

NAS 1630 - 1634
Pan Head Bolt
Torq-Set Recess
SPS Santa Ana, Calif.

NAS 1635
Screw, Pan Head
Cross Recess
SPS Santa Ana, Calif.

NAS 1703 thru NAS 1710
1/64 Oversize; Hi-Torque or Phillips
SPS Santa Ana, Calif.

NAS 2005 thru 2012
Pan Head Lock Bolt,
Titanium
SPS Jenkintown, Pa.

NAS 2105 thru 2112
100° Flush Head Lock Bolt,
Titanium
SPS Jenkintown, Pa.

NAS 2206 thru 2210
Bolt, Lock, Tension, Flat Head
Stump Type, Titanium
SPS Jenkintown, Pa.

NAS continued

NAS 2306 thru 2310
100° Flush Head Lock Bolt,
Titanium
SPS Jenkintown, Pa.

NAS 2406 thru 2412
Flat Head Lock Bolt
Titanium
SPS Jenkintown, Pa.

NAS 2506 thru 2512
100° Flush Head Lock Bolt
Titanium
SPS Jenkintown, Pa.

NAS 2606 thru 2612
Flat Head Lock Bolt,
Titanium
SPS Jenkintown, Pa.

NAS 2706 thru 2712
100° Flush Head Lock Bolt,
Titanium
SPS Jenkintown, Pa.

NAS 2803 thru 2810
100° Flush Head Bolt, Torq-Set
SPS Santa Ana, Calif.

NAS 2903 thru 2920
Hex Head Bolt, Oversize
SPS Santa Ana, Calif.

NAS 3003 thru 3020
Hex Head Bolt, Oversize
SPS Santa Ana, Calif.

NAS 4104 thru 4116
100° Flush Head Bolt, Alloy Steel
Tri-Wing Recess, Long Thread,
Self-Locking and Non-locking
SPS Santa Ana, Calif.

NAS 4204 thru 4216
100° Flush Head Bolt, A-286
Tri-Wing Recess, Long Thread
Self-locking and Non-locking
SPS Santa Ana, Calif.

NAS 4304 thru 4316
100° Flush Head Bolt, Titanium,
Tri-Wing Recess, Long Thread,
Self-locking and Non-locking
SPS Santa Ana, Calif.

NAS 4400 thru 4416
100° Flush Head Bolt, Alloy Steel,
Tri-Wing Recess, Short Thread,
Self-locking and Non-locking
SPS Santa Ana, Calif.

NAS 4500 thru 4516
100° Flush Head Bolt, A-286
Tri-Wing Recess, Short Thread,
Self-Locking and Non-locking
SPS Santa Ana, Calif.

NAS 4600 thru 4616
100° Flush Head Bolt, Titanium,
Tri-Wing Recess, Short Thread
Self-locking and Non-locking
SPS Santa Ana, Calif.

NAS 4703 thru 4716
100° Reduced Flush Head Bolt,
Tri-Wing Recess, Short Thread,
Alloy Steel, Non-locking
SPS Santa Ana, Calif.

NAS 4803 thru 4816
100° Reduced Flush Head Bolt,
Tri-Wing Recess, Short Thread,
A-286, Non-locking
SPS Santa Ana, Calif.

NAS 4903 thru 4916
100° Reduced Flush Head Bolt,
Tri-Wing Recess, Short Thread,
Titanium, Non-locking
SPS Santa Ana, Calif.

NAS 5000 thru 5006
Pan Head Bolt, Alloy Steel
Tri-Wing Recess, Short Thread
SPS Santa Ana, Calif.

NAS 5100 thru 5106
Pan Head Bolt, A-286
Tri-Wing Recess, Short Thread
SPS Santa Ana, Calif.

NAS 5200 thru 5206
Pan Head Bolt, Titanium
Tri-Wing Recess, Short Thread
SPS Santa Ana, Calif.

NAS 5300 thru 5306
Flat Fillister Head Machine Screw,
Tri-Wing Recess, Full Thread,
Alloy Steel
SPS Santa Ana, Calif.

NAS 5400 thru 5406
Flat Fillister Head Machine Screw,
Tri-Wing Recess, Full Thread,
A-286 CRES Steel
SPS Santa Ana, Calif.

NAS 5500 thru 5506
Flat Fillister Head Machine Screw,
Tri-Wing Recess, Full Thread,
Titanium
SPS Santa Ana, Calif.

NAS continued

NAS 5600 thru 5606
100° Flush Head Screw, Alloy Steel
Tri-Wing Recess, Full Thread,
Self-locking and Non-locking
SPS Santa Ana, Calif.

NAS 5700 thru 5706
100° Flush Head Screw, A-286
Tri-Wing Recess, Full Thread,
Self-locking and Non-locking
SPS Santa Ana, Calif.

NAS 5800 thru 5806
100° Flush Head Screw, Titanium
Tri-Wing Recess, Full Thread,
Self-locking and Non-locking
SPS Santa Ana, Calif.

NAS 6000 thru 6003
Hex Head Screw, A-286
Tri-Wing Recess, Full Thread,
Non-locking
SPS Santa Ana, Calif.

NAS 6100 thru 6103
Hex Head Screw, Titanium,
Tri-Wing Recess, Full Thread,
Non-locking
SPS Santa Ana, Calif.

NAS 6203 thru 6220
Hex Head Close Tolerance
Short Thread Bolt, Alloy Steel,
Self-locking and Non-locking
SPS Santa Ana, Calif.

NAS 6303 thru 6320
Hex Head Close Tolerance
Short Thread Bolt, A-286
Self-Locking and Non-locking
SPS Santa Ana, Calif.

NAS 6403 thru 6420
Hex Head Close Tolerance
Short Thread Bolt, Titanium,
Self-locking and Non-locking
SPS Santa Ana, Calif.

NAS 6604 thru 6620
Hex Head Close Tolerance
Long Thread Bolt, Alloy Steel,
Self-locking and Non-locking
SPS Santa Ana, Calif.

NAS 6704 thru 6720
Hex Head Close Tolerance
Long Thread Bolt, A-286,
Self-locking and Non-locking
SPS Santa Ana, Calif.

NAS 6804 thru 6820
Hex Head Close Tolerance
Long Thread Bolt, Titanium,
Self-locking and Non-locking
SPS Santa Ana, Calif.

AN

AN 3 thru AN 20
Hex Head Bolt
SPS Santa Ana, Calif.

AN 3C thru AN 20C
Hex Head Bolt S/S
SPS Santa Ana, Calif.

AN 3DD thru AN 20DD
Aluminum Hex Head Bolt
SPS Santa Ana, Calif.

AN 21 thru AN 36
Clevis Bolt
SPS Santa Ana, Calif.

AN 73 thru AN 81
Hex Head Bolt
SPS Santa Ana, Calif.

AN 173 thru AN 186
Hex Head Bolt, Close Tolerance Shank
Alloy Steel
SPS Santa Ana, Calif.

AN 173C thru AN 186C
Hex Head Bolt, Close Tolerance Shank
S/S
SPS Santa Ana, Calif.

AN 173DD thru 186DD
Hex Head Bolt, Close Tolerance Shank
Aluminum
SPS Santa Ana, Calif.

AN 361 - 362
Nutt-Shel Self-locking Plate Nut
Two Lug
SPS Santa Ana, Calif.

AN 363 and AN 365
Flexloc Nut
Full Height
Self-locking
SPS Jenkintown, Pa.

AN 364
Flexloc Nut
Thin
Self-locking
SPS Jenkintown, Pa.

AN 366
Nutt-Shel Self-locking Plate Nut
Two Lug, 250°F
SPS Santa Ana, Calif.

AN 373
Nutt-Shel Self-locking Plate Nut
Two Lug, 250°F
SPS Santa Ana, Calif.

AN 565
Unbrako Stds. Socket Set Screws
SPS Jenkintown, Pa.

AN 101001 thru AN 104600
Hex Head Bolt, Alloy Steel
SPS Santa Ana, Calif.

AN 104601 thru AN 108200
Hex Head Bolt, CRES Steel
SPS Santa Ana, Calif.

AN 148551 thru AN 149350
Internal Wrenching Bolt
SPS Jenkintown, Pa.

MS

MS 9033 thru MS 9039
12 Point Bolt A286 - 1200°
SPS Jenkintown, Pa.

MS 9047 and MS 9048
Spring Pin
SPS Jenkintown, Pa.

MS 9060 thru MS 9066
12 Point Bolt A286 - 1200° D.H.
SPS Jenkintown, Pa.

MS 9088 thru MS 9094
12 Point Bolt - Steel, D.H.
SPS Jenkintown, Pa.

MS 9110 thru MS 9113
12 Point Bolt, AMS 5731, Extende
Washer Head, Close Tolerance Sh
SPS Jenkintown, Pa.

MS continued

MS 9122
Slotted Hex Head Machine Screw, #10
SPS Santa Ana, Calif.

MS 9123
Slotted Hex Head Machine Screw, 1/4'
SPS Santa Ana, Calif.

MS 9146 thru MS 9152
12 Point Bolt - Steel - Cad. Plate
SPS Jenkintown, Pa.

MS 9157 thru MS 9163
12 Point Bolt - Steel
SPS Jenkintown, Pa.

MS 9169 thru MS 9175
12 Point Bolt - Steel, D.H.
SPS Jenkintown, Pa.

MS 9177 and MS 9178
12 Point Bolt A286 - 1200°
SPS Jenkintown, Pa.

MS 9183 thru MS 9186
12 Point Bolt - Steel
SPS Jenkintown, Pa.

MS 9187 and MS 9188
12 Point Bolt, A286 - 1200° D.H.
SPS Jenkintown, Pa.

MS 9189 and MS 9190
12 Point Machine Screw,
Alloy Steel
SPS Jenkintown, Pa.

MS 9191 and MS 9192
12 Point Machine Screw
Alloy Steel, Drilled Head
SPS Jenkintown, Pa.

MS 9206 thru MS 9223
12 Point Screw - Steel AMS 6304
Diffused Nickel - Cadmium Plated
SPS Jenkintown, Pa.

MS 9224
12 Point Bolt A286 - 1200°
SPS Jenkintown, Pa.

MS 9281 thru MS 9291
Bolt, Hexagon Head,
AMS 6322
SPS Santa Ana, Calif.

MS 9292 thru MS 9302
Bolt, Hexagon Head, Drilled,
1 Hole, AMS 6322
SPS Santa Ana, Calif.

MS 9316 and MS 9317
Slotted Hex Head Machine Screw
SPS Santa Ana, Calif.

MS 9397 thru MS 9402
Chamfered Tee Head Bolt,
AMS 6322, Cadmium Plated
SPS Santa Ana, Calif.

MS 9432 thru MS 9437
Bolt, Tee Head
AMS 5735, Chamfered
SPS Santa Ana, Calif.

MS 9438 and MS 9439
Screw, Hexagon Head, Drilled,
1 Hole, AMS 6304
SPS Santa Ana, Calif.

MS 9440 thru MS 9448
Bolt, Machine - Steel AMS 6304,
Diffused Nickel - Cadmium Plated,
Hexagon Head, Drilled.
SPS Santa Ana, Calif.

MS 9449 thru MS 9459
Screw, Machine - Steel AMS 6304,
Diffused Nickel - Cadmium Plated,
Hexagon Head
SPS Santa Ana, Calif.

MS 9487 thru MS 9497
Bolt, Hexagon Head,
AMS 5731, CRES
SPS Santa Ana, Calif.

MS 9498 thru MS 9508
Bolt, Hexagon Head, Drilled,
1 Hole, AMS 5731, CRES
SPS Santa Ana, Calif.

MS 9516 thru MS 9526
Hex Head Machine Screw,
AMS 6322, Cadmium Plated
SPS Santa Ana, Calif.

MS 9527 thru MS 9537
Hex Head Machine Screw,
Drilled Head, 1 Hole,
AMS 6322, Cadmium Plated
SPS Santa Ana, Calif.

MS 9554 thru MS 9562
12 Point Bolt, AMS 5731, Extended
Washer Head
SPS Jenkintown, Pa.

MS 9563 thru MS 9571
12 Point Bolt, AMS 5731, Extended
Washer Head, Drilled, 4 Holes
SPS Jenkintown, Pa.

MS continued

MS 9572 thru MS 9580
12 Point Bolt, AMS 5731, Extended
Washer Head, Drilled,
4 Holes, Silver Plated
SPS Jenkintown, Pa.

MS 9583 thru MS 9591
Bolt, Hexagon Head, Drilled,
6 Holes, AMS 5731, CRES
SPS Santa Ana, Calif.

MS 9676 thru MS 9679
12 Point Bolt, AMS 5731, Extended
Washer Head, Cupwasher Locked
SPS Jenkintown, Pa.

MS 9680 thru MS 9683
12 Point Bolt, AMS 6322, Extended
Washer Head, Cupwasher Locked
SPS Jenkintown, Pa.

MS 9694 thru MS 9702
12 Point Bolt, AMS 5708,
Extended Washer Head
SPS Jenkintown, Pa.

MS 9712 thru MS 9720
12 Point Bolt, AMS 5708, Extended
Washer Head, Drilled, 4 Holes, Silver Plated
SPS Jenkintown, Pa.

MS 9730 thru MS 9738
12 Point Bolt, AMS 5643, Extended
Washer Head, PD Shank
SPS Jenkintown, Pa.

MS 9739 thru MS 9747
12 Point Bolt, AMS 5643, Extended
Washer Head, Drilled, PD Shank
SPS Jenkintown, Pa.

MS 9748 thru MS 9756
12 Point Bolt, Titanium
Extended Washer Head, PD Shank
SPS Jenkintown, Pa.

MS 9757 thru MS 9765
12 Point Bolt, Titanium, Extended
Washer Head, Drilled, PD Shank
SPS Jenkintown, Pa.

MS 9883 thru MS 9891
12 Point Bolt, AMS 5616,
Extended Washer Head
SPS Jenkintown, Pa.

MS 16555
Dowel Pin
.0002" over nominal size
SPS Jenkintown, Pa.

MS 16556
Dowel Pin
.001" over nominal size
SPS Jenkintown

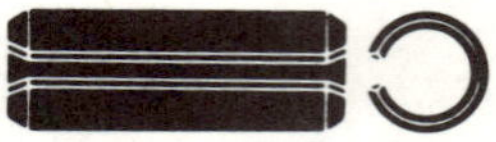

MS 16562
Spring Pin
Carbon Steel or S/S
SPS Jenkintown, Pa.

MS 16995 and MS 16996
Screw, Cap, Socket Head
Hexagon, Corrosion Resisting
Steel, 1960 Series
SPS Jenkintown, Pa.

MS 16997 and MS 16998
Screw, Cap, Socket Head
Hexagon, Alloy Steel, Cadmium
Plated, 1960 Series
SPS Jenkintown, Pa.

MS 18063 thru MS 18068
Screw, Set, Self-locking
Hexagonal Wrenching
Points: Cup, Flat, Cone
Alloy Steel, S/S
SPS Jenkintown, Pa.

MS 20004 thru MS 20024
Internal Wrenching Bolt
SPS Jenkintown, Pa.

MS 20033 thru MS 20046
Hex Head Bolt - Machine 1200°
SPS Santa Ana, Calif.

MS 20073 and MS 20074
Hex Head Bolt, D.H.
SPS Santa Ana, Calif.

MS 20364 (thin) and MS 20365
(Full Height)
Flexloc Nut - Self-locking
SPS Jenkintown, Pa.

MS 20500
Flexloc Nut - Self-locking 1200°F
SPS Jenkintown, Pa.

MS 20501
Nutt-Shel Self-locking Plate Nut
Two Lug, 1200°F
SPS Santa Ana, Calif.

MS 21040
Lightweight Sheet Metal Locknut
(96)
SPS Santa Ana, Calif.

MS 21095
Bolt, Self-locking, CRES. 250°F
Hexagon Head
SPS Santa Ana, Calif.

MS 21096
Bolt, Self-locking, Steel, 250°F
Pan Head, Cross Recessed
SPS Santa Ana, Calif.

MS 21097
Bolt, Self-locking, CRES. 250°F
Pan Head, Cross Recessed
SPS Santa Ana, Calif.

MS 21244
Castellated Nut, Hexagon,
Counterbored, Captive Washer, 450°
SPS Jenkintown, Pa.

MS 21250
12 Point Bolt
SPS Jenkintown, Pa.

MS 21262
Unbrako Std. Socket Head Cap Screw
1960 Series, Self-locking
SPS Jenkintown , Pa.

MS 21277 thru MS 21294
12 Point Bolt
A286
SPS Jenkintown, Pa.

MS 21295
Socket Head Cap Screw, Self-locking
S/S, 1960 Series
SPS Jenkintown, Pa.

MS 24667
82° Flat Head Socket Cap Screw,
Alloy Steel, Cadmium Plated
SPS Jenkintown, Pa.

MS 24671
82° Flat Head Socket Cap Screw,
Stainless Steel, Passivated
SPS Jenkintown, Pa.

MS 24673 thru MS 24678
Socket Head Cap Screw
Drilled Head, 1960 Series
SPS Jenkintown, Pa.

MS 27039
Pan Head Machine Screw,
Structural, Cross Recessed
SPS Santa, Ana, Calif.

MS 35455 thru MS 35460
Alloy Steel, 1936 Series
Unbrako Std. Socket Head Cap Screw
SPS Jenkintown, Pa.

MS 35461
Unbrako Std. Socket Head 1936 Series
Cap Screw - S/S
SPS Jenkintown, Pa.

MS 51017 thru MS 51052
Set Screw, Hexagon Socket
Points: Cup, Flat, Cone, Oval,
Half Dog
SPS Jenkintown, Pa.

MS 51963 thru MS 51966
Set Screw, Hexagon Socket
Points: Cup, Flat
SPS Jenkintown, Pa.

MS 51973 and MS 51974
Set Screw, Hexagon Socket,
Cone Point
SPS Jenkintown, Pa.

MS 51975
Screw, Shoulder-Socket Head,
Hexagon
SPS Jenkintown, Pa.

MS 51976 and MS 51977
Set Screw, Hexagon Socket,
Half-Dog Point
SPS Jenkintown, Pa.

MS 51981 and MS 51982
Set Screw, Hexagon Socket,
Oval Point
SPS Jenkintown, Pa.

MS 90415
Nut, Captive Washer, Steel
Self-locking, 450°
SPS Jenkintown, Pa.

MS 171401 thru MS 171900
Spring Pin
SPS Jenkintown, Pa.

BOLT HEAD MARKINGS

Specifications require that almost all bolts which are made to conform with standard drawings be identified by a specific marking on the bolt head. Shown here are the head markings on 20 different types of bolts. The initials "RM" and "RBM" identify the manufacturer of the bolt.

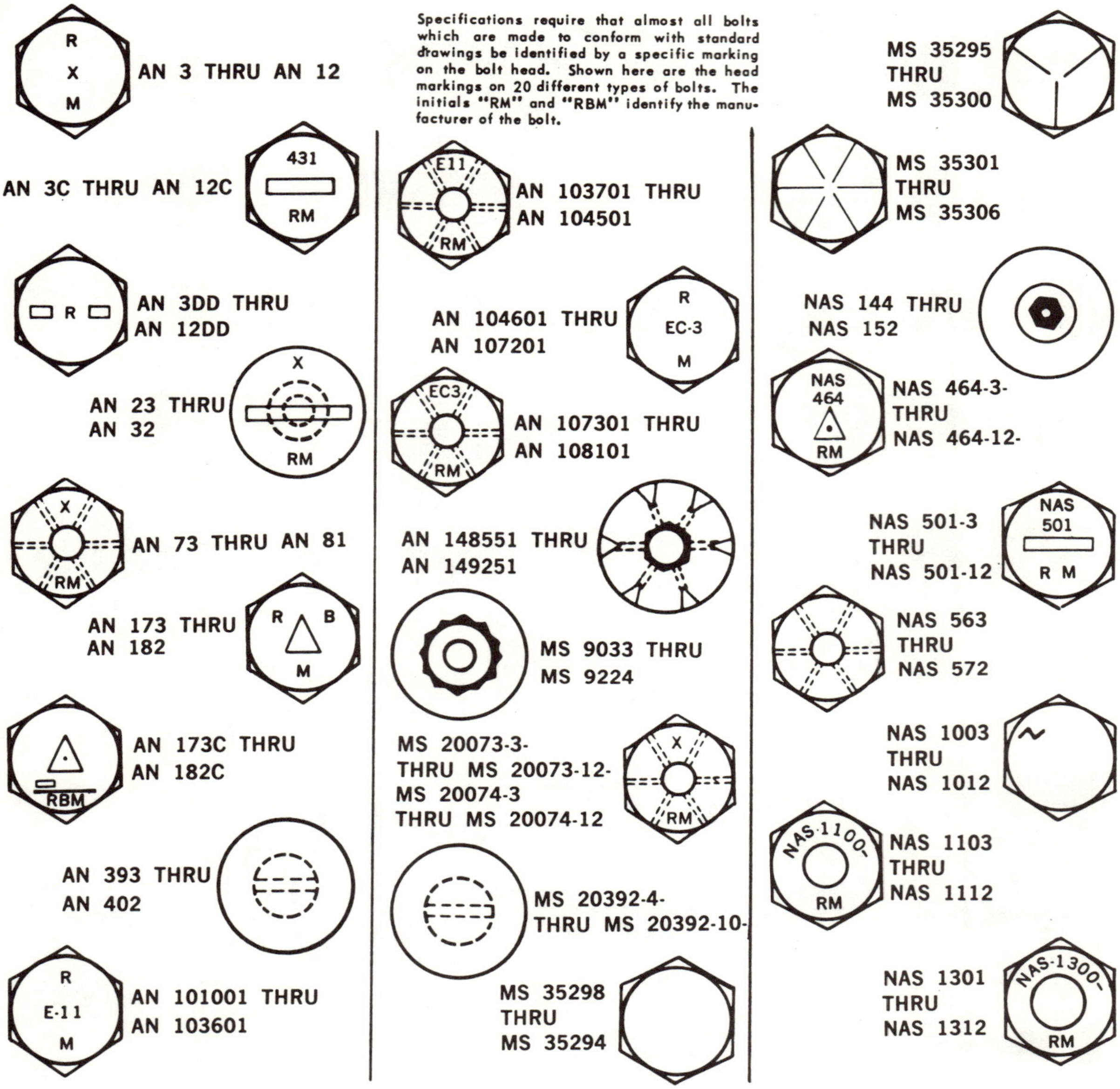

Simplified AN Chart

AN 3 THRU 20-BOLT-AIRCRAFT
AN 21 THRU 36-BOLT-CLEVIS
AN 60 THRU 65-BOLT-HEX HD.
AN 70 BOLT - CARRIAGE
AN 73 THRU 81-BOLT-DRLD. HD.
AN 83 THRU 96-BOLT-CLS. TOL.
AN 116 SHACKLE - SCREW PIN
AN 173 THRU 186-BOLT, CLS. TOL.
AN 310 NUT - CASTLE (AIR FRAME)
AN 315 NUT - PLAIN (AIR FRAME)
AN 316 NUT - CHECK
AN 320 NUT - CASTLE, SHEAR
AN 325 NUT - HEX, PLAIN (NF)
AN 330 NUT - CASTLE (NC)
AN 335 NUT - PLAIN HEX (USS)
AN 340 - LIGHT HEX (USS)
AN 341 NUT HEX, LIGHT, BRASS (ELEC)
AN 345 NUT HEX, LIGHT (SAE)
AN 350 NUT WING
AN 360 NUT - HEX, PLAIN (ENGINE)
AN 363 NUT - HEX, SELF LOCKING (HI-TEMP)
AN 364 NUT - HEX, SELF LOCKING, THIN
AN 365 NUT - HEX, SELF LOCKING
AN 380 PIN - COTTER
AN 381 PIN - COTTER, STAINLESS
AN 420 THRU 470 RIVETS, VARIOUS

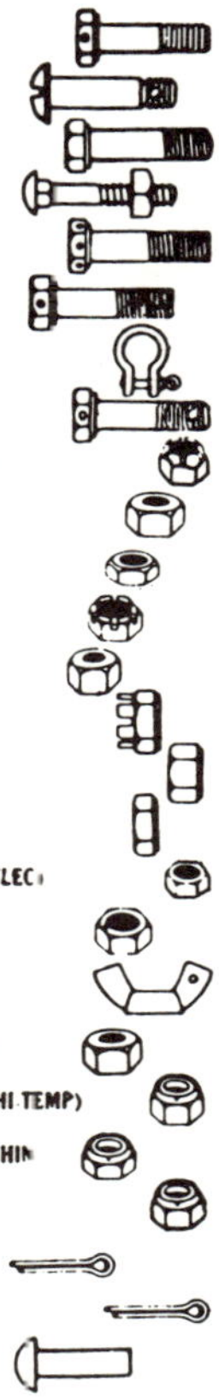

AN 500 SCREW - FILL. HD. (USS)
AN 501 SCREW - FILL. HD (SAE)
AN 502 SCREW - FIL. HD. MCH. ALLOY STL. (SAE)
AN 503 SCREW FIL. HD. MCH. ALLOY STL. (USS)
AN 504 SCREW - RD. HD. SELF TAP MCH.
AN 505 SCREW - 82° FL. HD. MCH. (USS)
AN 506 SCREW - 82° FL. HD. SELF TAP MCH.
AN 507 SCREW - 100° FL. HD MCH. (USS & SAE)
AN 508 SCREW - RD. HD. MCH. BRASS (ELEC)
AN 509 SCREW - 100° FL. HD. MCH. ALLOY STL. (SAE)
AN 510 SCREW - 82° FL. HD. MCH. (SAE)
AN 515 SCREW - RD. HD. MCH. (USS)
AN 520 SCREW - RD. HD. MCH. (SAE)
AN 525 SCREW - WASHER HD. MCH. ALLOY STL.
AN 526 SCREW - TRUSS HD. MCH. (USS & SAE)
AN 530 SCREW - RD. HD. TAPPING TYPE "B"
AN 531 SCREW - FL. HD. TAPPING TYPE "B"
AN 535 SCREW - RD. HD. DRIVE TYPE "U"
AN 540 SCREW - LAG
AN 545 SCREW - WOOD RD. HD.
AN 550 SCREW - WOOD, F. H.
AN 560 SCREW - SQ. HD. SET
AN 565 SCREW - HEADLESS SET
AN 932 PLUG - PIPE
AN 935 WASHER - SPRING LOCK
AN 936 WASHER - EX. & INT. TEETH LOCK
AN 940 WASHER - FLAT, BURR
AN 945 WASHER - FLAT PLAIN (COMMERCIAL)
AN 960 WASHER - FLAT, AIRCRAFT
AN 961 WASHER - FLAT, BRASS (ELEC)

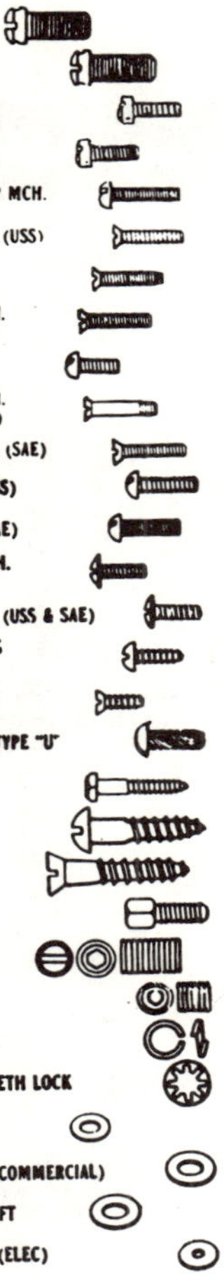

Identification Markings

AN and MS

Identification Marking	Designation of Product Standard	Description of Fastener	Nominal Size Range (in.)	Material	Procurement Specification	Is Mfr's Identification Mark Req'd?	REMARKS
E1	AN150401 thru AN150425	Nut—Hex, Check	.190 to .750	Alloy Steel AMS 6320 or AMS 6325 Cad. Plated	None	No	Marking stamped on side of nut.
	AN150426 thru AN150450	Nut—Hex, Shear, Slotted	.190 to .750				
E4	AN116901-116912 AN116913-116924 AN116925-116960 AN116961-117000 AN117001-117040 AN117041-117080	Screw—Oval, Fillister Screw—Oval, Fillister, Drilled Screw—Oval, Fillister Screw—Oval, Fillister Screw—Oval, Fillister Screw—Oval, Fillister, Drilled	.112 .112 .138 .138 .164 .164	Carbon Steel AMS 5061 Cad. Plated	None	No	Marking on top of head, raised or depressed.
E5	AN122576-112600	Washer—Plain	.094 to .164 I.D.	Carbon Steel AMS 5045 Cad. Plated	None	No	Rubber stamp identification
E9	MS9015 (ASG)	Plug, Machine Thread, "O" Ring Seal	.250 to .625	AMS 5024 Cad. Plated	None	No	Marking on top of head, raised or depressed.
E11	AN101001-101900 AN101901-102800 AN102801-103700 AN103701-104600 AN115401-115600 AN115601-115800 AN115801-116150	Bolt—Hex Head Bolt—Hex Head, Drilled Shank Bolt—Drilled Hex Head, 1 hole Bolt—Drilled Hex Head, 6 holes Screw—Flat Fillister Head Screw—Flat Fillster Head—Drilled Shank Screw—Flat Fillister, Drilled Head	.190 to .750 .190 to .750 .190 to .750 .190 to .750 .190 to .375 .190 to .375 .190 to .375	Alloy Steel AMS 6322 Cad. Plated	None	No	Marking on top of head, raised or depressed.
	AN121501-121525 AN121551-121575	Nut—Hex, Plain Nut—Hex, Castle	.190 to 1.000 .190 to 1.000				Marking stamped on side of nut.
	AN121601-121925	Pin, Straight, Headed	.125 to .500				Marking on top of head, raised or depressed.
	AN125951-126880	Stud—Straight	.250-20 x .250-28 to .625-11 x .625-18				Marking stamped into end of fine thread end.
	AN126881-128362	Stud—Straight, Necked	.375-16 x .375-24 to .625-11 x .625-18				Undersize or oversize limit of coarse thread stamped into other end.

AN and MS

Identification Marking	Designation of Product Standard	Description of Fastener	Nominal Size Range (In.)	Material	Procurement Specification	Is Mfr's Identification Mark Req'd?	REMARKS
E11	AN128363-129292	Stud—Straight, Drilled	.250-20 x .250-28 to .625-11 x .625-18	Alloy Steel AMS 6322 Cad. Plated	None	No	Marking stamped into end of fine thread end. Undersize or oversize limit of coarse thread stamped into other end.
	AN129293-130774	Stud—Straight, Drilled, Necked	.375-16 x .375-24 to .625-11 x .625-18				
	AN150501-152600	Stud—Stepped, 1.5 Dia. Engagement	.250-20 x .190-32 to .625-11 x .5625-18				Marking stamped into end of fine thread end. Undersize or oversize limit of coarse thread stamped into other end.
	AN152601-154700	Stud—Stepped, Drilled, 1.5 Dia. Engagement	.250-20 x .190-32 to .625-11 x .5625-18				
	AN154701-156800	Stud—Stepped, 2 Dia. Engagement	.250-20 x .190-32 to .625-11 x .5625-18				
	AN156801-158900	Stud—Stepped, Drilled, 2 Dia. Engagement	.250-20 x .190-32 to .625-11 x .5625-18				
	AN158901-160100	Stud—Stepped, Necked, 1.5 Dia. Engagement	.4375-14 x .375-24 to .625-11 x .5625-18				
	AN160101-161300	Stud—Stepped, Drilled, Necked, 1.5 Dia. Engagement	.4375-14 x .375-24 to .625-11 x .5625-18				
	AN161301-162500	Stud—Stepped, Necked, 2 Dia. Engagement	.4375-14 x .375-24 to .625-11 x .5625-18				
	AN162501-163700	Stud—Stepped, Drilled, Necked, 2 Dia. Engagement	.4375-14 x .375-24 to .625-11 x .5625-18				
	AN163701-165500	Stud—Stepped, Necked, Wrench Pad, 1.5 Dia. Engagement	.3125-18 x .250-28 to .625-11 x .5625-18				
	AN165501-167300	Stud—Stepped, Drilled, Necked, Wrench Pad, 1.5 Dia. Engagement	.3125-18 x .250-28 to .625-11 x .5625-18				
	AN167301-169100	Stud—Stepped, Necked, Wrench Pad, 2 Dia. Engagement	.3125-18 x .250-28 to .625-11 x .5625-18				
	AN169101-170900	Stud—Stepped, Drilled, Necked, Wrench Pad, 2 Dia. Engagement	.3125-18 x .250-28 to .625-11 x .5625-18				
	MS9088 (ASG) thru MS9094 (ASG)	Bolt—Machine, Drilled 12 Point Head	.190 to .5625				Marking on top of head, raised or depressed.
	MS9122 (ASG) and MS9123 (ASG)	Screw—Machine, Slotted, Hex Head	.190 to .250				
	MS9146 (ASG) thru MS9152 (ASG)	Bolt—Machine, 12 Point Head	.190 to .5625				
	MS9157 (ASG) thru MS9163 (ASG)	Bolt—Machine, 12 Point Head	.190 to .5625	Alloy Steel AMS 6322 Black Oxide Coating			
	MS9169 (ASG) thru MS9175 (ASG)	Bolt—Machine, Drilled 12 Point Head	.190 to .5625				
	MS9183 (ASG) thru MS9186 (ASG)	Bolt—Machine, Drilled 12 Point Head	.138 and .164	Alloy Steel AMS 6322 Cad. Plated			Marking on top of head, raised or depressed.
	MS9189 (ASG) thru MS9192 (ASG)	Bolt—Machine, 12 Point Head	.138 and .164	Alloy Steel AMS 6322 Black Oxide Coating			
E11	MS9303 (ASG)	Stud, Shouldered, Hexagon Wrenching	.190-32 x .190-32	Alloy Steel AMS 6322 Cad. Plated	None	No	Marking stamped on one side of hexagon.
	MS9305 (ASG)	Stud, Shouldered, Hexagon Wrenching	.250-28 x .250-28				
	MS9304 (ASG)	Stud, Shouldered and Stepped, Hexagon Wrenching	.190-32 x .250-28				
	MS9306 (ASG)	Stud, Shouldered and Stepped, Hexagon Wrenching	.250-28 x .3125-24				
	MS9307 (ASG)	Stud, Shouldered and Stepped, Hexagon Wrenching	.250-28 x .375-24				
	MS9308 (ASG)	Stud, Shouldered, Drilled, Hexagon Wrenching	.190-32 x .190-32				
	MS9310 (ASG)	Stud, Shouldered, Drilled, Hexagon Wrenching	.250-28 x .250-28				
	MS9309 (ASG)	Stud, Shouldered and Stepped, Drilled. Hexagon Wrenching	.190-32 x .250-28				
	MS9311 (ASG)	Stud, Shouldered and Stepped, Drilled. Hexagon Wrenching	.250-28 x .3125-24				
	MS9312 (ASG)	Stud, Shouldered and Stepped, Drilled. Hexagon Wrenching	.250-28 x .375-24				
E12	MS9015 (ASG)	Plug, Machine Thread—"O" Ring Seal	.750 to 2.500	Aluminum Alloy AMS 4120	None	No	Marking on top of plug, raised or depressed.
	MS9099 (ASG)	Nut, Hex, Boss Connection	.3125 to 2.500				Marking stamped on one side of nut.
	MS9200 (ASG)	Nut, Plain, Hex, Boss Connection	.3125 to 2.500				
E23	MS172201 (ASG)—MS172235 (ASG)	Washer, Key—Single Bearing Retainer	.3937 to 4.3307 Nom. Brg. Bore Dia.	Alloy Steel AMS 6350 Phosphate Coated	None	Yes	Marking on top face of washer.
	MS172271 (ASG)—MS172320 (ASG)	Washer, Key—Single	.500 to 2.500 Shaft Thread Dia.				
	AN122576-122600	Washer—Plain	.190 to 1.000 I.D.	AMS 6350 Cad. Plated	None	No	Rubber stamp identification
E30	MS172236 (ASG)—MS172270 (ASG)	Nut—Spanner, Bearing Retainer Aeronautical	.3937 to 4.3307 Nom. Brg. Bore Dia.	Alloy Steel AMS 6381 or AMS 6382 Phosphate Coated	None	No	Marking stamped on top face of nut.
	MS172321 (ASG)—MS172370 (ASG)	Nut—Spanner, Aeronautical	.500 to 2.500				
E37	MS9316 (ASG) and MS9317 (ASG)	Screw, Machine—Slotted Hex Head	.190-32 and .250-28	Alloy Steel AMS 6304 Nickel-Cad. Coating	None	No	Marking on top of head, raised or depressed.
EC2	AN121526-121550	Nut—Plain	.190 to 1.000	Corrosion Resistant Steel AMS 5628	None	No	Marking stamped on one side of nut.
	AN121576-121600	Nut—Castle	.190 to 1.000				
EC3	AN104601-105401	Bolt—Hex Head	.190 to .750	Corrosion Resistant Steel In Accordance With AMS 7472	None	No	Marking on top of head, raised or depressed
	AN105501-106400	Bolt—Hex Head, Drilled Shank	.190 to .750				
	AN106401-107300	Bolt—Drilled Hex Head, 1 hole	.190 to .750				
	AN107301-108200	Bolt—Drilled Hex Head, 6 holes	.190 to .750				

AN and MS

Identification Marking	Designation of Product Standard	Description of Fastener	Nominal Size Range (in.)	Material	Procurement Specification	Is Mfr's Identification Mark Req'd?	REMARKS
EC10	MS9100 (ASG)	Nut—Hex, Boss Connection	.3125 to 2.500	Corrosion Resistant Steel AMS 5639	None	No	Marking stamped on one side of nut.
EH5	MS9201 (ASG)	Nut—Plain, Hex, Boss Connection	.3125 to 2.500	Corrosion Resistant Steel AMS 5646 Silver Plated	None	No	Marking stamped on one side of nut.
	MS9404 (ASG)	Plug—Machine Thread	.250 to .750	Corrosion Resistant Steel AMS 5646	None	Yes	Marking on top of plug.
EH19	MS9033 (ASG)—MS9038 (ASG)	Bolt, Machine, 12 Point Head, Heat Resistant	.190 to .500	Steel Alloy AMS 5735	None	No	Marking on top of head, raised or depressed.
	MS9060 (ASG)—MS9066 (ASG)	Bolt, Machine, 12 Point Drilled Head	.190 to .5625				
	MS9177 (ASG) and MS9178 (ASG)	Screw, Machine, 12 Point Head	.138 to .164				
	MS9187 (ASG) and MS9188 (ASG)	Screw, Machine, 12 Point Drilled Head	.138 and .164				
	MS9224 (ASG)	Bolt, Machine, 12 Point Head, Heat Resistant	.5625				
	MS9360 (ASG)	Nut, Plain, Hex, Drilled	.112 to .164	Steel Alloy AMS 5735 Silver Plated			Marking stamped on one side of nut.
—	AN3 thru AN20	Bolt—Machine, Aircraft (Including bolts undrilled, and bolts with drilled head and/or shank)	.190 to 1.250	Corrosion Resistant Steel	MIL-B-6812	Optional	Marking on top of head, raised or depressed.
	AN504	Screw, Tapping, Thread Cutting, Round Head, Machine Screw Thread	.138 to .250		MIL-S-6033	No	Depressed marking on top of head. **Note:** Some screws are marked with two dashes. Only one dash need be visible after slotting.
	AN505	Screw—Machine, Flat Head, 82 Degree, Coarse Thread	.125 to .375		FF-S-92	No	
	AN507	Screw—Machine, Flat Head, 100 Degree	.138 to .375		FF-S-92	No	
	AN509	Screw—Machine, Flat Head, 100 Degree, Structural	.164 to .5625		MIL-S-7839	No	
	AN510	Screw—Machine, Flat Head, 82 Degree, Fine Thread	.125 to .250		FF-S-92	No	
	AN515	Screw—Machine, Round Head, Coarse Thread	.125 to .375		FF-S-92	No	
	AN520	Screw—Machine, Round Head, Fine Thread	.125 to .250		FF-S-92	No	
	AN526	Screw—Machine, Truss Head	.138 to .250		FF-S-92	No	
	AN530	Screw, Tapping, Thread Forming or Thread Cutting, Round Head, Spaced Thread	.138 to .250		MIL-S-6033	No	
	AN531	Screw, Tapping, Thread Forming or Thread Cutting, 82 Degree Flat Head, Spaced Thread	.138 to .250		MIL-S-6033	No	
	MS20427	Rivet, Solid—100 Degree Countersunk Head	.062 to .375		—	No	Depressed marking on top of head. **Note:** Some screws are marked with two dashes. Only one need be visible after slotting.
	MS21207 (ASG)	Screw, Tapping, 100 Degree Flat Head, Spaced Thread	.138 to .250		MIL-S-6033	Optional	
	MS24693	Screw, Machine, Flat Countersunk, 100 Degree, Cross Recessed, UNC-2A and UNF-2A	.138 to .375		FF-S-92	No	
	MS24694	Screw, Machine, Flat Countersunk Head, 100 Degree, Structural, Cross Recessed, UNC-3A and UNF-3A	.164 to .5625		MIL-S-7839	No	
	MS35249	Screw, Machine, Flat Countersunk Head, Slotted, UNC-2A	.138 to .750		FF-S-92	No	
	MS35250	Screw, Machine, Flat Countersunk Head, Slotted, UNF-2A	.138 to .750		FF-S-92	No	
X	AN3 thru AN20	Bolt—Machine, Aircraft	.190 to 1.250	Non-Corrosion Resistant Steel	MIL-B-6812	Optional	Marking on top of head, raised or depressed.
	MS20073 (ASG)	Bolt—Machine, Aircraft, Drilled Head, Fine Thread	.190 to .750		MIL-B-6812	No	
	MS20074 (ASG)	Bolt—Machine, Aircraft, Drilled Head, Coarse Thread	.190 to .750		MIL-B-6812	No	
X X	AN21 thru AN36	Bolt—Clevis	.138 to 1.000	Non-Corrosion Resistant Steel	MIL-B-6812	Optional	Marking on top of head, raised or depressed. **Note:** Only one "X" need be visible after slotting.
	AN502	Screw—Machine, Drilled Fillister Head, Fine Thread	.190 to .3125		MIL-S-7839	No	
	AN503	Screw—Machine, Drilled Fillister Head, Coarse Thread	.138 to .3125		MIL-S-7839	No	
	AN509	Screw—Machine, Flat Head, 100 Degree, Structural	.164 to .5625		MIL-S-7839	No	
=	AN509	Screw—Machine, Flat Head, 100 Degree, Structural	.164 to .5625	Bronze	MIL-S-7839	No	Marking on top of head.
— —	AN3 thru AN20	Bolt—Machine, Aircraft	.190 to 1.250	Aluminum Alloy	MIL-B-6812	Optional	Marking on top of head, raised or depressed.
△ (with x)	AN173 thru AN186	Bolt—Machine, Close Tolerance, Aircraft	.190 to 1.000	Non-Corrosion Resistant Steel	MIL-B-6812	Optional	Marking on top of head, raised or depressed.
-△-	AN173 thru AN186	Bolt—Machine, Close Tolerance, Aircraft	.190 to 1.000	Aluminum Alloy	MIL-B-6812	Optional	Marking on top of head, raised or depressed.
△	AN173 thru AN186	Bolt—Machine, Close Tolerance, Aircraft	.190 to 1.000	Corrosion Resistant Steel	MIL-B-6812	Optional	Marking on top of head, raised or depressed.
△	MS20427	Rivet, Solid—100 Degree Countersunk Head	.062 to .375	Carbon Steel Cad. Plated	None	No	Marking indented into center of top of head.
	MS20613	Rivet, Solid—Universal Head	.062 to .375				
	AN125551 thru AN125700	Rivet, Universal Head	.062 to .375				
+	MS20426	Rivet, Solid, Countersunk 100 Degree Precision Head	.062 to .375	Aluminum Alloy 5056H32	MIL-R-5674	No	Raised on top of head.
— —	MS20426	Rivet, Solid, Countersunk 100 Degree Precision Head	.062 to .375	Aluminum Alloy 2024-T4	MIL-R-5 4	No	Raised on top of head.
•	MS20426	Rivet, Solid, Countersunk 100 Degree Precision Head	.062 to .375	Aluminum Alloy 2017-T4	MIL-R-56 4	No	Raised on top of head.
•	MS20426	Rivet, Solid, Countersunk 100 Degree Precision Head	.062 to .375	Aluminum Alloy 2117-T4	MIL-R-5 4	No	Indented into top of head.
+	MS20602	Rivet, Blind, Chemically Expanded, Protruding Head	.125 to .1875	Aluminum Alloy 5056	MIL-R-7885	No	Indented into center of head.
	MS20603	Rivet, Blind, Chemically Expanded, 100 Degree Flush Head	.125 to .1875				

AN and MS

Identification Marking	Designation of Product Standard	Description of Fastener	Nominal Size Range (in.)	Material	Procurement Specification	Is Mfr's Identification Mark Req'd?	REMARKS
ORANGE COLOR	MS20600	Rivet, Blind, Structural Pull Stem, Self Plugging, Protruding Head	.125 to .250	Aluminum Alloy 5056	MIL-R-7885	No	Parts are dyed orange. Color used is No. 32246 of Fed. Std. No. 595.
	MS20601	Rivet, Blind, Self Plugging, 100 Degree Flush Head	.125 to .250		MIL-R-7885		
	MS20604 (ASG)	Rivet, Blind, Non Structural, Universal Head	.094 to .250		MIL-R-8814		
	MS20605 (ASG)	Rivet, Blind, Non Structural, 100 Degree Flush Head	.094 to .250		MIL-R-8814		
	MS20606 (ASG)	Rivet, Blind, Non Structural Modified Truss Head	.125 to .250		MIL-R-8814		
• •	MS20615	Rivet, Solid, Universal Head	.062 to .250	Nickel-Copper Alloy (Monel)	None	No	Indented into center of head.
B16	MS16187	Studs, Continuous Thread, High Temperature Service (875°-1000°F.)	.250 to 3.000	Alloy Steel	MIL-S-1222	Yes	Marking on one end of stud.
J	MS17293 (WEP)—MS17296 (WEP)	Stud—Stepped, 1.5 Dia. Engagement (MIL-S-8879 Thread)	.250-20 x .190-32 to .4375-14 x .375-24	AMS 6322 Cad. Plated	None	No	Marking stamped into end of fine thread end. Oversize limit of coarse. thread stamped into other end.
	MS17297 (WEP) and MS17298 (WEP)	Stud—Stepped, Drilled, 1.5 Dia. Engagement (MIL-S-8879 Thread)	.3125-18 x .250-28 and .375-16 x .3125-24				
	MS17299 (WEP)—MS17301 (WEP)	Stud—Straight (MIL-S-8879 Thread)	.190-24 x .190-32 to .3125-18 x .3125-24				
	MS17302 (WEP) and MS17303 (WEP)	Stud—Straight, Drilled, (MIL-S-8879 Thread)	.250-28 x .250-20 and .3125-18 x .3125-24				
•K• or **NI-CU K**	MS18116 (Ships)	Bolt, Bolt-Stud, Stud, Stud-Bolt; Nickel-Copper-Aluminum Alloy (K-Monel); Mechanical Properties and Marking	.250 to 4.000	Nickel-Copper-Aluminum Alloy	MIL-B-857	Yes	Marking on top of head of bolt and on end of stud.
1200	MS20033 thru MS20046	Bolt, Machine, Hex Head, 1200° F.	.190 to 1.000	Corrosion and Heat Resistant Steel	MIL-B-7874	No	Marking on top of head.
12	MS20500 (ASG)	Nut, Self Locking, Hexagon, 1200°F.	.190 to 1.000	Corrosion and Heat Resistant Steel	MIL-N-7873	No	Marking on one side of nut.
	MS20501 (ASG)	Nut, Self Locking, Plate, 2 Lug, 1200°F.	.138 to .375				Marking on top of plate.
S12	20503 (U.S. Air Force Std.)	Nut—Self Locking, Free Spinning, 1200°F.	.190 to 1.000	Corrosion and Heat Resistant Steel	MIL-N-8056	No	Marking on one side of nut.
B	MS21044 (ASG)	Nuts, Self Locking, Hexagon—250°F.	.112 to 1.250	Copper Base Alloy	MIL-N-25027	No	Marking on top or one side of nut.
	MS21262 (ASG)	Screws, Self Locking, 250°F, Cylindrical Head, Hex Wrenching Socket, 160 KSI Ftu	.164 to .625	Alloy Steel	FF-S-86 and MIL-F-18240	No	Circle of 6 dots. Raised or depressed on top surface of head.
	MS21295 (ASG)	Screws, Self Locking, 250°F, Cylindrical Head, Hex Wrenching Socket, 80 KSI Ftu	.164 to .625	Corrosion Resistant Steel			
B	MS27039	Screw, Machine—Pan Head, Structural, Cross Recessed	.164 to .500	High Strength Bronze	MIL-S-7839	No	Marking on top of head.
C	MS27039	Screw, Machine—Pan Head, Structural, Cross Recessed	.164 to .500	Corrosion Resistant Steel	MIL-S-7839	No	Marking on top of head.
—	MS27039	Screw, Machine—Pan Head, Structural, Cross Recessed	.164 to .500	Alloy Steel	MIL-S-7839	No	Marking on top of head.
	MS35307	Screw, Cap, Hex Head, UNC-2A	.250 to 1.250	Corrosion Resistant Steel	FF-S-85	No	Marking on top of head. Raised preferred.
	MS35308	Screw, Cap, Hex Head, UNF-2A	.250 to 1.250				
	MS51099	Screw, Cap, Drilled Hex Head UNC-2A	.250 to 1.000				
	MS51100	Screw, Cap, Drilled Hex Head, UNF-2A	.250 to 1.000				
	MS51109	Screw, Cap, Hex Head, Drilled Shank UNC-2A	.250 to 1.000				
	MS51110	Screw, Cap, Hex Head, Drilled Shank	.250 to 1.000				
	MS35297	Screw, Cap, Hex Head, UNC-2A Cad. or Zinc Plated	.250 to 2.500	Medium Carbon Steel, Heat Treated (Mechanical Properties are Equivalent to SAE Grade 5)	FF-S-85	No	Marking on top of head. Raised preferred.
	MS35299	Screw, Cap, Hex Head, UNC-2A Phosphate Finish	.250 to 2.500				
	MS51095	Screw, Cap, Drilled Hex Head, UNC-2A, Cad. Plated	.250 to 1.000				
	MS51096	Screw, Cap, Drilled Hex Head, UNF-2A, Cad. Plated	.250 to 1.000				
	MS51105	Screw, Cap, Hex Head, Drilled Shank, UNC-2A, Cad. Plated	.250 to 1.000				
	MS51106	Screw, Cap, Hex Head, Drilled Shank, UNF-2A, Cad. Plated	.250 to 1.000				
	MS90726	Screw, Cap, Hex Head, UNF-2A Cad. Plated	.250 to 1.500				
	MS51107	Screw, Cap, Hex Head, Drilled Shank, UNC-2A	.250 to 1.000	Alloy Steel (Mechanical Properties are Equivalent to SAE Grade 8)	FF-S-85	No	Marking on top of [redacted] Raised prefer[redacted].
	MS51108	Screw, Cap, Hex Head, Drilled Shank, UNF-2A	.250 to 1.000				
	MS90727	Screw, Cap, Hex Head, UNF-2A Cad. Plated	.250 to 1.500				
	MS90728	Screw, Cap, Hex Head, UNC-2A Cad. Plated	.250 to 2.500				

AN and MS

The following AN and MS items are marked with their own part numbers. AN and MS part numbers are the number of the applicable AN or MS standard followed by supplementary numerals and letters designating product size, material, and coating. The first numerals give the number of the standard and reference to this document will permit completion of identification of the part.

Identification Marking	Designation of Product Standard	Description of Fastener	Nominal Size Range (in.)	Material	Procurement Specification	Is Mfr's Identification Mark Req'd?	REMARKS
	AN42 thru AN49	Bolt, Eye	.190 to .5625	As Required	MIL-B-6812	No	Marking on side of head.
	AN5023	Nut, Propeller Shaft Thread Shipping	1.375 to 6.000	As Required	None	No	Marking idented on one face of nut.
	MS9197 (ASG)	Nut, Tube Coupling	.5625 to 3.500	AMS 5646 Silver Plated	None	Yes	Marking on one side of nut.
	MS9198 (ASG)	Nut, Tube Coupling	.5625 to 3.500	AMS 5646	None	Yes	Marking on one side of nut.
	MS9199 (ASG)	Nut, Tube Coupling	.750 to 3.500	Aluminum AMS 4121	None	Yes	Marking on one side of nut.
	MS9356 (ASG)	Nut, Plain Hex	.112 to 1.000	AMS 5735	None	Yes	Marking on side of nut.
	MS9357 (ASG)	Nut, Plain, Hex	.112 to 1.000	AMS 5735 Silver Plated	None	Yes	Marking on side of nut.
	MS9358 (ASG)	Nut, Castellated, Hex	.190 to 1.000	AMS 5735	None	Yes	Marking on side of nut.
	MS9359 (ASG)	Nut, Castellated, Hex	.190 to 1.000	AMS 5735 Silver Plated	None	Yes	Marking on side of nut.
	MS9360 (ASG)	Nut, Plain, Hex, Drilled	.190 to 1.000	AMS 5735 Silver Plated	None	Yes	Marking on side of nut.
	MS9361 (ASG)	Nut, Plain, Hex, Check	.190 to 1.000	AMS 5735	None	Yes	Marking on side of nut.
	MS9362 (ASG)	Nut, Plain, Hex, Check	.190 to 1.000	AMS 5735 Silver Plated	None	Yes	Marking on side of nut.
THESE ITEMS MARKED WITH PART NUMBERS	MS9363 (ASG)	Nut, Slotted, Hex, Shear	.190 to 1.000	AMS 5735	None	Yes	Marking on side of nut.
	MS9364 (ASG)	Nut, Slotted, Hex, Shear	.190 to 1.000	AMS 5735 Silver Plated	None	Yes	Marking on side of nut.
	MS9397 (ASG) thru MS9402 (ASG)	Bolt, Tee Head, Chamfered, UNJF-3A	.190 to .500	AMS 6322 Cad. Plated	None	Yes	Marking on top of head.
	MS9404 (ASG)	Plug, Machine Thread	.750 to 2.500	AMS 5646	None	Yes	Marking on top of head.
	MS9432 (ASG) thru MS9437 (ASG)	Bolt, Tee Head, Chamfered, UNJF-3A	.190 to .500	AMS 5735	None	Yes	Marking on top of head.
	MS9440 (ASG) thru MS9448 (ASG)	Bolt, Machine, Drilled Hex Head, UNJF-3A	.190 to .750	AMS 6304 Diffused Nickel Cadmium Plated	None	Yes	Marking on top of head.
	MS9449 (ASG) and MS9450 (ASG)	Screw, Machine, Hex Head, UNJF-3A	.138 and .164	AMS 6304 Diffused Nickel Cadmium Plated	None	Yes	Marking on top of head.
	MS9451 (ASG) thru MS9459 (ASG)	Bolt, Machine, Hex Head, UNJF-3A	.190 to .750	AMS 6304 Diffused Nickel Cadmium Plated	None	Yes	Marking on top of head.
	MS17354 (WEP)	Nut—Plain, Hex, Boss Connection, (MIL-S-8879 Thread)	.3125 to 2.500	Aluminum AMS 4121	None	Yes	Marking indented on one side of nut.
	MS17731 (ASG)	Fastener, Rotary, Quick-Operating, Flush Head, Floating Type	.375	Steel Cad. Plated	MIL-F-22978	Yes	Marking on top of head.
	MS17732 (ASG)	Fastener, Rotary, Quick-Operating, Protruding Head, Floating Type	.375	Steel Cad. Plated	MIL-F-22978	Yes	Marking on top of head.
	MS20004 thru MS20024	Bolt, Internal Wrenching	.250 to 1.500	Alloy Steel Cad. Plated	MIL-B-7838	No	Marking on side or top of head.
	MS21025	Nut, Castellated, Bearing Retainer	1.000 to 4.4375	Alloy Steel Cad. Plated	MIL-N-6034	No	Marking on side or top of head.
	MS21250 (ASG)	Bolt, 12 Point, External Wrenching	.250 to 1.500	Alloy Steel Cad. Plated	MIL-B-8831	No	Marking indented on top of head.
	MS21277 (ASG) and MS21278 (ASG)	Screw, Machine, 12 Point Head	.138 and .164	AMS 5735	None	Yes	Marking on top of head, raised or depressed.
	MS21279 (ASG) thru MS21285 (ASG)	Bolt, Machine, 12 Point Head	.190 to .5625	AMS 5735	None	Yes	Marking on top of head, raised or depressed.
	MS21286 (ASG) and MS21287 (ASG)	Screw, Machine, Drilled 12 Point Head	.138 and .164	AMS 5735	None	Yes	Marking on top of head, raised or depressed.
THESE ITEMS MARKED WITH PART NUMBERS	MS21288 (ASG) thru MS21294 (ASG)	Bolt, Machine, Drilled 12 Point Head	.190 to .5625	AMS 5735	None	Yes	Marking on top of head, raised or depressed.
	MS24592 (USAF)	Nut, Tube Coupling	.375 to 2.500	Steel or Aluminum Alloy	MIL-A-5070	Yes	Marking on side of nut.
	MS25080 (ASG)	Terminal Stud	.138 to .375	Brass Cad. or Silver Plated	None	No	Marking on face of wrenching flange.
	MS27073 (USAF)	Nut, Tube Coupling—Swivel	.375 to 1.625	Corrosion Resistant Steel or Aluminum Alloy	MIL-F-27272	Yes	Marking on side of nut.
	MS39078 (ORD)	Nut, Coupling, Electrical Conduit	.500 to 3.000	Aluminum Alloy	None	No	Marking on top or side of nut.
	MS39081 (ORD)	Nut, Coupling, Electrical Conduit, Plain Hexagon	.500 to 3.000	Aluminum Alloy	None	No	Marking on top or side of nut.

AMS

Code Marking	AMS Material Specification	Title of Specification	Alloy	Similar Specification
E1	6320	Bars and Forgings—0.55 Ni, 0.5 Cr, 0.25 Mo (0.33-0.38 C)	8735	MIL-S-6098 Cond. C
E1	6325	Bars and Forgings—0.55 Ni, 0.5 Cr, 0.25 Mo (0.38-0.43 C) Heat Treated 105,000 TS	8740	MIL-S-6049
E4	5061	Bars and Wire—0.08-0.20 C	—	—
E5	5045	Sheet and Strip—0.25 Max. C, Cold Rolled, Hard Temper	1020	QQ-S-636, Cond. 1
E9	5024	Bars, Forgings and Mechanical Tubing—0.32-0.39 C, 1.5 Mn, Free Cutting	1137	QQ-S-633 Comp 1137
E11	6322	Bars and Forgings—0.55 Ni, 0.5 Cr, 0.25 Mo (0.38-0.43 C)	8740	MIL-S-6049 Cond. C
E12	4120	Bars, Rods, Wire, Rolled—4.5 Cu, 1.5 Mg, 0.6 Mn	2024-T4	QQ-A-268, Temp. T4
E23	6350	Plate, Sheet and Strip—Annealed—0.95 Cr, 0.2 Mo (0.28-0.33 C)	4130	MIL-S-18729
E30	6381	Tubing, Mechanical—0.95 Cr, 0.2 Mo (0.38-0.42 C)	4140	—
E30	6382	Bars and Forgings—0.95 Cr, 0.2 Mo (0.38-0.43 C)	4140	MIL-S-5626
E37	6304	Bars, Forgings and Mechanical Tubing—1 Cr, 0.55 Mo, 0.3 V (0.40-0.50 C)	17-22A	—
EC2	5628	Bars and Forgings—16 Cr, 2 Ni	51431	QQ-S-763 Class 431
EC10	5639	Bars, Forgings, Mechanical Tubing and Rings—19 Cr, 9 Ni	30304	QQ-S-763 Class 304
EH5	5646	Bars, Forgings, Mechanical Tubing and Rings—18 Cr, 11 Ni (Cb+Ta)	30347	QQ-S-763 Class 347
EH19	5735	Bars, Forgings, Mechanical Tubing and Rings—15 Cr, 26 Ni, 1.3 Mo, 2.1 Ti, 0.3 V	A-286	—

MIL

Identification Marking	Reference Specification	Description of Fastener	Nominal Size Range (in.)	Material	Is Mfr's Identification Symbol Required?	REMARKS
	MIL-B-857A (Ships)	Bolts and Studs	¼ to 1½	Medium Carbon Steel, Heat Treated	No	Marking on heads of bolts and on nut end of studs. Properties equivalent to SAE Grade 5.
	MIL-B-857A (Ships)	Bolts and Studs	¼ to 1½	Alloy Steel	No	Marking on heads of bolts and on nut end of studs. Properties equivalent to SAE Grade 8.
H	MIL-S-1222F (Ships)	Nuts for High Temperature Service	¼ to 2½	Carbon Steel	Yes	Marking die stamped on top face of nut.
4	MIL-S-1222F (Ships)	Nuts for High Temperature Service	¼ to 2½	Alloy Steel	Yes	Marking die stamped on top face of nut.
3	MIL-R-1150A	Rivets, Solid	.063 to .750	Alum. Alloy 3S-F	No	Marking raised on end of rivet shank.
5	MIL-R-1150A	Rivets, Solid	.063 to 1.250	Alum. Alloy 53S-T61	No	Marking raised on end of rivet shank.
6	MIL-R-1150A	Rivets, Solid	.063 to 1.250	Alum. Alloy 61S-T6	No	Marking raised on end of rivet shank.
+	MIL-R-1223A	Rivets, Solid	.250 to 1.250	Carbon Steel	Optional	Marking raised at center of head.
	MIL-R-1223A	Rivets, Solid	.250 to 1.250	Carbon Steel	Optional	Marking raised on top of head at outer periphery.

NAS

Identification Marking	Designation of Product Standard	Description of Fastener	Nominal Size Range (in.)	Material	Procurement Specification	Is Mfr's Identification Mark Req'd?	REMARKS
THESE ITEMS MARKED WITH PART NUMBERS	NAS144 thru NAS158	Bolt—Internal Wrenching	.250 to 1.125	Alloy Steel Cad. Plated	NAS159	Yes	Marking on side of head, raised or depressed.
	NAS172	Bolt—Internal Wrenching	1.250	Alloy Steel Cad. Plated	NAS159	Yes	Marking on side of head, raised or depressed.
	NAS174	Bolt—Internal Wrenching	1.375	Alloy Steel Cad. Plated	NAS159	Yes	Marking on side of head, raised or depressed.
	NAS176	Bolt—Internal Wrenching	1.500	Alloy Steel Cad. Plated	NAS159	Yes	Marking on side of head, raised or depressed.
	NAS333 thru NAS340	Bolt—100 Degree, Close Tolerance, High Strength	.190 to .625	Alloy Steel	NAS498 and MIL-B-7838	No	Marking on top of head, raised or depressed.
	NAS428	Bolt—Adjusting Crowned Hex Head	.190 to .375	Alloy Steel	MIL-B-6812	Yes	Marking on top of head, raised or depressed.
	NAS464	Bolt—Shear, Close Tolerance	.190 to 1.000	Alloy Steel	NAS498	No	Marking on top of head, raised or depressed.
	NAS467	Bolt—Hex Head, Electrical	.250 to .500	Brass Per QQ-N-35, Comp. 1	None	No	Marking on top of head, raised or depressed.
	NAS501	Bolt—Hex Head	.190 to 1.250	Corrosion Resistant Steel (Type 321 or 347)	None	No	Marking on top of head, raised or depressed.
	NAS517	Screw—100 Degree, Close Tolerance, Flat Head, 160,000 psi.	.164 to .500	Alloy Steel Cad. Plated	MIL-S-7839 and NAS498	No	Marking on top of head, indented only.
	NAS560	Screw—Machine, Non-Magnetic, High Temperature, Structural, 100 Degree Flush Head	.164 to .5625	Corrosion Resistant Steel	Noted on Std.	Optional	Marking on top of head, indented only.
	NAS563 thru NAS572	Bolt—Hex Head, Threaded Full Length	.190 to .750	Alloy Steel Cad. Plated	MIL-B-6812	Optional	Marking on top of head, raised or depressed.
	NAS593	Nut—Rigid Tube Connector	.4375 to 3.000	Aluminum Alloy	NAS597	No	Marking on one flat of nut.
	NAS596	Nut—Rigid Tube Connector	.4375 to 4.500	Corrosion Resistant Steel	NAS597	No	Marking on one flat of nut.
	NAS600 thru NAS606	Screw, Machine—Aircraft, Pan Head, Phillips Recess, Threaded Full Length	.112 to .375	Alloy Steel	MIL-S-7839	Optional	Marking on top of head, raised or depressed.
	NAS623	Screw, Machine—Aircraft, Pan Head, Phillips Recess, Short Thread	.164 to .375	Alloy Steel	NAS498	Yes	Marking on top of head, raised or depressed.
	NAS624 thru NAS644	Bolt—12 Point, External Wrenching, 180,000 psi.	.250 to 1.500	Alloy Steel	NAS496	Yes	Marking on top of head, raised or depressed.
	NAS653 thru NAS658	Bolt—Hex Head, Close Tolerance, Short Thread	.190 to .500	Titanium Alloy	NAS621	Yes	Marking on top of head, raised or depressed.
	NAS663 thru NAS668	Bolt, 100 Degree Flat Head, Close Tolerance	.190 to .500	Titanium Alloy	NAS621	Yes	Marking on top of head, depressed only.
	NAS673 thru NAS678	Bolt—Hex Head, Close Tolerance	.190 to .500	Titanium Alloy	NAS621	Yes	Marking on top of head, raised or depressed.
	NAS1003 thru NAS1020	Bolt—Machine, Hex Head	.190 to 1.250	Corrosion Resistant Steel	None	Optional	Marking on top of head, raised or depressed.
	NAS1083 thru NAS1088	Bolt—100 Degree Flat Head, Close Tolerance, Short Thread,	.190 to .500	Titanium Alloy	NAS621	Yes	Marking on top of head, depressed only.

NAS

Identification Marking	Designation of Product Standard	Description of Fastener	Nominal Size Range (in.)	Material	Procurement Specification	Is Mfr's Identification Mark Req'd?	REMARKS
THESE ITEMS MARKED WITH PART NUMBERS	NAS1101	Screw, Flat Fillister Head Torq-Set, Full Thread	.190 to .375	Alloy Steel or Corrosion Resistant Steel	NAS498 and AMS7478	Yes	Marking on top of head, raised or depressed.
	NAS1102	Screw, 100 Degree Flat Head, Torq-Set, Full Thread	.190 to .375	Alloy Steel or Corrosion Resistant Steel	NAS498 and AMS7478	Yes	Marking on top of head, raised or depressed.
	NAS1103 thru NAS1120	Bolt, Hex Head, Close Tolerance, Short Thread, 160,000 psi.	.190 to 1.250	Alloy Steel Cad. Plated	NAS498	Yes	Marking on top of head, raised or depressed.
	NAS1121 thru NAS1128	Bolt, Flat Fillister Head, Torq-Set, Close Tolerance, Short Thread	.190 to .500	Alloy Steel or Corrosion Resistant Steel	NAS498 and AMS7478	Yes	Marking on top of head, raised or depressed.
	NAS1131 thru NAS1138	Bolt, Pan Head, Torq-Set, Close Tolerance, Short Thread	.190 to .500	Alloy Steel, Corrosion Resistant Steel, or Titanium Alloy	NAS498, AMS7478, or NAS621	Yes	Marking on top of head, raised or depressed.
	NAS1141 thru NAS1148	Bolt, Brazier Head, Torq-Set, Close Tolerance, Short Thread	.190 to .500	Alloy Steel, or Corrosion Resistant Steel	NAS498 or AMS7478	Yes	Marking on top of head, raised or depressed.
	NAS1151 thru NAS1158	Bolt, 100 Degree Flat Head, Torq-Set, Close Tolerance, Short Thread	.190 to .500	Alloy Steel, Corrosion Resistant Steel, or Titanium Alloy	NAS498, AMS7478, or NAS621	Yes	Marking on top of head, raised or depressed.
	NAS1161 thru NAS1168	Bolt, Self Locking, 100 Degree Flat Head, Torq-Set, Close Tolerance	.190 to .500	Alloy Steel Cad. Plated	NAS498 and MIL-F-18240	Yes	Marking on top of head, raised or depressed. Marking includes circle of six (minimum) dots.
	NAS1171 thru NAS1178	Bolt, Self Locking, Pan Head, Torq-Set, Close Tolerance	.190 to .500	Alloy Steel Cad. Plated	NAS498 and MIL-F-18240	Yes	Marking on top of head, raised or depressed. Marking includes circle of six (minimum) dots.
	NAS1181 thru NAS1188	Bolt, Self Locking, Flat Fillister Head, Torq-Set, Close Tolerance	.190 to .500	Alloy Steel or Corrosion Resistant Steel	NAS498, or AMS7478 and MIL-F-18240	Yes	Marking on top of head, raised or depressed. Marking includes circle of six (minimum) dots.
	NAS1189	Bolt, Self Locking, 100 Degree Flat Head, Full Thread	.190 to .375	Alloy Steel or Corrosion Resistant Steel	NAS498, or AMS7478 and MIL-F-18240	Yes	Marking on top of head, raised or depressed. Marking includes circle of six (minimum) dots.
	NAS1190	Bolt, Self Locking, Pan Head, Full Thread	.190 to .375	Alloy Steel or Corrosion Resistant Steel	NAS498, or AMS7478 and MIL-F-18240	Yes	Marking on top of head, raised or depressed. Marking includes circle of six (minimum) dots.
	NAS1191	Bolt, Self Locking, Flat Fillister Head, Full Thread	.190 to .375	Alloy Steel or Corrosion Resistant Steel	NAS498, or AMS7478 and MIL-F-18240	Yes	Marking on top of head, raised or depressed. Marking includes circle of six (minimum) dots.
	NAS1202 thru NAS1210	Bolt, 100 Degree Flat Head, Close Tolerance, Phillips Recess, 160,000 psi.	.164 to .625	Alloy Steel Cad. Plated	NAS498	Yes	Marking on top of head, raised or depressed. Marking includes circle of six (minimum) dots.
THESE ITEMS MARKED WITH PART NUMBERS	NAS1216	Screw, Pan Head, Hi-Torque Recess, Full Thread	.138 to .375	Various Steel Alloys Cad. Plated	NAS498	Yes	Marking on top of head, raised or depressed.
	NAS1217	Screw, Pan Head, Hi-Torque Recess, Short Thread	.164 to .375	Various Steel Alloys Cad. Plated	NAS498	Yes	Marking on top of head, raised or depressed.
	NAS1218	Screw, Pan Head, Hi-Torque Recess, Long Thread	.138 to .375	Various Steel Alloys Cad. Plated	NAS498	Yes	Marking on top of head, raised or depressed.
	NAS1219	Screw, 100 Degree Flat Head, Hi-Torque Recess, Full Thread	.138 to .375	Various Steel Alloys Cad. Plated	NAS498	Yes	Marking on top of head, raised or depressed.
	NAS1220	Screw—100 Degree Flat Head, Hi-Torque Recess, Short Thread	.164 to .375	Various Steel Alloys Cad. Plated	NAS498	Yes	Marking on top of head, raised or depressed.
	NAS1221	Screw—100 Degree, Flat Head, Hi-Torque Recess, Long Thread	.164 to .375	Various Steel Alloys Cad. Plated	NAS498	Yes	Marking on top of head, raised or depressed.
	NAS1223 thru NAS1235	Bolt, Self Locking, Hex Head, Close Tolerance	.190 to 1.250	Alloy Steel, or Corrosion Resistant Steel	MIL-B-7838, AMS7478, and MIL-F-18240	Yes	Marking includes circle of six (minimum) dots.
	NAS1243 thru NAS1250	Bolt, .0156 Oversize Shank, 100 Degree Flat Head, Hi-Torque Recess, Close Tolerance, Short Thread	.190 to .625	Alloy Steel Cad. Plated	NAS498	Yes	Marking on top of head, raised or depressed.
	NAS1253 thru NAS1260	Bolt, .0312 Oversize Shank, 100 Degree Flat Head, Hi-Torque Recess, Close Tolerance, Short Thread	.190 to .625	Alloy Steel Cad. Plated	NAS498	Yes	Marking on top of head, raised or depressed.
	NAS1261 thru NAS1265	Bolt, Hex Head, Close Tolerance, Short Thread	.5625 to 1.000	Titanium Alloy	NAS621	Yes	Marking on top of head, raised or depressed.
	NAS1266 thru NAS1270	Bolt, Hex Head, Close Tolerance	.5625 to 1.000	Titanium Alloy	NAS621	Yes	Marking on top of head, raised or depressed.
	NAS1271 thru NAS1280	Bolt, 12 Point, External Wrenching	.250 to 1.000	Titanium Alloy	NAS621	Yes	Marking on top of head, raised or depressed.
	NAS1292 thru NAS1296	Pin—Threaded, Shear, 100 Degree Flat Head	.164 to .375	AMS6485 or AMS5712	NAS1289 or NAS1290	Yes	Marking on top of head, depressed only. "NAS" may be omitted.
	NAS1297	Bolt—Hex Head Shoulder	.138 to .625	Alloy Steel Cad. Plated	None	Yes	Marking on top of head, raised or depressed.

NAS

Identification Marking	Designation of Product Standard	Description of Fastener	Nominal Size Range (in.)	Material	Procurement Specification	Is Mfr's Identification Mark Req'd?	REMARKS
	NAS1298	Screw—Brazier Head, Shoulder	.138 to .4375	Alloy Steel Cad. Plated	None	Yes	Marking on top of head, raised or depressed.
	NAS1299	Screw—100 Degree Flat Head, Shoulder	.138 to .4375	Alloy Steel Cad. Plated	None	Yes	Marking on top of head, raised or depressed.
	NAS1303 thru NAS1320	Bolt—Hex Head, Close Tolerance, 160,000 psi.	.190 to 1.250	Alloy Steel Cad. Plated	MIL-B-7838	Yes	Marking on top of head, raised or depressed.
	NAS1322 thru NAS1326	Pin—Threaded, Shear, Protruding Head	.164 to .375	AMS6485 or AMS5712	NAS1289 or NAS1290	Yes	Marking on top of head, depressed only. "NAS" may be omitted.
	NAS1402 thru NAS1406	Screw, Machine, Pan Head, Phillips Recess, 160,000 psi.	.164 to .375	Alloy Steel Cad. Plated	MIL-B-7838	Yes	Marking on top of head, raised or depressed.
	NAS1503 thru NAS1510	Bolt, 100 Degree Flat Head, Close Tolerance, Short Thread, 160,000 psi.	.190 to .625	Alloy Steel Cad. Plated	NAS498	Optional	Marking on top of head, depressed only.
	NAS1578	Bolt, Shear, Pan Head, 1200°F	.190 to .500	U-212 per NAS1597	NAS1589	Yes	Marking on top of head, raised or depressed.
	NAS1579	Screw, Pan Head, Full Thread, 1200°F	.190 to .375	U-212 per NAS1597	NAS1589	Yes	Marking on top of head, raised or depressed.
	NAS1580	Bolt, Tension, Flush Head, 1200°F	.190 to .625	U-212 per NAS1597	NAS1589	Yes	Marking on top of head, raised or depressed.
	NAS1581	Bolt, Shear, Reduced, 100 Degree Flush Head, 1200°F	.190 to .625	U-212 per NAS1597	NAS1589	Yes	Marking on top of head, raised or depressed.
THESE ITEMS MARKED WITH PART NUMBERS	NAS1582	Screw, Flush Head, Full Thread, 1200°F	.190 to .375	U-212 per NAS1597	NAS1589	Yes	Marking on top of head, raised or depressed.
	NAS1583	Pin, 100 Degree Flat Head, Hi Shear Rivet, 1200°F	.125 to .375	Inconel-X (AMS5667)	NAS1590	Yes	Marking on top of head, raised or depressed.
	NAS1584	Pin, Flat Head, Hi Shear Rivet, 1200°F	.125 to .375	Inconel-X (AMS5667)	NAS1590	Yes	Marking on top of head, raised or depressed.
	NAS1586	Bolt, Tension, 12 Point External Wrenching, 1200°F	.250 to 1.250	U-212 per NAS1597	NAS1589	Yes	Marking on top of head, raised or depressed.
	NAS1587	Washer, Plain and Countersunk, 1200°F	.190 to 1.250 I.D.	Corrosion Resistant Steel	None	No	Marking indented into one face.
	NAS1588	Bolt, Shear, Hex Head, 1200°F	.190 to 1.000	U-212 per NAS1597	NAS1589	Yes	Marking on top of head, raised or depressed.
	NAS1603 thru NAS1610	Bolt, .0312 Oversize Shank, 100 Degree Flat Head, 160,000 psi.	.190 to .625	Alloy Steel Cad. Plated	NAS498	Yes	Marking depressed on top of head.
	NAS1620 thru NAS1628	Screw, 100 Degree Flat Head, Torq-Set	.138 to .500	Alloy Steel, or Corrosion Resistant Steel	NAS498 or AMS7478	Yes	Marking depressed on top of head.
	NAS1630 thru NAS1634	Screw, Pan Head, Torq-Set	.138 to .250	Alloy Steel, or Corrosion Resistant Steel	MIL-S-7839 or AMS7478	Yes	Marking on top of head, raised or depressed.
	NAS1703 thru NAS1710	Bolt, .0516 Oversize Shank, 100 Degree Flat Head, 160,000 psi.	.190 to .625	Alloy Steel Cad. Plated	NAS498	Yes	Marking depressed on top of head.
	NAS1806 thru NAS1816	Rivet—Hi Shear, Flat Head, Interference Fit	.1875 to .500	Titanium Alloy	NAS621	Yes	Marking depressed on top of head. "NAS" may be omitted.
	NAS1906 thru NAS1916	Rivet—Hi Shear, 100 Degree Flat Head, Interference Fit	.1875 to .500	Titanium Alloy	NAS621	Yes	Marking depressed on top of head. "NAS" may be omitted.
	NAS2006 thru NAS2010	Bolt—Lock, Tension, Protruding Head, Pull Type	.1875 to .3125	Titanium Alloy	NAS621	Yes	Marking depressed on top of head. "NAS" may be omitted.
	NAS2106 thru NAS2110	Bolt—Lock, Tension, 100 Degree Flat Head, Pull Type	.1875 to .3125	Titanium Alloy	NAS621	Yes	Marking depressed on top of head. "NAS" may be omitted.
	NAS2206 thru NAS2210	Bolt—Lock, Tension, Protruding Head, Stump Type	.1875 to .3125	Titanium Alloy	NAS621	Yes	Marking depressed on top of head. "NAS" may be omitted.
	NAS2306 thru NAS2310	Bolt—Lock, Tension, 100 Degree Flat Head, Stump Type	.1875 to .3125	Titanium Alloy	NAS621	Yes	Marking depressed on top of head. "NAS" may be omitted.
THESE ITEMS MARKED WITH PART NUMBERS	NAS2406 thru NAS2412	Bolt—Lock, Shear, Protruding Head, Pull Type	.1875 to .375	Titanium Alloy	NAS621	Yes	Marking depressed on top of head. "NAS" may be omitted.
	NAS2506 thru NAS2512	Bolt—Lock, Shear, 100 Degree Flat Head, Pull Type	.1875 to .375	Titanium Alloy	NAS621	Yes	Marking depressed on top of head. "NAS" may be omitted.
	NAS2606 thru NAS2612	Bolt—Lock, Shear, Protruding Head, Stump Type	.1875 to .375	Titanium Alloy	NAS621	Yes	Marking depressed on top of head. "NAS" may be omitted.
	NAS2706 thru NAS2712	Bolt—Lock, Shear, 100 Degree Flat Head, Stump Type	.1875 to .375	Titanium Alloy	NAS621	Yes	Marking depressed on top of head. "NAS" may be omitted.
	NAS2803 thru NAS2810	Bolt, 100 Degree Flat Head, Torq-Set, Close Tolerance, 180,000 psi.	.190 to .625	Alloy Steel Cad. Plated	NAS498	Yes	Marking on top of head, raised or depressed.
	NAS2903 thru NAS2920	Bolt, .0156 Oversize Shank, Hex Head, Close Tolerance, 160,000 psi.	.190 to 1.250	Alloy Steel Cad. Plated	MIL-B-7838	Yes	Marking on top of head, raised or depressed.
	NAS3003 thru NAS3020	Bolt, .0312 Oversize Shank, Hex Head, Close Tolerance, 160,000 psi.	.190 to 1.250	Alloy Steel Cad. Plated	MIL-B-7838	Yes	Marking on top of head, raised or depressed.
□	NAS525	Rivet—Hi Shear, 100 Degree Flat Head, Close Tolerance	.125 to .625	Various Steel Alloys	NAS498	Optional	Depressed marking at center of head.
□	NAS529	Rivet—Hi Shear, Flat Head Close Tolerance	.125 to .625	Various Steel Alloys	NAS498	Optional	Depressed marking at center of head.
-_	NAS662	Screw, Machine, 100 Degree Flat Head, Plain and Self Locking	.060 to .099	Corrosion Resistant Steel	FF-S-92	Yes	Marking on top of head, raised or depressed. Locking screws also marked with 6 dots.
=	NAS662	Screw, Machine, 100 Degree Flat Head, Plain and Self Locking	.060 to .099	Brass	FF-S-92	Yes	Marking on top of head, raised or depressed. Locking screws also marked with 6 dots.
C	NAS679	Nut, Self Locking, Hex, Low Height, 550°F, 800°F	.112 to .4375	Corrosion Resistant Steel	MIL-N-25027	No	Marking on side of nut.
C	NAS1021	Nut, Self Locking, Hex, Regular Height, 800°F	.112 to 1.250	As Required	MIL-N-20527	No	Marking on side of nut.
C	NAS1022	Nut, Self Locking, Hex, Shear, 800°F	.138 to 1.250	As Required	MIL-N-25027	No	Marking on side of nut.
C	NAS1291	Nut—Self Locking, Hex, Low Height, Light Weight, 450°F, 800°F	.086 to .375	Corrosion Resistant Steel	MIL-N-25027	No	Marking on side of nut.

NAS

Identification Marking	Designation of Product Standard	Description of Fastener	Nominal Size Range (in.)	Material	Procurement Specification	Is Mfr's Identification Mark Req'd?	REMARKS
—	NAS1065	Stud—High Temperature, Electrical Terminal Block	.138 to .375	Corrosion Resistant Steel	None	No	Marking depressed on top of head.
X	NAS1096	Screw, Hex Head, Cross Recess, Full Thread	.164 and .190	Alloy Steel Cad. Plated	MIL-S-7839	No	Marking on top of head, raised or depressed.
⬠	NAS1054	Rivet—Hi Shear, Protruding Head	.125 to .375	Various Steel Alloys	NAS498	Optional	Marking depressed at center of head.
	NAS1055	Rivet—Hi Shear, 100 Degree Flat Head, Close Tolerance	.125 to .375	Various Steel Alloys	NAS498	Optional	Marking depressed at center of head.
•	NAS1097	Rivet—Solid, 100 Degree Flat Head	.094 to .250	Alum. Alloy 2117-T4	MIL-R-5674	No	Marking depressed at center of head.
	NAS1241	Rivet—Solid, 100 Degree Flat Head	.094 to .375	Alum. Alloy 2117-T4	MIL-R-5674	No	Marking depressed at center of head.
	NAS1242	Rivet—Solid, Universal Head	.094 to .375	Alum. Alloy 2117-T4	MIL-R-5674	No	Marking depressed at center of head.
	NAS1321	Rivet, Slug, Sleeveless	.125 to .375	Alum. Alloy 2117-T4	MIL-R-5674	No	Marking depressed at center of head.
•	NAS1097	Rivet—Solid, 100 Degree Flat Head	.094 to .250	Alum. Alloy 2017-T4	MIL-R-5674	No	Marking raised at center of head.
	NAS1241	Rivet—Solid, 100 Degree Flat Head	.094 to .375	Alum. Alloy 2017-T4	MIL-R-5674	No	Marking raised at center of head.
	NAS1242	Rivet—Solid, Universal Head	.094 to .375	Alum. Alloy 2017-T4	MIL-R-5674	No	Marking raised at center of head.
+	NAS1097	Rivet—Solid, 100 Degree Flat Head	.094 to .250	Alum. Alloy 5056-H32	MIL-R-5674	No	Marking raised at center of head.
	NAS1241	Rivet—Solid, 100 Degree Flat Head	.094 to .375	Alum. Alloy 5056-H32	MIL-R-5674	No	Marking raised at center of head.
	NAS1242	Rivet—Solid, Universal Head	.094 to .375	Alum. Alloy 5056-H32	MIL-R-5674	No	Marking raised at center of head.
— —	NAS1097	Rivet—Solid, 100 Degree Flat Head	.094 to .250	Alum. Alloy 2024-T4	MIL-R-5674	No	Marking raised at center of head.
	NAS1241	Rivet—Solid, 100 Degree Flat Head	.094 to .375	Alum. Alloy 2024-T4	MIL-R-5674	No	Marking raised at center of head.
	NAS1242	Rivet—Solid, Universal Head	.094 to .375	Alum. Alloy 2024-T4	MIL-R-5674	No	Marking raised at center of head.
	NAS1321	Rivet, Slug, Sleeveless	.125 to .375	Alum. Alloy 2024-T4	MIL-R-5674	No	Marking raised at center of head.
•	NAS1198	Rivet—Solid, Universal Head	.062 to .250	Alloy Steel A286	None	No	Raised dot on top of head, not at center.
	NAS1199	Rivet—Solid, 100 Degree Flat Head	.062 to .250	Alloy Steel A286	None	No	Raised dot on top of head, not at center.
	NAS1200	Rivet—Solid, 100 Degree Flat Head, Shear	.094 to .250	Alloy Steel A286	None	No	Raised dot on top of head, not at center.
•••••• or ••••••	NAS1351	Screw, Cap, Socket Head, Self Locking, Fine Thread	.060 to 1.000	Alloy Steel and Corrosion Resistant Steel	FF-S-86	No	Marking on top or side of head, raised or depressed.
	NAS1352	Screw, Cap, Socket Head, Self Locking, Coarse Thread	.073 to 1.500	Alloy Steel and Corrosion Resistant Steel	FF-S-86	No	Marking on top or side of head, raised or depressed.
	NAS1408	Nut, Self Locking, Hex, Standard Height, Non-Airborne Use, Coarse Thread	.250 to 1.500	Steel, and Corrosion Resistant Steel	MIL-N-6034 and MIL-N-25027	No	Marking on top of nut, at corners, raised.
	NAS1409	Nut, Self Locking, Hex, Standard Height, Non Airborne Use, Fine Thread	.250 to 1.500	Steel, and Corrosion Resistant Steel	MIL-N-6034 and MIL-N-25027	No	Marking on top of nut, at corners, raised.
	NAS1635	Screw, Pan Head, Self Locking, Full Thread, Cross Recess	.138 to .375	Corrosion Resistant Steel	None	No	Marking on top of head.
M	NAS1398	Rivet—Blind, Protruding Head, Locked Spindle	.125 to .250	Monel	NAS1400	No	Marking on top of head.
	NAS1399	Rivet—Blind, 100 Degree Flat Head, Locked Spindle	.125 to .250	Monel	NAS1400	No	Marking on top of head.
C	NAS1398	Rivet—Blind, Protruding Head, Locked Spindle	.125 to .250	Alloy Steel A286	NAS1400	No	Marking on top of head.
	NAS1399	Rivet—Blind, 100 Degree Flat Head, Locked Spindle	.125 to .250	Alloy Steel A286	NAS1400	No	Marking on top of head.
H	NAS1414 thru NAS1422	Pin, Swage Locking, 100 Degree Shear Head, Stump Type	.125 to .375	Alloy Steel	NAS1413	Yes	Marking depressed at center of head.
	NAS1424 thru NAS1432	Pin, Swage Locking, Protruding Head, Shear, Stump Type	.125 to .375	Alloy Steel	NAS1413	Yes	Marking depressed at center of head.
	NAS1436 thru NAS1442	Pin, Swage Locking, 100 Degree Shear Head, Pull Type	.1875 to .375	Alloy Steel	NAS1413	Yes	Marking depressed at center of head.
	NAS1446 thru NAS1452	Pin, Swage Locking, Protruding Head, Shear, Pull Type	.1875 to .375	Alloy Steel	NAS1413	Yes	Marking depressed at center of head.
—	NAS1456 thru NAS1462	Pin, Swage Locking, 100 Degree Flat Head, Tension, Pull Type	.1875 to .375	Alloy Steel	NAS1413	Yes	Marking depressed at center of head.
	NAS1465 thru NAS1472	Pin, Swage Locking, Protruding Head, Tension, Pull Type	.156 to .375	Alloy Steel	NAS1413	Yes	Marking depressed at center of head.
	NAS1486 thru NAS1492	Pin, Swage Locking, 100 Degree Flat Head, Tension, Stump Type	.1875 to .375	Alloy Steel	NAS1413	Yes	Marking depressed at center of head.
	NAS1496 thru NAS1502	Pin, Swage Locking, Protruding Head, Tension, Stump Type	.1875 to .375	Alloy Steel	NAS1413	Yes	Marking depressed at center of head.
— —	NAS1475 thru NAS1482	Pin, Swage Locking, 100 Degree Flat Head, Tension, Pull Type	.156 to .375	Alloy Steel	NAS1413	Yes	Marking depressed at center of head.
∴	NAS1516 thru NAS1522	Pin, Swage Locking, 100 Degree Flat Head, Tension, Pull Type	.1875 to .375	Alum. Alloy 7075-T6	NAS1413	Yes	3 dots raised at center of head. Mfr's mark depressed.
	NAS1546 thru NAS1552	Pin, Swage Locking, 100 Degree Flat Head, Tension, Stump Type	.1875 to .375	Alum. Alloy 7075-T6	NAS1413	Yes	3 dots raised at center of head. Mfr's mark depressed.
••	NAS1525 thru NAS1532	Pin, Swage Locking, Protruding Head, Tension, Pull Type	.156 to .375	Alum. Alloy 7075-T6	NAS1413	Yes	2 dots raised at center of head. Mfr's mark depressed.
	NAS1535 thru NAS1542	Pin, Swage Locking, 100 Degree Flat Head, Tension, Pull Type	.156 to .375	Alum. Alloy 7075-T6	NAS1413	Yes	2 dots raised at center of head. Mfr's mark depressed.
	NAS1556 thru NAS1562	Pin, Swage Locking, Protruding Head, Tension, Stump Type	.1875 to .375	Alum. Alloy 7075-T6	NAS1413	Yes	2 dots raised at center of head. Mfr's mark depressed.

TYPES OF BOLT AND SCREW HEADS

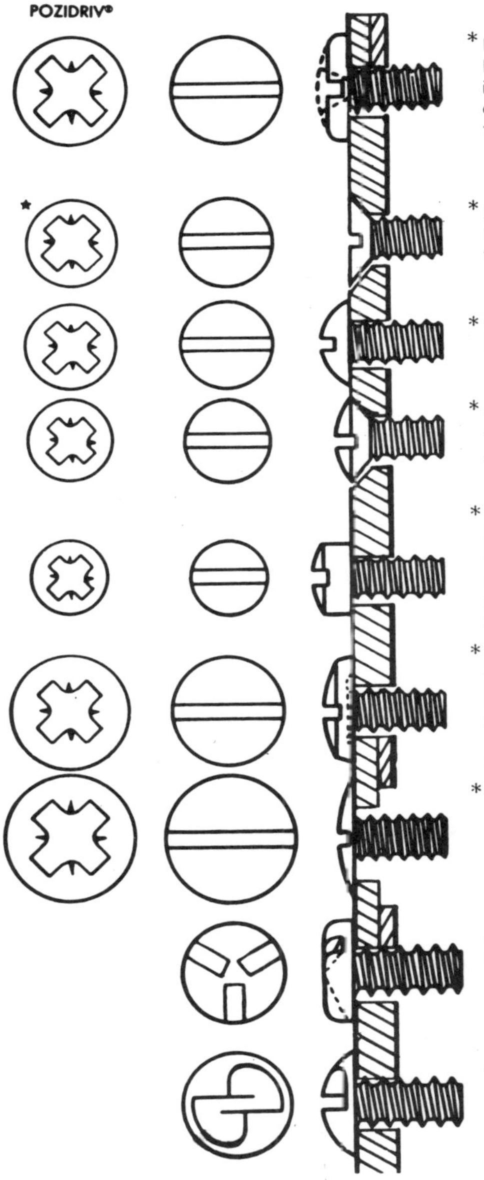

* **PAN HEAD:** Recommended for new designs to replace round, truss and binding heads. Provides a low large diameter head, but with characteristically high outer edge along the outer periphery of the head where driving action is most effective for high tightening torques. Slightly different head contour where supplied with recessed head. See dotted line.

* **FLAT HEAD:** Supplied to standard dimensions with an 80° to 82° angle to be used where finished surfaces require a flush fastening unit. The countersunk portion offers good centering possibilites.

* **ROUND HEAD:** Not recommended for new design (see pan head). This head was the most universally used design in the past.

* **OVAL HEAD:** Fully specified as "oval countersunk", this head is identical to the standard flat head, but possesses, in addition, a rounded, neat appearing upper surface for attractiveness of design.

* **FILLISTER HEAD:** The standard oval fillister head has a smaller diameter than the round head, but is higher with a correspondingly deeper slot. The smaller diameter head increases the pressure applied on the smaller area and can be assembled close to flanges and raised surfaces. Headed in counterbored dies to insure concentricity, they may be used successfully in counterbored holes.

* **BINDING HEAD (Straight Side):** Most generally used in electrical and radio work because of its identifying undercut beneath the head, which binds and eliminates fraying of stranded wire. Offers an attractively designed, medium-low head with ordinarily sufficient bearing surface. Not ordinarily recommended as a Phillips Recessed head—see Pan Head for better functional design.

* **TRUSS HEAD:** Also known as oven head, stove head, and oval binding head. A low, neat appearing, large diameter head having excellent design qualities, and as illustrated can be used to cover larger diameter clearance holes in sheet metal when additional play in assembly tolerance is required. Suggest pan head as a substitute.

HOLT HEAD (PATENTED): Provides the perfect tamper-proof assembly. Theft-proof—decorative, yet inexpensive. Special drivers available for field removal or power driven assembly machines.

ONE-WAY HEAD: This ingenious, tamper-proof type of head, once assembled cannot be removed, yet is driven with a standard screw driver. Manufactured with amazing economy in productive quantities, this simple design can frequently solve costly assembly problems.

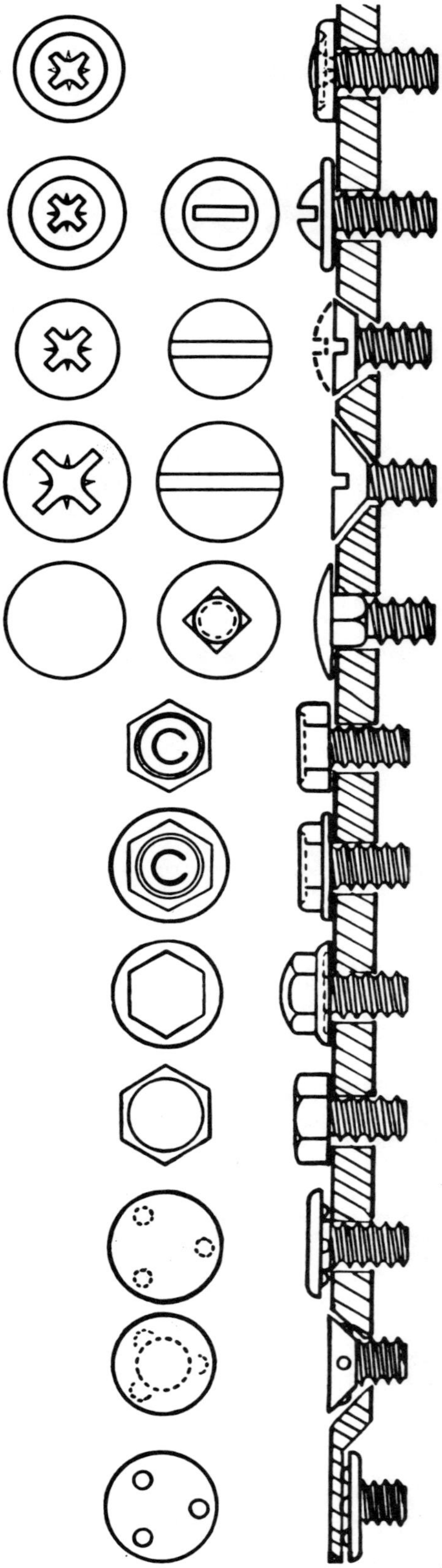

PHILLIPS FINISHING WASHER HEAD: Designed as a neat appearance product for the electronic and appliance trade with all thread styles.

WASHER HEAD: This design has the finished appearance of a conventional round head plus washer, and was originally created to provide extra large bearing surface under the head. The modern "truss" head (carried in stock) normally answers this purpose. When a larger head diameter is required this washer head design is recommended.

* **FLAT AND OVAL HEADS (UNDERCUT):** This is the standard flat or oval head 80° to 82° countersunk screw which has the lower one-third of the countersunk portion removed to facilitate production of extremely short lengths. As illustrated, it will fit a standard counterbored hole and is particularly adaptable to flush assemblies in thin stock.

FLAT HEAD (100° COUNTERSUNK): This special flat head screw has been developed for applications requiring flush surfaces, and is recommended for use in soft materials, to distribute pressure over a larger and less angular surface. Very well adapted for use with thin aluminum, soft plastics, etc.

SQUARE SHOULDER SCREWS: An adaptation of the standard carriage bolt design. Possesses a truss head on a square shank which resists rotation when located or driven into place. This square shoulder may also be staked into place as a permanent fastener. A great many varieties in all screw diameters are available in productive quantities.

* **INDENTED HEXAGON:** An inexpensive wrench head fastener made to standard hexagon head dimensions. The hex is completely cold upset in a counterbored die and possesses an identifying depression in the top surface of the head.

* **INDENTED HEXAGON WASHER HEAD:** Produced in the same manner as the standard indented hexagon head but with a washer section at the base of the head to protect the finish of the assembly from wrench disfigurement.

ACORN HEAD (FULL UNDERCUT): A very neat appearing trim screw for appliance application—excellent wrench surfaces.

* **HEXAGON HEAD (TRIMMED):** This is the standard type of wrench-applied hexagon head, characterized by clean, sharp corners trimmed to close tolerances. Recommended for general commercial applications, it is available in all standard patterns and in all thread diameters.

* **HEAD STYLES (WELDING SCREWS):** The welding screw has been developed to provide a strong, permanent threaded fastener which becomes an integral part of the assembly. It utilizes the principle of projection welding by means of multiple lugs applied to various head surfaces.
Type "U" (Underlug) for general application. Assembles easily into prelocated holes and fully utilizes head strength. Top surfaces of head designed for efficient welding anode contact.

Flat Head styles take full advantage of the self-centering feature of the countersunk portion and provide a smooth, flush outer surface.

Type "T" (Overlug) for applications requiring smooth, finished outer surfaces. Under surface of head is designed for perfect electrode contact.

* **Most head styles also available as a Recessed or Clutch.**

TYPES OF SCREW POINTS

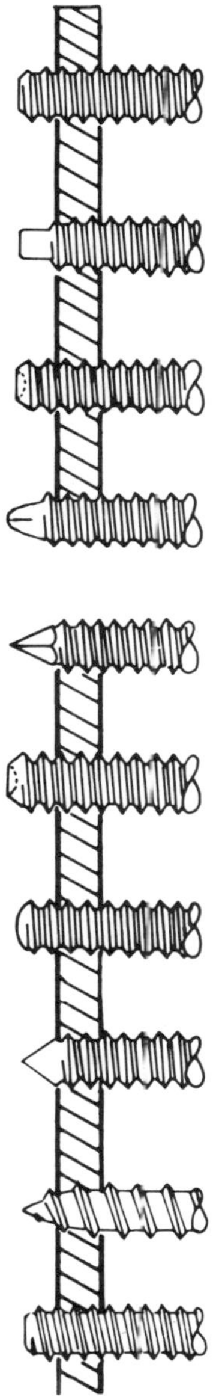

ASA	MANUFACTURER	FEDERAL
A	A	A
B	B or Z	B

DIE POINT: One of the least expensive pointing operations applied at the time of heading. This operation provides an end chamfer starting with a diameter smaller than the root diamter of the thread. The minimum reduction of the point is approximately 10% below the maximum minor diameter with an included angle of 40° to 50°.

DOG POINT: A straight pointed section reduced in diameter slightly below the root diameter of the thread, usually extending in length about two-thirds the diameter of the thread. Recommended for ease in starting, to insure against stripping fine threaded products, and to increase efficiency along production lines.

ROLLED POINT: An efficient method of producing pointed long studs or long screws with an end chamfer similar to the Die Point. The last thread and a half is slightly cupped by the thread roll-over operation.

PINCH POINT (ROUNDED): An inexpensive method of applying a 40°, 60° or 90° lead-in point having a slightly rounded contour but with pinch-off marks on its surface. Used for aligning several sheets or assembling several parts requiring pilot action.

NAIL POINT (PINCHED): Usually supplied with an approximate 45° included angle having a sharp point and slightly squared surface. Used for impinging or locking against wood or other soft material. Other degrees of included angle and sharpness also available.

CUPPED POINT: A special cup section supplied on the end of the threaded member having a depression in the end to reduce the area in contact with the surface which increases its holding and locking power under pressure.

ROUND POINT: A dome-like rounded surface applied to the end of a threaded member in order to offer pressure without disfigurement. Used for adjusting members where friction without cutting action is desirable.

CONE POINT: A precision forming operation to provide any required included angle. Offers a smooth surface, accurate length, and a sharp point which can be produced to any desired contour to fit your particular requirements.

TYPE A POINT: A thread forming screw for use in thin metal .015 to .050 thick. Used with drilled, punched or nested holes in sheet metal, resin impregnated plywood, asbestos combinations, among others. Not recommended for new design.

TYPE B POINT: A thread forming screw for use in heavier metal .050 to .200 thick. Larger root diameter with finer thread pitch for light and heavy sheet metal non-ferrous castings, plastics, impregnated plywoods, asbestos combinations, and other materials.

ASA	MANUFACTURER	FEDERAL	
AB	AB	—	**TYPE AB POINT:** A thread forming screw combining locating type point of Type A with thread size and pitch of Type B. Normal limitations of Type B apply.
C	C	C	**TYPE C POINT:** A thread forming screw with either coarse or fine pitch machine screw thread and blunt tapered point. Eliminates chips and permits replacement with standard screw in the field. Higher driving torque required. Usable in heavy sheet metal and die castings.
U	U	U	**TYPE U POINT:** A thread forming screw with high Helix thread for driving or hammering into sheet metal, castings, fiber or plastics for permanent, quick assemblies.
F	F	CF	**TYPE F POINT:** A thread cutting screw with machine screw thread with blunt tapered point, having multi-cutting edges and chip cavities. For heavy gauge sheet metal, aluminum, zinc and lead die castings, cast iron, brass and plastic.
BF	FZ	BF	**TYPE FZ POINT:** A thread cutting screw with a tapping screw thread with blunt tapered point and multi-cutting edges and chip cavities. For plastics, die castings, metal clad and resin impregnated plywoods, and asbestos.
D	I	CS or 1	**TYPE 1 POINT:** A thread cutting screw with single flute for general use. Produces a fine standard machine screw thread for field replacement.
17	17	-	**TYPE 17 POINT:** A thread cutting screw for wood with a coarse tapping screw thread and a special long sharp point fluted to capture chips.
T	23	CG	**TYPE 23 POINT:** A thread cutting screw in the fine thread series offering maximum thread cutting area and excellent chip clearing, with minimum tightening torques.
BT	25	BG	**TYPE 25 POINT:** A thread cutting screw similar to Type 23 point except with coarse Type B thread. For plastics and other soft materials with large chip clearing and cutting edges.

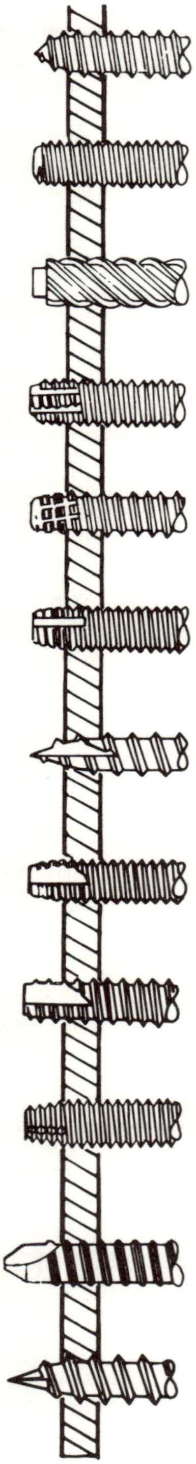

SWAGEFORM®—A new tapping screw with 3 rows of lobes on point equally spaced at 120° sections. Low "on torque" and high break away torque features this thread swaging screw.

TEKS®—A self-drilling tapping screw with special drilling points—lengths—diameters that will drill through ¼" metal. Eliminates all hole preparation—drills faster than a drill. No punching, drilling or tapping required. Reduces die costs. TEKS are used with conventional driving equipment.

SELF-DRILLING TAPPING SCREW—RAPIDRIL: Produces more secure sheet metal assemblies faster . . . used as self-drilling screw or driving thru pre-punched holes. Can be used with or without pilot holes. Positive rake "forward cutting edge" drills straight thru sheet metal at peak speed. Perfectly mated threads increase strip and back out pressures.

Types of Threads

Because of the necessity for standardization, there have gradually developed three main systems of screw threads. These are the Unified and American Standard Form, the British Whitworth Form, and the European Metric Form.

There are many variations of these basic types, as well as specialized forms. Some of these are illustrated on these two pages.

Unified AND AMERICAN STD. THREAD FORM (COARSE)

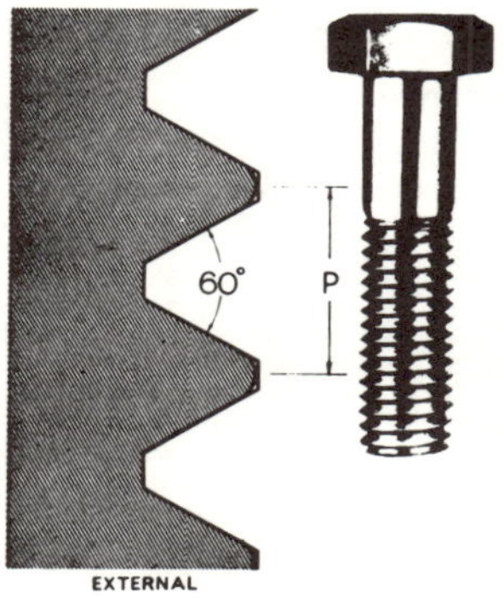

This thread form is used for most of the Bolts, Nuts, Screws and miscellaneous threaded products manufactured in the United States, Canada, and the United Kingdom.

Although the "American National Standard" thread system (Classes 1, 2, 3, 4 and 5 without an "A" or "B" suffix) is still a recognized standard for screw threads, it has been largely supplanted by the "Unified and American" system.

The "Unified and American" thread form—whether coarse or fine thread series—segregates external (bolt) and internal (nut) threads into "A" and "B" dimensions. It provides a more liberal allowance with no sacrifice of strength between mating threads which simplifies plating problems.

Unified AND AMERICAN STD. THREAD FORM (FINE)

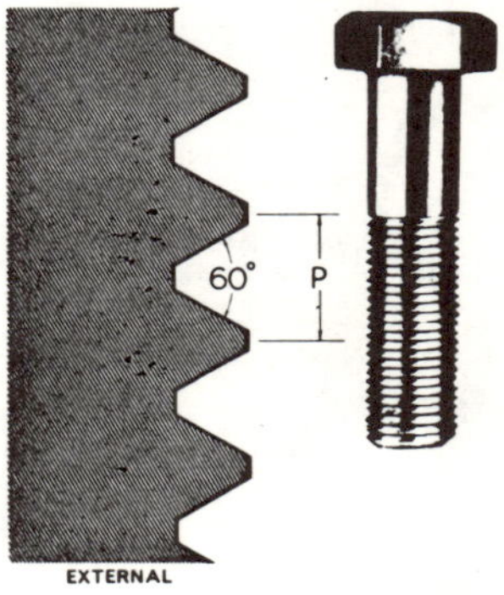

The Fine Thread series employs the same thread form as the Coarse Thread series but is distinguished by the greater number of threads per inch applied to a specific diameter. It is recommended for general use in automotive and aircraft work and where wall thickness requires a fine thread.

The basic coarse and fine thread series are supplemented by four other series which have limited popularity. These are the Extra-Fine Thread series (NEF) and three constant pitch series (8N, 12N, and 16N).

For the Class 2A, 2B, 3A and 3B coarse and fine thread specifications of the "Unified and American" system, see ASA B.1.1-1960 on pages 8, 9, 10, and 11.

AMERICAN STANDARD Wood Screw THREAD FORM

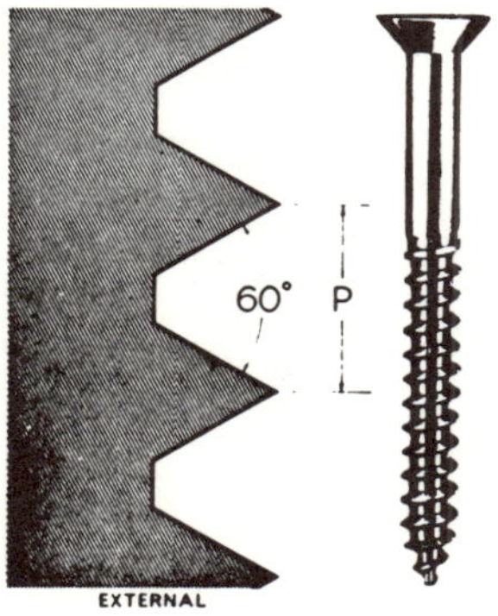

Interchangeability of parts as provided for in the various Screw Thread Forms is not a prerequisite for Wood Screw thread forms since the threads of this product cut into the work piece and do not mate with tapped holes. The thread form of Wood Screw illustrated is in accordance with American Standard B18.6.1-1956.

In general, the American Standard specifications for Wood Screws has evolved from constant industry practice in the United States.

AMERICAN STANDARD Tapping Screw THREAD FORM

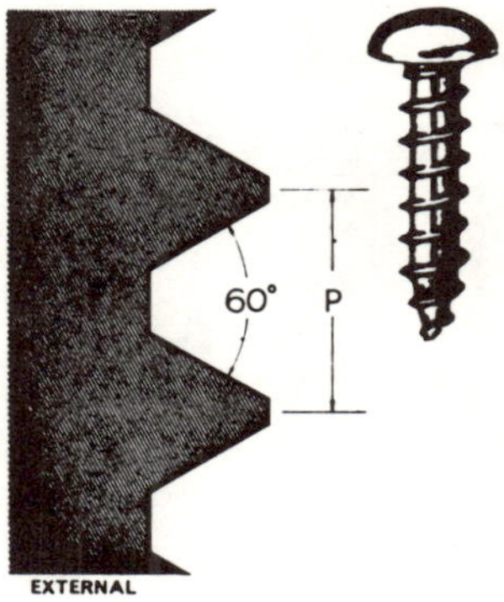

Similar to Wood Screws, the interchangeability of parts as provided for in the various Screw Thread Forms and Screw Thread standards is not a prerequisite for Tapping Screw thread forms. The thread form of Tapping Screws as illustrated is in accordance with American Standard B18.6.4-1958.

In general, the American Standard specifications for Tapping Screws has evolved from constant industry practice in the United States.

AMERICAN STANDARD Acme THREAD FORM

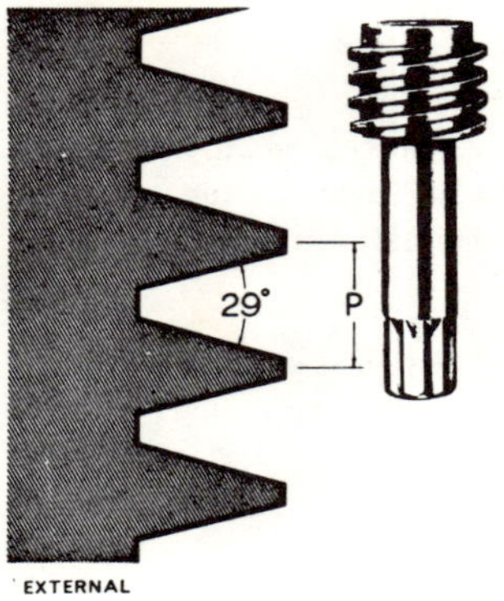

This thread form was developed to replace the square thread and is used chiefly for the purpose of producing traversing motions. ASA B1.5-1952 standardizes Acme Threads into two general classes: General Purpose and Centralizing Acme Threads.

The General Purpose threads have been classified into three classes which have clearances upon all diameters for free movement.

The Centralizing Threads are grouped into five classes which have a more limited clearance at the major diameter between external and internal threads.

AMERICAN STANDARD Stub Acme THREAD FORM

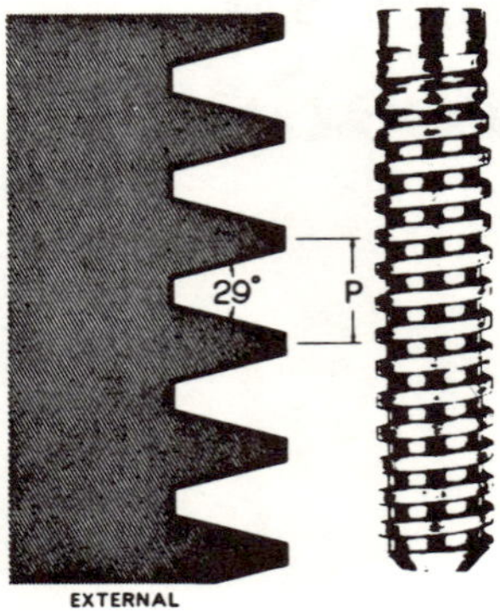

This screw thread is intended for use in those applications where a coarse pitched thread of shallow depth is required because of mechanical or metallurgical considerations. The thread is based upon the Standard Acme Thread except that the basic height of the thread is shortened. Only one class of thread, for general usage, is provided in the standard ASA B1.8-1952.

BRITISH STANDARD Whitworth THREAD FORM

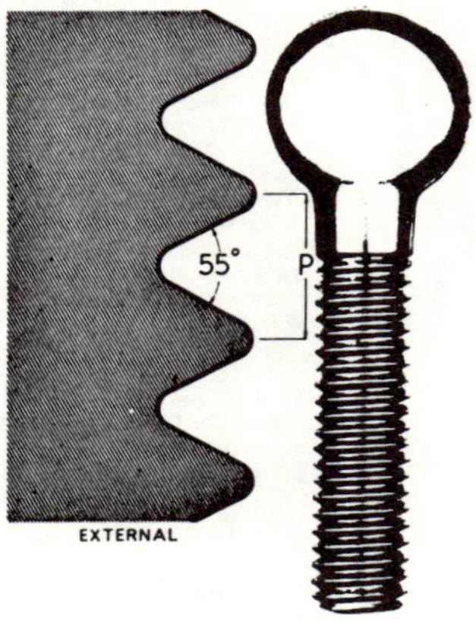

The crest and root of this thread form are rounded. This form is in general use in the United Kingdom.

This thread form is also used in the British Standard Fine Screw Threads (BSF), British Standard Pipe Threads (BSP), and British Standard Conduit Threads (BSC).

AMERICAN STANDARD Buttress THREAD FORM

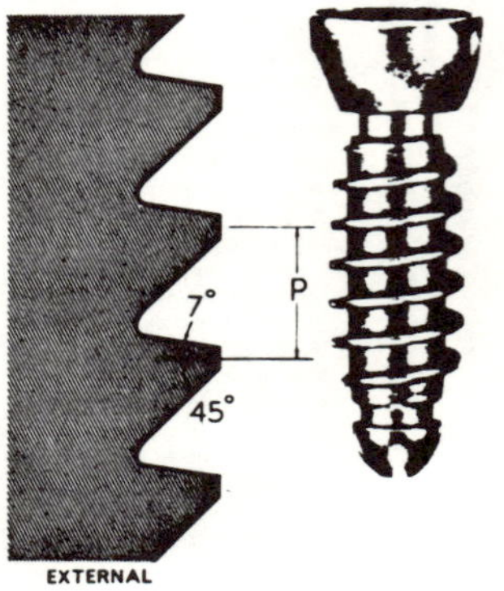

This thread form is capable of withstanding high stresses in the direction parallel to its axis. The past variations in the thread form have been standardized in the United States by ASA B1.9-1953.

The crests and roots of the internal and external threads may be truncated or rounded. The British Standard for the Buttress Thread is similar to the ASA Standard except that the crest and root are truncated or rounded to form a less deep engagement of the mating parts.

EUROPEAN Metric THREAD FORM

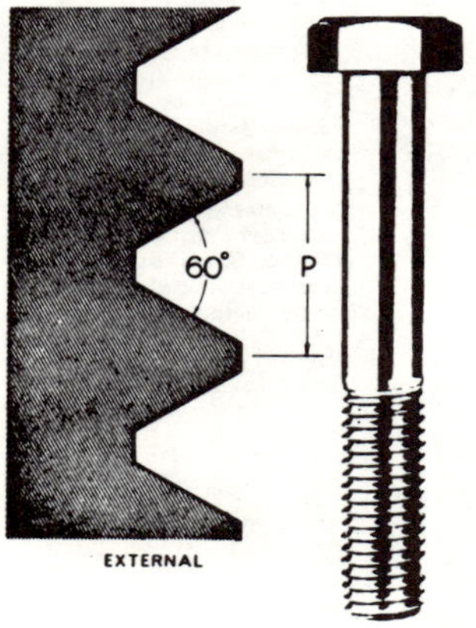

This thread form is similar to the Unified American Standard excepting the depth of the thread which is greater. This system has been adopted by most European countries using the metric system, by Japan, and by the U.S.S.R. either wholly or in part.

British Association STANDARD THREAD FORM

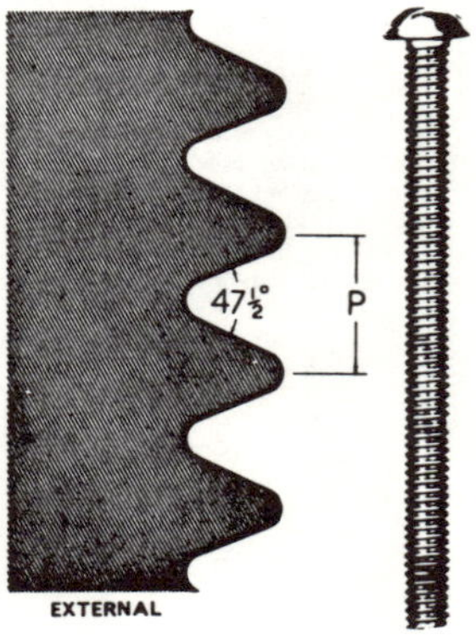

This screw thread system is recommended by the British Standards Institution for use in preference to the BSW and BSF systems for all screws smaller than ¼ inch.

This system was orginated for use in clock, watchmaking, and instrument work.

(Courtesy The H. M. Harper Co.)

SCREW THREAD CLASSES
UNIFIED SCREW THREAD SERIES

Historical

Experience with threads throughout World War II indicated that a definite need existed for the adoption of an international standard to be established on screw threads. November 18, 1948, the United States, Great Britain and Canada reached an accord to establish a Unified Screw Thread Standard Series to be known as the UNIFIED COARSE (UNC) coarse series and UNIFIED FINE (UNF) fine series. In addition to letters "A" and "B" were added to the thread callout nomenclature indicating "A" would be external and "B" would be of internal thread. Thus if a 1/4-20 class 2 fit screw was to be ordered, the designation would be 1/4-20 UNC-2A; or if a 1/4-20 class 2 nut was desired it would be 1/4-20 UNC-2B nut.

It is noted that upon the adoption of the UNIFIED SCREW THREAD SERIES on November 18, 1948, that the application of symbols UNC and UNF applied to sizes 1/4 and larger. Sizes below 1/4 although having adopted the general characteristics of the UNIFIED SCREW THREAD SERIES could not have the application of the symbols UNC and UNF for it was not approved officially for such sizes. Hence, the callout for a 4-40 screw was properly designated as a 4-40 NC-2A screw.

February 18, 1960, Sectional Committee B1 of the American Standards Association held their meeting on the continued standardization and unification of screw threads, and as a result they approved American Standard ASA B1.1-1960 which authorized the application of the UNIFIED SCREW THREAD SERIES for screws, bolts, nuts and other threaded parts from sizes 0-80 through 6 inches. As a consequence, the symbols of UNIFIED COARSE (UNC) and UNIFIED FINE (UNF), as of February 18, 1960, was officially approved for adoption. Many existing procurement documents are still described by the old AMERICAN NATIONAL THREAD SERIES with the numerical class callout. They remain that way until revision for other factors is required.

Interchangeability

UNIFIED and AMERICAN NATIONAL threads have substantially the same thread form. Both have the same diameter and pitch and are mechanically interchangeable. The principal difference between UNIFIED and AMERICAN NATIONAL thread series is in the application of tolerances and the allowance applied to 2A threads.

The main question, particularly from an inspection point of view, is that in revising hardware procurement documents from the thread number class callout to the UNIFIED series number and letter callout, are such parts interchangeable? Can a former class 1/4-20 NC-2 screw be substituted by a 1/4-20 UNC-2A screw? The answer is yes. To eliminate doubt, 1A/1B, 2A/2B, 3A/3B may be inserted in lieu of the older classes, 1, 2, and 3.

It may be further emphasized that in the future it will become increasingly difficult to obtain hardware items to the AMERICAN NATIONAL SCREW THREAD series number designation because industry has adopted the UNIFIED SCREW THREAD SERIES.

In reference to the UNIFIED SCREW THREAD SERIES thread, it may be said that any class of internal thread (B) may be assembled with any class of external thread (A). However, a Class 3A bolt in a 1B tapped hole can be applied, but such an assembly might be of looser fit than interchanging a 3A fit with a 3B. The 2/2B assembly (assuming that the external thread is not plated) will be a freer assembly than Class 2. The initial clearance and somewhat greater tolerance for the 2B thread accounts for that. There is however, much less difference in a 3A/3B assembly as compared to Class 3, inasmuch as there is no initial clearance.

A breakdown of the UNIFIED SCREW THREAD SERIES classes is as follows:

Classes 1A and 1B screw threads replace AMERICAN NATIONAL Class 1 for new design. They are used on threaded components where quick and easy assembly is necessary and where a liberal allowance is required for ready assembly.

Classes 2A and 2B threads are the most commonly used thread standards for general applications including bolts, screws, nuts and threaded fasteners. The allowance of 2A may be utilized to accommodate plated finishes or other coatings where applicable. The minimum diameter of class 2B whether or not plated or coated is basic and does not contribute to allowance or clearance at maximum metal limits.

Classes 3A and 3B screw threads are to be used if closer tolerances are desired than that provided by 2A and 2B. The maximum diameters of Class 3A (external) and minimum diameter of 3B (internal) whether or not plated are basic and give no allowance or clearance for assembly of metal parts.

Classes 2 and 3 of AMERICAN NATIONAL are available for continued use in existing design. Transition from AMERICAN NATIONAL SCREW THREAD SERIES to UNIFIED SCREW THREAD SERIES has been substantially completed. However, Classes 2 and 3 of the AMERICAN NATIONAL series in their maximum external and minimum internal diameters whether or not plated or coated are basic and afford no allowance or clearance for assembly of material components.

NOTE: Gaging practice for minimum Class 3 external thread is the same as for Class 3A thread; Class 2 external thread is the same for Class 2A; and gaging for maximum of Classes 2 and 3 internal threads is the same for Classes 2B and 3B.

Informative Data

COARSE-THREAD SERIES: (UNC, formerly NC) Generally utilized for bulk production of bolts, screws, nuts and other general engineering applications. It is generally used in lower tensile strength materials. It is applicable for rapid assembly or disassembly or if corrosion or slight damage is possible.

FINE-THREAD SERIES: (UNF, formerly NF) Suitable for production of bolts, screws, nuts and other applications where coarse series is not applicable. External threads of this series have greater tensile stress area than comparable sizes of coarse series. It is used where length of engagement is short or where wall thickness demands a finer pitch.

EXTRA-FINE SERIES: (UNEF, formerly NEF) Use this series where very fine pitches of thread are desirable for short engagement lengths and for thin wall tubes, nuts, ferrules or couplings.

CONSTANT PITCH SERIES: These thread series offer a comprehensive range of diameter/pitch combinations where the threads in the coarse (UNC) fine (UNF) and extra fine (UNEF) series do not meet particular requirements of design. Threads from CONSTANT-PITCH SERIES are in three general classifications as follows:

8 THREAD SERIES (8UN, formerly 8N) Consisting of a uniform pitch for large diameters or as a compromise between coarse and fine thread series. It is frequently used as a substitute for coarse thread series for diameters larger than 1 inch.

12 THREAD SERIES (12UN, formerly 12N) is a uniform pitch series for large diameters requiring threads of medium fine thread. It is now used mostly for continuation of fine thread series for diameters larger than 1-11/16 inches.

16 THREAD SERIES (16UN, formerly 16N): The 16 thread series is a uniform pitch series for large diameters requiring fine pitch threads. It serves for a continuation of extra fine thread series for diameters larger than 1-11/16 inches.

NOTE: Whenever a thread of the CONSTANT-PITCH Series appears in the UNC, UNF or UNEF series, the symbols and tolerances for limits of size of those standard series are applicable.

SCREW THREAD FORMS

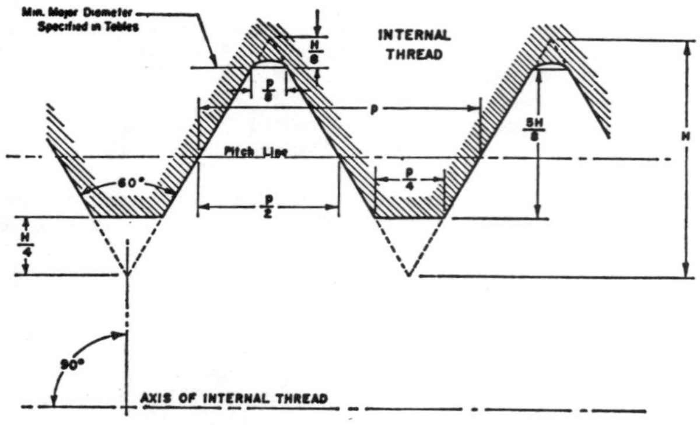

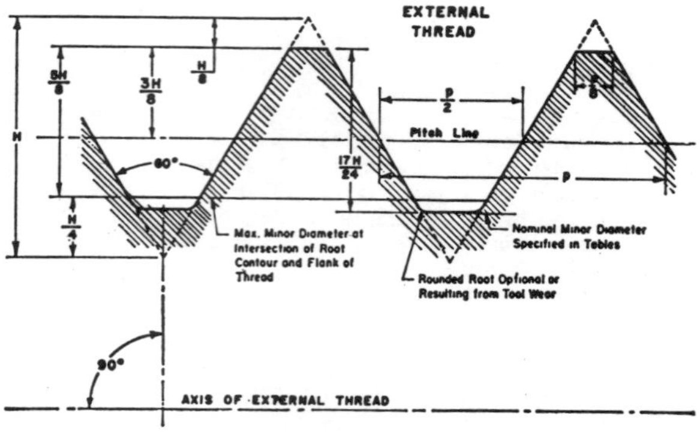

Unified Screw Thread Design Forms

(Maximum Material Condition)

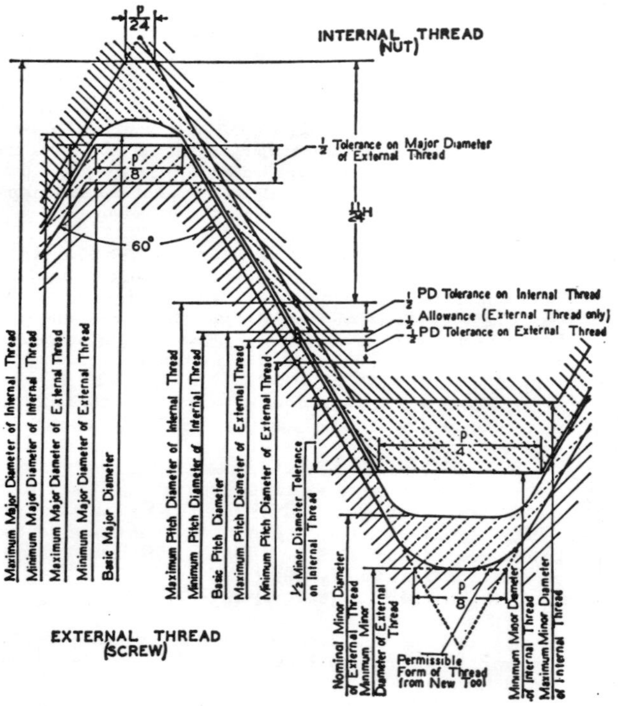

Tolerances, Allowances, and Crest Clearances for Classes 1A, 2A, 1B, and 2B
(Note: "Nominal minor diameter of screw" is that specified in tables)

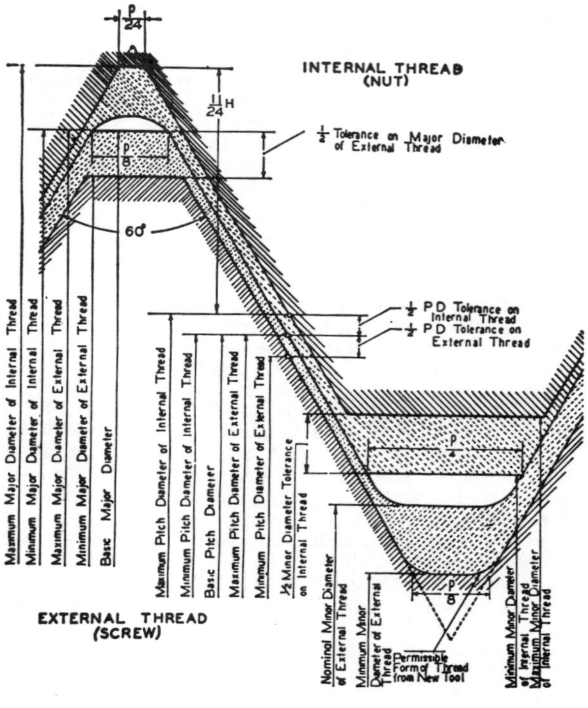

Tolerances and Crest Clearances for Classes 3A and 3B
(Note: "Nominal minor diameter of screw" is that specified in tables)

Extracted from National Bureau of Standards Handbook H28 (1957), Part I (*Screw-Thread Standards for Federal Services*), which is in general agreement with *American Standard Unified Screw Threads* (ASA B1.1-1960).

MODIFIED THREAD SERIES: It is sometimes necessary to modify the limits of the major diameter of an external thread or the minor diameter of an internal thread from the limits established by standards series and special threads for a specific purpose without change in the diameter or pitch limits. Such threads should read 1/4-20 UNC-2A MOD. Such application as the foregoing is for high temerature, high strength combinations. Coarse thread series is recommended from 1/4 to 1 inch and the 8 thread series in sizes over 1 inch.

SELECTED COMBINATIONS (UNS): Special combinations of diameter and pitch with pitch allowance based on a length of thread engagement 9 times the pitch. The pitch diameters limits are applicable to a length of engagement from 5 to 15 times the pitch.

Other Thread Classes

Thread Forms and Series Symbols

ACME Acme Thread
Two general applications for Acme thread used chiefly for purpose of producing traversing motions on machines and tools. Acme thread is divided into three general classes, 2G, 3G, and 4G; and five centralizing classes: 2C, 3C, 4C, 5C and 6C.
Classes—2G, 3G, 4G
2C, 3C, 4C, 5C, 6C

STUB ACME **Stub Acme Thread**
Used for unusual applications pertinent to Acme thread but where a coarse pitch or shallow depth is required. Stub Acme thread should never be used without consultation with the manufacturer. No class callout but it corresponds to 2G of the Acme.

60° STUB 60° Stud Thread
The angle between the flanks of the thread is 60°. Threads are truncated top and bottom; have a basic height of 0.433p, a basic thickness of 0.500p. As an example a callout for a 60° Stub would be 1-⅛-9 SPECIAL FORM 60° Thread.

***NC** American National Coarse Thread
Generally utilized for the bulk production of bolts, screws and nuts.
Classes—1, 2, 3, 4, 5

***NF** American National Fine Thread
Generally utilized for bulk production of bolts, screws and nuts for application where coarse series not applicable. Gives greater tensile strength than coarse series.
Classes—1, 2, 3, 4, 5

***NEF**
Used where finer pitches of threads are desired, applicable to thin wall tubes, nuts, ferrules or couplings.
Classes—2, 3

***8N**
Uniform pitch for large diameters, generally used as substitute for coarse thread series for diameters over 1 inch.
Classes—2, 3

***12N**
Uniform pitch series for large diameters requiring threads of medium fine pitch. Used for diameters larger than 1-½ inch.
Classes—2, 3

CLASS 4 THREADS

Formerly designated for thread work requiring a fine snug fit or slight interference fit as illustrated in Handbook H28-1944. Class 4 threads have not been reactivated for new design.

CLASS 5 THREADS

Interference fit threads for a wrench fit. Used principally for studs and their mating tapped holes. This is considered a trial Standard of the American Standards. See National Bureau of Standards Handbook H28 Part III 1957.

NC5 HF
NC CSF Designations for External Class 5 fit
NCR ONF
NC5 IF
NCR INF Designations for Internal Class 5 fit

***16N**
Uniform pitch series for large diameters requiring fine thread. Use in diameter larger than 2 inch.
Classes—2, 3

***NS**
Special combinations of diameter and pitch superseded by Unified Screw Threads.
Classes 2, 3

* Superseded by UNIFIED SCREW THREAD SERIES

N-BUTT National Buttress threads used where high stresses in direction parallel to axis is required.
Classes 1, 2

NGO American National Gas Outlet Thread
NGS American National Gas Straight Thread
SGT American National Special Gas Taper Thread
NH American National Hose Coupling and Fire Hose Coupling Thread
NPSH American National Hose Coupling Thread (iron pipe sizes)
AMO American Standard Microscope objective and nosepierce thread, American Standard surveyed mounting thread
UNC Unified coarse thread see NC for usage description
Classes—1A, 2A, 3A
1B, 2B, 3B
UNF Unified fine thread see NF for usage description
Classes—1A, 2A, 3A
1B, 2B, 3B
UNEF Unified extra fine see NEF
Classes—1A, 2A, 3A
1B, 2B, 3B
UNM Unified Miniature Thread
4UN, 6UN, 8UN
12UN, 16UN, 20UN Unified Constant Pitch Series
28UN, 32UN
UNS Special threads of Unified form
Classes—2A, 2B
27UNS 27 thread Unified Special thread series for use on thin wall tubing.

AMERICAN STANDARD PIPE THREADS

NPT American Standard taper pipe thread per general use
NPTR American Standard taper pipe thread per railing joints
NPSL American Standard straight pipe thread for loose fitting mechanical joints with locknuts.
NPSM American Standard straight pipe thread for free-fitting mechanical joints for fixtures.
ANPT Aeronautical taper pipe-thread.
NPSC American Standard straight pipe thread in pipe couplings.

DRYSEAL AMERICAN STANDARD PIPE THREADS

DRYSEAL F-PTF Dryseal American Standard fine taper pipe thread.
DRYSEAL NPSF Dryseal American Standard fuel internal straight pipe thread
DRYSEAL NPSI Dryseal American Standard intermediate internal straight pipe thread.
DRYSEAL NPSF Dryseal American Standard taper pipe thread
DRYSEAL PTF-SAE SHORT Dryseal SAE short taper pipe thread
DRYSEAL PTF-SPL SHORT Dryseal American Standard special short pipe thread
DRYSEAL PTF-SPL EXTRA Dryseal American Standard special extra short pipe thread
DRYSEAL SPL-PTF Dryseal American Standard special diameter-pitch combination pipe thread

(Courtesy Laboratory for Electronics, Inc.)

Facts About Corrosion

Corrosion is the wearing away or alteration of a metal or alloy either by direct chemical attack or by electrochemical reaction. There are several basic types, described below, which may occur singly or in combination.

TYPES OF CORROSION:

1. ATMOSPHERIC:

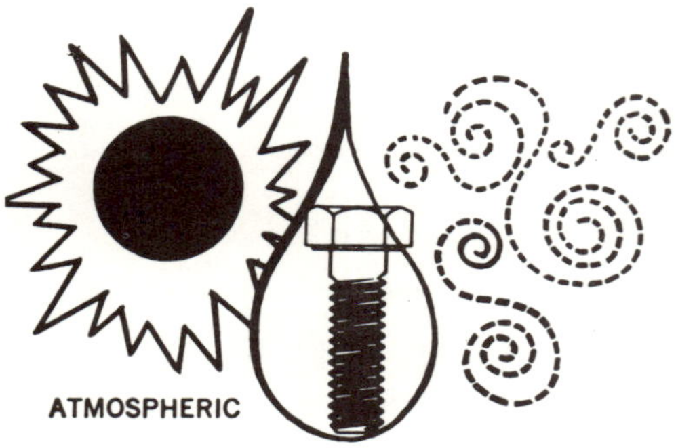

ATMOSPHERIC

This refers to the effect of corrosive agents present in the atmosphere, such as oxygen, carbon dioxide, water vapor, and sulfur and chlorine compounds. The severity of this type of attack is directly related to the amount of water vapor and sulphur and chlorine compounds present.

2. GALVANIC:

This is caused by the coupling of unlike metals in an electrolyte which results in an electrical transfer of particles from one metal to another. A reference to the Galvanic Series will show the relative position of various metals and alloys. In general, depending upon the electrolyte, the further apart on the list, the greater the degree of attack, with the less noble metal corroding sacrificially and protecting the more noble. Many so-called specialized types of corrosion, such as contact corrosion, crevice corrosion, deposit attack, and impingement attack, are forms of galvanic corrosion caused by localized galvanic cells of different potentials.

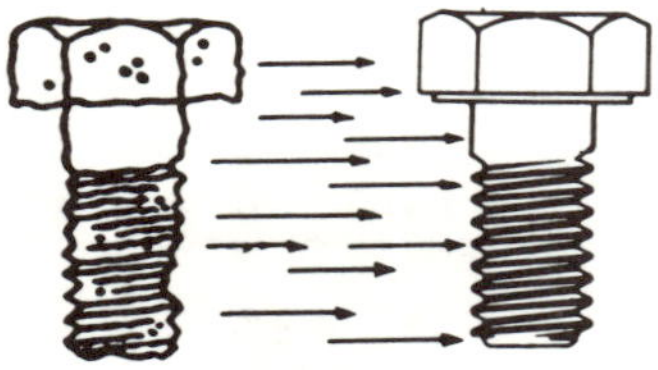

GALVANIC

The chart shows a representative sample of metals and alloys arranged according to their relative potentials in a specified environment. Any metal will have a tendency to corrode when it is in contact with another metal in a lower position in the series in the presence of an electrolyte.

GALVANIC SERIES CHART

ANODE (Least Noble)

Magnesium
Magnesium alloys

Zinc

Aluminum 1100

Cadmium

Aluminum 2024-T4

Steel or Iron
Cast Iron

Chromium-iron (active

Ni-Resist

Type 304 Stainless (active)
Type 316 Stainless (active)

Lead-tin solders
Lead
Tin

Nickel (active)
Inconel (active)

Brasses
Copper
Bronzes
Copper-nickel alloys
Monel

Silver solder

Nickel (passive)
Inconel (passive)

Chromium-iron (passive)
Type 304 Stainless (passive)
Type 316 Stainless (passive)

Silver

Graphite
Gold
Platinum

CATHODE (Most Noble)

PITTING

3. PITTING:

Even passive (corrosion-resistant) metals will corrode in specialized environments. Extreme temperature, acids or other factors will cause a localized breakdown of the material. The weakened area then becomes the anode and loses metal locally to the passive or cathodic area. The result is pitting.

4. CONCENTRATION CELL CORROSION:

This is sometimes called "Deposit attack" or "crevice corrosion." It refers to the tendency of corrosion to build up more rapidly in the cracks and crevices of an assembly.

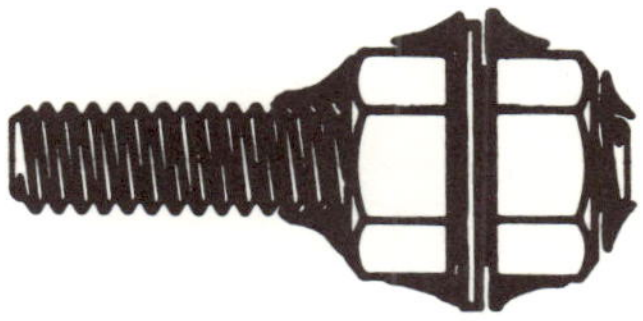

CONCENTRATION CELL CORROSION

5. INTERGRANULAR CORROSION:

This is preferential corrosion at the grain boundaries of a metal.

6. STRESS CORROSION:

Stresses in metal caused either by internal or external pressures such as a bolt and nut assembly will create paths within the grain of the metal which tend to corrode more readily under certain conditions. Corrosion is accelerated along these boundaries resulting in premature failure.

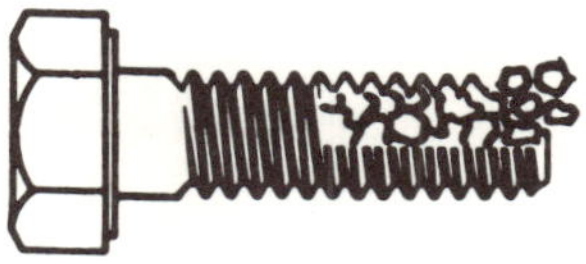

INTERGRANULAR CORROSION

7. CORROSION FATIGUE:

This is similar to stress corrosion except that the stresses are cyclic. Repeated loading and unloading usually causes accelerated rates of corrosion over those found in static stress corrosion.

8. STRAY CURRENTS:

Fabricated metal sections either buried or in a solution sometimes come in contact with electric currents caused by extraneous influences such as generators or motors. This frequently causes an accelerated disintegration of the metal.

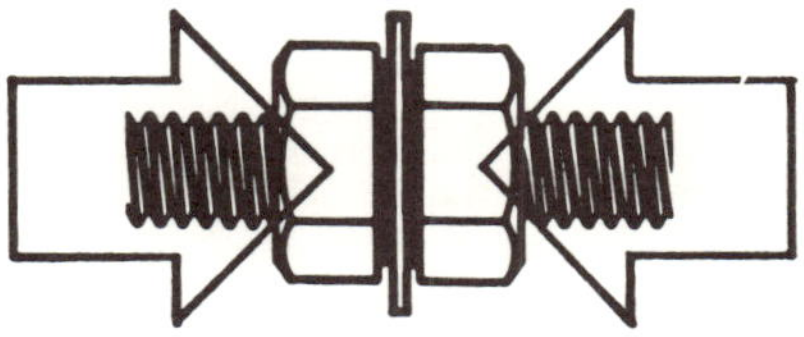

STRESS CORROSION

To Minimize Corrosion:

A. GENERALLY

1. **Select the material most likely to resist the corrosive environment to which it will be subjected. Refer to the Corrosion Guide.**
2. **Avoid irregular stresses in design. Even high stresses in bolted assemblies do not necessarily impair corrosion resistance as long as they are uniform.**
3. **If practical, add suitable chemical inhibitors to a corrosive solution.**

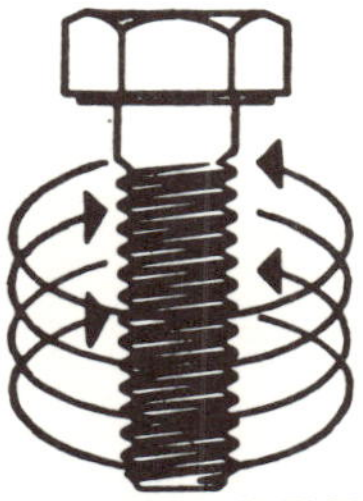

CORROSION FATIGUE

B. WHERE GALVANIC CORROSION IS INVOLVED

1. **If possible, use the same or similar metals in an assembly, where an electrolyte is present.**
2. **Avoid making combinations where the area of the less noble material is relatively small.**
3. **When dissimilar metals are used together in the presence of an electrolyte, separate them with a dielectric; insulate, paint, or coat.**
4. **The galvanic process can be used to advantage. Couple to the item to be protected pieces of less noble metal which are not functional and can thus corrode sacrificially.**

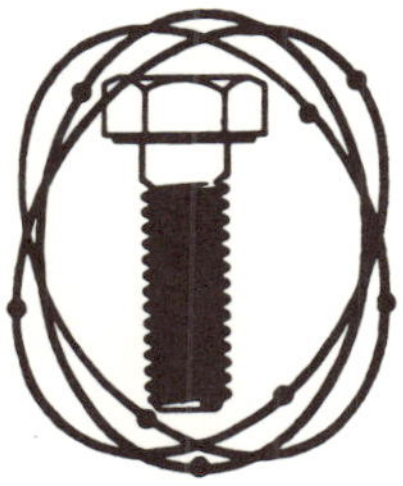

STRAY CURRENTS

Torque Law

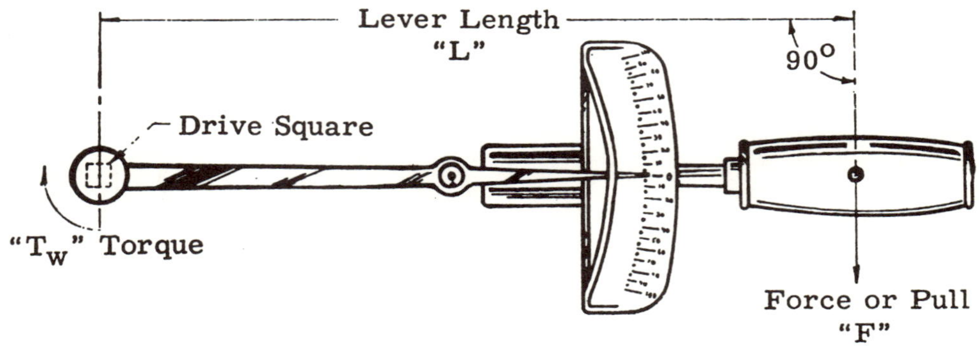

Torque is based on the fundamental law of the lever . . . that is . . . DISTANCE times FORCE equals the moment or TORQUE about a point. If this law is fully understood, all problems of torque relative to torque wrenches and adapters can be solved.
Drawing number 11 represents a torque wrench. The distance between the centerline of the drive square and centerline of the force being applied at the handle is the lever length of the torque wrench. This is designated as "L."
The arrow pointing downward at the handle of the torque wrench represents a force or pull which is exerted by the operator. The force or pull is designated as "F."
The curved arrow about the drive square of the torque wrench represents the torque which is being applied to the nut or cap screw. This is also the torque which is indicated on the scale of the wrench and will be represented as "T_w". In order to calculate the torque at the drive square, it is necessary to multiply the lever length "L" times the force "F" . . . this is shown in the following formula:

$T_w = L \times F$
T_w = Torque
L = Lever Length
F = Force

There is one precaution which must be observed. In using the formula, the lever length must be measured at 90° to the direction of the force. This relationship is shown in Drawing No. 11 by the 90° angle. If this precaution is not taken, an erroneous answer will result.

(Courtesy P. A. Sturtevant Co.)

TORQUE CONVERSION CHART

Inch Grams	Inch Ounces	Inch Pounds	Foot Pounds	Centimeter-Kilopond	Meter-Kilopond	Joules
7.09	1/4					
14.17	1/2					
21.26	3/4					
28.35	1					0.0070615
113.40	4	1/4				
226.80	8	1/2				
453.60	16	1		1.15		0.112985
	96	6	1/2	6.91		
	192	12	1	13.82	0.138	1.35582
	384	24	2	27.65	0.276	
	576	36	3	41.47	0.415	
	768	48	4	55.30	0.553	
	960	60	5	69.20	0.692	
		72	6	82.95	0.829	
		84	7	96.77	0.967	
		96	8	110.60	1.106	
		108	9	124.42	1.244	
		120	10	138.25	1.382	

Fastener Glossary

A

AMS specifications Aeronautical Material Specifications are material and process specifications for aircraft components conforming to established engineering and metallurgical practices in the aircraft industries. They are developed by the SAE Aeronautics Committee.

AN specifications Dimensional standards for aircraft fasteners developed by the Aeronautical Standards Group. See Air Force-Navy Aeronautical Standard Drawings.

ASTM specifications Standards developed by the American Society for Testing Materials.

acorn die A form of threading die for use in screw machines. The cutting portion resembles an acorn.

acorn nut See CAP NUT.

age hardening A process of aging that increases hardness and strength and ordinarily decreases ductility. Age hardening usually follows rapid cooling or cold working.

air force-navy aeronautical standard drawings Dimensional standards for aircraft fasteners developed by the Aeronautical Standards Group. All drawings are prefixed by AN.

aircraft quality Aircraft quality as applied to fasteners implies that they are to be used in highly stressed applications and that they are to be produced under closely controlled, special, and, in most cases, restricted methods of manufacture and inspection.

allowance An allowance is an intentional difference between the maximum material limits of mating parts. It is the minimum clearance (positive allowance) or maximum interference (negative allowance) between such parts.

alloy steel An alloy steel is a steel containing elements, other than carbon, which have been added to obtain definite mechanical or physical properties, such as higher strength at elevated temperatures, toughness, etc.

american standards Dimensional standards for fasteners, etc., issued by the American Standard Association.

angle of head In countersunk heads, the included angles of the conical underportion or bearing surface, usually 82° or 100°.

angularity Angularity is the angle between the axes of two surfaces of a fastener.

annealed A fastener is considered in the annealed state when it has been heated and cooled to make it soft—that is, free of hardness caused by working or previous heat treatment.

anodizing Anodizing is the formation of an oxide film on the surface by means of an anodic treatment. This is commonly used on aluminum.

B

barrel nut An internally threaded screw having a slotted head.

basic size The basic size is that size from which the limits of size are derived by the application of allowances and tolerances.

bearing surface The bearing surface is the supporting or locating surface of a fastener with respect to the part which it fastens (mates). The loading of a fastener is usually through the bearing surface.

bevel A surface not at right angles to the rest of the piece.

blinding head Rounded top surface and slightly tapered sides. The bearing surface is flat, with annular undercut optional.

binding post A binding post is a special type of subassembly used for clamping or holding electrical conductors in a rigid position. It commonly consists of a screw having a collar head or body with one or more clamping screws.

blank A blank is a fastener in some intermediate stage of manufacture.

blind rivet A rivet designed for use where only one side of the work is accessible.

blunt start The removal of the partial thread at the entering end of thread.

body The body of a threaded fastener is the unthreaded portion of the shank.

bolt A bolt is an externally threaded fastener (See: Mechanical Fasteners section for descriptions of bolt types).

bolt blank A headed rod or bar intended for a subsequent threading operation.

bolt end A bolt end is a headless rod threaded at one end and assembled with a square nut; it is designed to be welded or otherwise fastened to a part or structure.

bound body (body-bound or fitted) A bound body is a body which has a definite interference or extremely small clearance with its mating hole.

bow or camber Bow or camber is the amount that a side of a surface of a fastener deviates from being straight.

brinell hardness test A method of determining the hardness of metallic materials by applying a known load to the surface of the material through a hardened steel ball of known diameter. The diameter (or depth) of the resulting impression in the metal is measured. The Brinell hardness number is taken as the quotient of the applied load divided by the area of the surface impression, which is assumed to be spherical.

broaching Broaching is the process of removing metal by pushing or pulling a cutting tool, called a broach, along the surface.

burnishing Burnishing is the process of producing a smooth surface by rubbing or rolling a tool against the surface.

burr A burr is a small amount of material extended out from the edge of a hole, shoulder, etc. as the result of a machining operation.

button die Small adjustable dies for screw machines and similar work, supported in a ring holder.

C

carbon steel Carbon steel is a steel which does not contain any substantial amounts of alloying materials other than carbon.

case hardened A case hardened fastener is a fastener of ferrous material having a surface which has been harder than the core.

ceiling hook A ceiling hook is a fastener similar to an open eye bolt except that it has lag threads.

chamfer (v) To bevel a sharp external edge. (n) A beveled edge.

chamfer angle The chamfer angle is the angle of the chamfer measured from the normal to the axis of the fastener and is generally specified in conjunction with either a length or a diameter.

chamfer point A chamfer point is a truncated cone point, the end of which is approximately flat and perpendicular to the fastener axis. Three points on threaded fasteners generally have point included angles of 45 to 90 degrees and a point diameter equal to or slightly less than the minor diameter of the thread. This point is intended to facilitate entry of fasteners into holes at assembly.

chasing threads Cutting screw threads by moving a tool along the axis of the work to be threaded.

chip A chip is a small fragment of metal removed from a surface by cutting with a tool.

class of thread Classes of threads are distinguished from each other by the amount of tolerance or tolerance and allowance specified. Classes 1A, 2A, and 3A apply to external threads, and Classes 1B, 2B and 3B apply to internal threads. Classes 2 and 3 apply to both external and internal threads.

clevis U-shaped shackle for connecting a rod to a pin.

coating Coating is the application of some material such as a metal, organic compound, etc., to the surface of a fastener.

coining Coining is the process of forming metal by forcing it to flow while confined within closed dies.

cold heading Forcing metal to flow cold into dies to form thicker sections and more or less intricate shapes. The operation is performed in specialized machines where the metal, in the form of a wire or bar stock, may be upset or headed in certain sections to a larger size and, if desired, may be extruded in other sections to a smaller diameter than the stock wire.

cold heading stock Cold heading stock is material produced under closely controlled manufacturing and inspection methods so as to be suitable for heading and to be free from those defects causing fractures during heading.

cold swaging process A method of reducing or forming steel or other material while cold, by drawing to a point or reducing the diameter, as may be required.

cold work To deform metal stock by hammering, forming, drawing, etc., while the metal is at ordinary room temperature.

collar A collar is a raised ring or flange of material on the head or shank of a fastener.

commercial fastener A commercial fastener is a fastener manufactured to published standards and stocked by manufacturers or distributors. The material, dimensions and finish of commercial fasteners conform to the quality level generally recognized by manufacturers and users as commercial quality.

comparator A device for inspecting screw threads and outlines by comparing them with a greatly enlarged standard chart.

complete thread The length of complete thread is the length of that cross section of a threaded length having full form at both crest and root. Where there is a chamfer at the start of the thread not exceeding two pitches in length, it is included within the length of the complete thread. The thread length on the drawing shall be the gaging length or the length of threads having full form, i.e., the partial threads shall be outside or beyond the length, specified. When designing threaded products, it is necessary to take cognizance of: (1) Such permissible length of chamfer and (2) The first two threads which by virtue of HI-LO gaging practice may exceed the product limits and which may be included within the length of complete thread. However, where the application is such as to require a minimum or maximum number, or length, of complete threads the specification shall so state. Similar specification is required for definite length of engagement.

compression fastener A compression fastener is a fastener the primary function of which is to resist forces which tend to compress it.

concentric, concentricity Two surfaces of a fastener are concentric when they have a common center or axis. Concentricity is the term used to describe this condition.

cone point A cone point is a sharp conical point designed to perform perforating or aligning functions at assembly.

copper steel When any minimum copper content is specified, the steel is classed as copper steel. The copper is added to enhance corrosion resistance of the steel.

corrosion Gradual chemical or electrochemical attack on a metal by atmosphere, moisture or other agents.

counterbore (v) To enlarge a hole to a given depth. (n) 1. The cylindrical enlargement of the end of a drilled or bored hole. 2. A cutting tool for counterboring, having a piloted end of the size of the drilled hole.

counterboring Counterboring is the process of enlarging for part of its depth a hole previously formed and to provide a shoulder at the bottom of the enlarged hole. Special tools called counterbores are generally used for this operation.

countersink A countersink is an internal chamfer.

countersinking Countersinking is the process of beveling or flaring the end of a hole. Holes in which countersunk head type fasteners are to be used must be countersunk to provide a mating bearing surface.

countersunk head A head, the underside of which is beveled to fit a flaring hole. The bearing surface of other types of heads is generally perpendicular to the body axis.

crest That surface of the thread which joins the flanks of the thread and is farthest from the cylinder or cone from which the thread projects.

crest clearance As in a thread assembly, the distance, measured perpendicular to the axis, between the crest of a thread and the root of its mating thread.

crest truncation The crest truncation of a thread is the distance measured perpendicular to the axis, between the sharp crest (or crest apex) and the cylinder or cone which bounds the crest.

cross drilled A cross drilled fastener is a fastener having one or more holes in the head or shank at right angles to, and normally intersecting, the axis of the fastener.

cup point A point in the form of a cone, commonly having an included angle of 90°, with a conical depression in the end commonly having an included angle of 118°. The contact area is a circular ridge which has considerable holding power with slight penetration; applied to set screws generally.

cut thread A cut thread is a thread produced by removing material from the surface with a form cutting tool.

D

decarburized A fastener has a decarburized surface when the carbon content of the surface is lower than the carbon content of the core.

design size The design size is that size from which the limits of size are derived by the application of tolerances. When there is no allowance the design size is the same as the basic size.

die 1. One of a pair of hardened metal blocks for forming, impressing, or cutting out a desired shape. 2. (thread). A tool for cutting external threads. Opposite of tap.

die chaser The separate cutting tools used in die heads, which actually cut the screw threads.

dog point A cylindrical extension, or pilot, of diameter smaller than the minor diameter of the thread, commonly equal to about D/2 in length, with a conical section between it and the thread; usually used as a pilot in assembling or as the end of a set screw projecting into a fairly deep hole or clot.

driver head A head, on a bolt or screw, designed for driving the fastener by means of a tool other than a wrench, such as a screwdriver.

drill (v) To sink a hole with a drill, usually a twist drill. (n) A pointed cutting tool rotated under pressure.

drilling Drilling is the process of forming holes by means of specialized cutting tools called drills.

drop forging Forming metal, usually under impact, by compression within dies designed to produce the required shape.

E

eccentric, eccentricity Two surfaces of a fastener are eccentric when they do not have the same center or axis. The amount by which the centers or axes are displaced from each other is called eccentricity. This is not to be confused with Total I Indicator Reading (TIR). (See Runout).

effective thread The effective (or useful) thread includes the complete thread and that portion of the incomplete thread having crests not fully formed.

electro-galvanizing Electro-galvanizing is the process of coating metal with zinc by electroplating.

embossing Embossing is the process of raising a boss or protuberance on the surface.

endurance limit The endurance limit is the maximum stress that a fastener can withstand without failure for a specified number of stress cycles (also called fatigue limit).

externally relieved body An externally relieved body is a body on which the diameter of the entire body or a portion thereof is reduced to less than the minimum pitch diameter of the thread.

extruding Extruding is the process of reducing the size of some feature or diameter by forcing it through a die.

eyelet An eyelet is a flanged tubular fastener designed for securing by curling or splaying the tubular end.

F

facing Facing is a machining operation on the end, flat face or shoulder of a fastener.

fastener A fastener is a mechanical device for holding two or more bodies in definite positions with respect to each other.

fatigue strength The stress to which a metal can be subjected for a specified number of cyclic changes of stress.

fillet A fillet is the concave junction at two intersecting surfaces of a fastener.

fillister head Rounded top surface, cylindrical sides, and a flat bearing surface.

fin A fin is a form of key under the head of a fastener which serves to keep the fastener from turning during assembly and use.

fin neck A fin neck is a style of neck consisting of two or more fins under and integral with the head.

finish The term finish is commonly applied to the condition of the surface of a fastener as a result of chemical or organic treatment subsequent to fabrication.

The term finish is also applied to some types of fasteners to indicate the condition of the surface as a result of mechanical operations and the degree of precision. See: Finished, Semi-Finished and Unfinished Fasteners.

finished fastener A finished fastener is a fastener made to close tolerances and having surfaces other than the threads and bearing surface finished to provide a general high grade appearance. (See: Semi-Finished Fastener, Unfinished Fastener).

fit Fit is the general term used to signify the range of tightness which may result from the application of a specific combination of allowances and tolerances in the design of mating parts.

flash Flash is the thin fin of metal along the sides or around the edges of a forged or upset section. It is caused when metal flows out between the edges of the forging dies.

flash plating Flash plating is a very thin deposit of metal, sufficient to give a solid color.

flat head Flat top surface and a conical bearing surface.

following flank The flank of a thread opposite to the leading flank.

forging Forging is the process of forming a product by hammering or pressing. When the material is forged below the recrystallization temperature it is said to be cold forged. When worked above the recrystallization temperature it is said to be hot forged.

form of thread The profile of a thread in an axial plane for a length of one pitch.

full annealing Annealing a ferrous alloy by austenitizing and then cooling slowly through the transformation range.

full or nominal diameter body A full or nominal diameter body is a body the diameter of which is generally within the dimensional limits of the major diameter of the thread. Sometimes referred to as "full size body."

G

galvanize To treat with a bath of lead and zinc to prevent rusting.

galvanizing Galvanizing is the process of coating metal with zinc by hot dipping.

gimlet point A gimlet point is a threaded cone point usually having a point angle of 45 to 50 degrees. It is used on thread forming screws such as Type A tapping screws, wood screws, lag bolts, etc.

grains Grains are the individual crystals of the material.

grinding Grinding is the process of removing material from the surface by the cutting action of a bonded abrasive wheel.

grip In general, the grip of a fastener is the thickness of material or parts which the fastener is designed to secure when assembled.

grommet A grommet is a large eyelet type fastener designed for securing by curling the tubular end over a formed washer to provide strength in holes through resilient materials.

ground thread A ground thread finished on the flanks by a grinding operation.

H

half dog point The same as a dog point but half as long; used on short screws for the same purposes as the dog point, but in shallower hole or slot.

hardenability In a ferrous alloy, the property that determines the depth and distribution of hardness induced by quenching.

hardening A method of heat treating metals by heating to a temperature within, or above, the critical range, holding at that temperature for a given time, and then cooling rapidly, usually by quenching in oil or water.

hardness Resistance to plastic deformation by indentation, penetration, scratching or bending.

head The head of a fastener is the enlarged shape preformed on one end of a headed fastener to provide a bearing surface.

head angle The head angle is the included angle of the bearing surface of the head.

head diameter The head diameter is the diameter at the largest periphery of the head.

head eccentricity Head eccentricity is the amount that the head of a fastener is eccentric with the fastener body or shank.

head height For a flat bearing surface head, the head height is the overall distance, measured parallel to the fastener axis, from the extreme top to the bearing surface. For a conical bearing surface head the head height is the overall distance, measured in a line parallel to the fastener axis from the extreme top to the intersection of the bearing surface with the extended thread major diameter cylinder on a threaded fastener or with the shank on an unthreaded fastener. For flat and oval undercut heads, it is this distance measured to the intersection of the bearing surface with the undercut. For oval heads and undercut oval heads, the overall distance is referred to as total head height.

head length For rectangular or irregular shaped heads, the head length is the distance along the longest axis of the head, measured in a plane perpendicular to the axis of the fastener.

head taper Head taper is the angle formed by the side or sides of the head and the axis of the fastener. This is not applicable to conventional countersunk heads and should not be confused with head angle.

head width The head width is the distance across opposite flats of hexagon, square or twelve-point heads measured in a plane perpendicular to the fastener axis. For rectangular or irregular shaped heads the head width is the distance along the narrowest axis of the head measured in a like manner.

headed fastener A headed fastener is a fastener having one end enlarged or preformed.

headed and threaded rod A headed and threaded rod is a fastener similar to a machine screw except that it has a very much greater length. It has a round, truss, or flat head and an end threaded for a nut.

header A header is a specialized form of horizontal press.

header point A header point is a chamfered point normally produced during the heading operation. The screw blank is chamfered before threads are rolled. It is applied to machine screws in certain sizes and lengths.

heading Heading is a manufacturing process involving the use of a header. This process may or may not involve upsetting or extruding. A part made from wire below the recrystallization temperature is said to be cold headed whereas parts made from wire above the recrystallization temperature are said to be hot headed.

headless A headless threaded fastener is a fastener normally having a slot, recess, or socket in one end.

headless fastener A headless fastener is a fastener, either threaded or unthreaded which does not have either end enlarged.

heat treating An operation invoving the heating and cooling of a metal to obtain desirable conditions or properties.

height of thread The distance, measured perpendicular to the axis, between the major and minor cylinders or cones, respectively.

hexagon head Flat top surface with hexagonal sides and with a flat bearing surface.

high strength fastener A high strength fastener is a fastener having high tensile and shear strengths attained through combinations of materials, work-hardening, and heat treatment.

hot forming Working operations such as bending and drawing sheet and plate, forging, pressing, and heading, performed on metal heated to temperatures above room temperature.

I

immunize To remove small particles of iron or grit from the surface of stainless steel by pickling in an acid solution.

impact test A test to determine the energy absorbed in fracturing a test bar at high velocity. The test may be in tension or in bending, or it may properly be a notch test if a notch is present, creating multiaxial stresses.

incomplete thread This is also known as the vanish or washout thread. On straight threads, the incomplete thread is that portion at the end having roots not fully formed by the lead or chamfer on threading tools.

On the taper threads, the crest at the end may also be not fully formed due to the intersection of the major cone of an external thread or the minor cone of an internal thread with the cylindrical surface of the work.

included angle of thread The angle between the flanks of the thread measured in an axial plane.

inclusions Inclusions are particles of non-metallic impurities contained in material.

interference fit A thread fit having limits of size so prescribed that an interference always results when mating parts are assembled.

internal thread A thread on the internal surface of a hollow cylinder or cone.

internal wrenching bolt An internal wrenching bolt is a bolt having a large cylindrical head with flat top, flat bearing surface, and hexagon socket. The bolt is designed for use in high strength, high temperature applications such as steam turbines. A bolt designated as an Internal Wrenching Bolt is used in aricraft. It is similar in design to the one previously described except that the head is tapered.

internally relieved body An internally relieved body is a body which has an axial hole drilled through a portion of the body.

internally relieved body bolt This bolt has an axial hole drilled through the head and through a portion of its body.

international metric thread system A thread form similar to the American standard, excepting the depth which is greater. There is a clearance between the root and mating crest fixed at a maximum of 1/16 the height of the fundamental triangle or 0.054 x pitch. A rounded root profile is recommended. The angle in the plane of the axis is 60 degrees and the crest has a flat like the American standard equal to 0.125 x pitch.

J

jam nut (1.) A second nut forced or jammed against the main nut to prevent loosening. (2.) A thin nut.

K

knurling Knurling is the process of producing a roughened surface by means of a specialized forming tool called a knurl.

L

laps Laps are surface defects caused by folding over fins or sharp corners into the surface of the material.

lead error A variation in the distance between the threads of a screw.

left-hand thread A thread is a left-hand thread, if, when viewed axially, it winds in a counterclockwise and receding direction. All left-hand threads are designated LH.

length The length of a headed fastener is the distance from the intersection of the largest diameter of the head with the bearing surface to the extreme point, measured in a line parallel to the axis of the fastener. Exceptions: The length of a shoulder screw and a socket head shoulder screw is the length of the shoulder. The length of a flat top countersunk head tubular rivet (with chamfered top) is measured from the intersection of the bearing surface with the shank diameter to the extreme point. The length of a headless fastener is the distance from one extreme point to the other, measured in a line parallel to the axis of the fastener.

length of thread engagement The length of thread engagement of two mating threads is the distance between the extreme points of contact on the pitch cylinders or cones, measured parallel to the axis.

lentil head screw Obsolete term for oval head screw.

limits of size The limits of size (commonly referred to as "Limits") are the applicable maximum and minimum sizes.

lock nut There are two basically different types of lock nuts; (1) a prevailing torque type which resists relative bolt-nut movement with or without an axially applied load to the bolt-nut combination, and (2) a free-running type which exhibits a locking ability when there is an axial load applied to the base of the nut. The "locking" or stopping action of the nut is accomplished by thread deformation, or clamping, or by the addition of non-metallic inserts. The free-running type usually has a design feature which adds to the elastic elongation of the bolt-nut combination.

M

machining Machining is the process of forming the surface by cutting away material.

major diameter On a straight thread, the diameter of the coaxial cylinder which would pass through the crests of an external thread or the roots of an internal thread.

master gage A thread-plug gage which represents the physical dimensions of the nominal or basic size of the part. It clearly establishes the minimum size of the threaded hole and the maximum size of the screw at the point at which interference between mating parts begins.

maximum material condition The maximum material condition of a feature of a fastener means the maximum amount of material permitted by the tolerance shown for that feature (abbreviated MMC).

maximum material limit The maximum limit of size of an external dimension or the minimum limit of size of an internal dimension.

mechanical properties Mechanical properties are those properties which involve a relationship between strain and stress.

milled from the bar Milled from the bar fasteners machined from bar stock on a lathe, screw machine, etc.

miniature screw A miniature screw is a screw less than 0.06 in. in diameter, having a slotted head, and threaded for assembly with a preformed internal thread.

minor diameter On a straight thread, the minor diameter is the diameter of the coaxial cylinder which would pass through the roots of an external thread.

N

NAS drawings and specifications Dimensional and material standards for aircraft fasteners developed by the National Aircraft Standards Committee. All drawings and specifications are prefixed by NAS.

nail point A nail point is a sharp pyramidal point generally having a point angle of 30 to 45 degrees, and is produced by a pinching operation. It is designed for piercing wood or other resilient materials.

neck Neck is used to define: (1) a specialized form of a portion of the body of fasteners near the head to perform a definite function, such as preventing rotation, etc.; and (2) a reduced diameter of a portion of the shank of a fastener which is required for design or manufacturing reasons.

needle point A needle point is a cone point of long length intended to perform a piercing function.

nitriding A surface hardening process used on ferrous metals by heating the metal in contact with ammonia gas or other nitrogenous material.

nominal size The nominal size is the designation used for the purpose of general identification.

non-ferrous metal Metals or alloys without an appreciable amount of iron. Examples are aluminum, brass, copper, etc.

nonstandard fastener A nonstandard fastener is a fastener which differs in size, length, material, or finish from established and published standards.

normalize To remove internal stresses by heating a metal piece to its critical temperature and allowing it to cool very slowly.

nut A nut is a block or sleeve having an internal thread designed to assemble with the external thread on a bolt, screw, or other threaded part. It may serve as a fastening means, an adjusting means, a means for transmitting motion, or a means for transmitting power with a large mechanical advantage and non-reversible motion. (See Mechanical Fasteners section for descriptions of Nut types.)

nut thickness Nut thickness is the overall distance from the top of the nut to the bearing surface, measured parallel to the axis of the nut.

O

oiled Oiled is the term denoting the application of a suitable corrosion retarding oil to a fastener.

oval binding head screw Obsolete term for a truss head screw.

oval head Rounded top surface and a conical bearing surface with an included angle usually of 82°.

oven head screw Obsolete term for truss head screw.

oval point An oval point is a radiused point, sometimes referred to as a "Crowned End" or "Round Point."

P

pan head Flat top surface rounded into cylindrical sides, and a flat bearing surface. The recessed pan head has a rounded top surface blending into cylindrical sides and a flat bearing surface.

passivating Passivating is the process of dissolving ferrous particles and surface impurities from stainless steel by chemical means (normally a nitric acid dip) and to produce a passive film on the surface. The purpose is to improve the corrosion resistance of the surface.

physical properties Physical properties are the properties defining the basic characteristics of the material or fastener.

pickling Pickling is the process of removing surface oxides or impurities by chemical or electrochemical means.

pilot point A pilot point is a cylindrical point having a diameter. It is designed to facilitate the alignment and starting of such fasteners as drive screws and groove pins into holes at assembly.

pin A pin is a straight cylindrical or tapered fastener, with or without a head, designed to perform a semi-permanent attaching or locating function (See Mechanical Fastener Section for descriptions of pin types.)

pinch point A pinch point is a short sharp cone point, usually having a point angle of 45 degrees, formed by a pinching operation. This point is normally limited to diameters of ¼ inch or smaller and is applied to metal drive screws, and Type BP tapping screws.

pipe screw thread American Standard pipe threads are tapered 1 inch in 16, or ¾ inch per foot. They are 60 degree threads, of National form with flat or rounded top and bottom.

pitch The distance; measured parallel to fastener axis, between corresponding points on adjacent thread forms in the same axial plane and on the same side of the axis.

pitch diameter On a straight thread, the diameter of the coaxial cylinder, the surface of which would pass through the thread profiles at such points as to make the width of the groove equal to one-half of the basic pitch. On a perfect thread this occurs at the point where the widths of the thread and groove are equal.

pitch line A generator of the cylinder or cone of a screw thread specified in the definition of pitch diameter.

pits Pits are sharp depressions on the surface of a raw material or fastener.

plain Plain as applied to finish of fasteners is used to indicate that the fastener has had no supplementary surface treatment, such as plating, coating, etc., other than being oiled.

plain point The term plain point is applied to the end of a fastener cut approximately flat and perpendicular to the fastener axis. These points on threaded fasteners may be slightly concave especially when threads are rolled.

plain sheared point The end of a fastener cut approximately flat and square to the axis, without chamfer.

plating Plating is the application of a metallic deposit on the surface of the fastener by electrolysis, impact or other suitable means.

plating build up Plating build up is the term used to describe the disposition of more plating on edges or corners than on the other surfaces of the fastener.

point The point of a fastener is the configuration of the end of the shank of a headed fastener or of each end of a headless fastener.

point angle The point angle is the included angle of the point.

point diameter The point diameter is the diameter of the point measured at the extreme end of the fastener. It may sometimes be designated as "Chamfer Diameter" or "Pilot Diameter" on respective point types.

point length The point length is the length of the pointed portion of the fastener, measured parallel to the axis of the fastener from the extreme end. It may sometimes be designated as "Chamfer Length" or "Pilot Length" on respective point types.

point radius The point radius is the spherical radius on an oval or spherical point.

pointing Pointing is a secondary machine operation consisting of cutting points on fastener blanks which were not pointed during the heading operation.

polishing Polishing is the process of producing a smooth surface by rubbing with fine abrasive wheels, belts or compounds.

precipitation hardening Hardening caused by the precipitation of a constituent from a supersaturated solid solution.

process annealing Heating a ferrous alloy to a temperature close to, but below, the lower limit of the transformation range and then cooling, in order to soften the alloy for further cold working.

proof load A proof load is a specified test required for a fastener to indicate that it is suitable for the purpose intended.

proof test A proof test is any specified test required for a fastener to indicate that it is suitable for the purpose intended.

punching Punching is the process of trimming or removing material with dies in a press.

Q

quality Quality denotes the suitability of a fastener for the purpose for which it is intended. Quality should not be confused with precision or workmanship as it is possible that precision parts of good workmanship and finish can be of poor quality if they fail to perform the function for which they are intended. Also for countless applications good quality parts do not require precision or fine finish to serve satisfactorily.

quench hardening Hardening a ferrous alloy by austenitizing and then cooling rapidly enough so that some or all of the austenite transforms to martensite.

quenching Rapid cooling. When applicable, the following more specific terms should be used: direct quenchinch, fog quenching, hot quenching, interrupted quenching, selective quenching, spray quenching, and time quenching.

R

ream To finish a drilled or punched hole very accurately with a rotating fluted tool of the required diameter.

reamer Tool used for enlarging holes previously formed by drilling or boring.

recess depth The recess depth is the distance measured parallel to the fastener axis from the intersection of the head surface with the maximum diameter of the recess to the bottom of the recess.

recess diameter The recess diameter is the diameter measured in a plane perpendicular to the axis of the fastener over the intersection of the outermost extremities of the recess with the head surface.

recess eccentricity Recess eccentricity is the amount that a recess in a recessed head is eccentric with the shank of the fastener.

recess width The recess width is the distance measured in a plane perpendicular to the axis of the fastener across the intersection of the sides or wings of the recess with the head surface.

recessed head A recessed head is a head having a specially formed indentation or recess centered on its top surface. Two common forms of recessed heads are the "Cross Recess" and "Clutch Recess."

reduced diameter body A reduced diameter body is a body the diameter of which may range from the minimum pitch diameter to the minimum major diameter of the thread and is common on screws having rolled heads. Sometimes referred to as "undersize body," which term is not recommended.

reference dimension A reference dimension on a fastener is a dimension without tolerance used for information purposes only.

relief The amount one plane surface of a piece is set below or above another plane, usually for clearance or for economy in machining.

ribbed neck A ribbed neck is a style of neck consisting of longitudinal ribs around the shank adjacent to the underside of the head.

ribs Ribs are small ridges of material usually formed longitudinally around the shank.

right-hand thread A thread is a right-hand thread if, when viewed axially, it winds in a clockwise and receding direction. All threads are right-hand threads unless otherwise designated.

rivet A rivet is a headed metal fastener of malleable material used to join parts of structures and machines by inserting the shank through the aligned holes in each piece and forming a head on the headless end by upsetting. (See Mechanical Fasteners Section for descriptions of rivet types.)

riveting burr A riveting burr is a small plain washer which is assembled with a small rivet before peening the end to provide a large area of contact on the part.

rockwell hardness test A measure of hardness by determining the depth of penetration of a penetrator into the specimen under certain fixed conditions of test. The penetrator may be either a steel ball or a diamond sphero-conical penetrator. The hardness number is related to the depth of indentation and the higher the number the harder the material.

roll threading Applying a thread to a bolt or screw by rolling the piece between two grooved die plates, one of which is in motion, or between rotating grooved circular rolls.

rolled point The point frequently produced by the cupping of the last 1 to 1½ threads by the thread rolling pressure. This type of point is not produced intentionally but is considered as an alternative form of the plain sheared point.

rolled thread A rolled thread is a thread produced by the action of a form tool which when pressed into the surface of a blank displaces material radially.

root That surface of the thread which joins the flanks of adjacent thread forms and is identical with or immediately adjacent to the cylinder or cone from which the thread projects.

root truncation The root truncation of a thread is the distance measured perpendicular to the axis, between the sharp root (or root apex) and the cylinder or cone which bounds the root.

round head Semi-elliptical top surface and flat bearing surface.

runout Runout is a term frequently used interchangeable with eccentricity but which normally refers to the amount which the outside surface of one component of a fastener runs out with rspect to the outside surface of another component. As such it includes eccentricity, angularity and bow. The amount of runout is usually expressed in terms of Total Indicator Reading (TIR).

S

SAE specifications Standards developed by the Society of Automotive Engineers, Inc.

SAE standard screw threads The SAE Screw Thread Standard, as revised in 1954, conforms with the Unified and American Standard (ASA B1.1-1949).

scale Scale is an oxide of iron sometimes formed on the surfaces of hot headed or forged fasteners.

screw dowel A screw dowel is a dowel pin provided with a straight or tapered thread.

screw stock Screw stock is metal in the form of wire or rod which is used for making screw machine parts. Usually it is of a free machining type of material.

screw thread chasers Cutting tools having teeth spaced to match the pitch of the threads to be cut.

screws A screw is an externally threaded fastener. (See Mechanical Fasteners section for descriptions of screw types.)

semi-finished fastener A semi-finished fastener is a fastener made to the same basic dimensions as a finished fastener but having greater tolerances on most dimensions and only the bearing surface and threads finished. (See Finished Fastener, Unfinished Fastener.)

sems This is a preassembled screw and washer unit in which the washer is retained free to rotate under the screw head by the rolled thread. These units expediate assembly operations and assure the presence of a washer in each assembly. They are generally available in various combinations of head styles and washer types.

shank The shank is that portion of a headed fastener which lies between the head and the extreme point.

shaving Shaving is a cutting operation in which thin layers of material are removed from the outer surfaces of the product.

shear fastener A shear fastener is a fastener whose primary function is to resist forces which tend to shear it.

shear strength The stress required to produce fracture when impressed vertically upon the cross-section of a material. Expressed in psi.

shim A thin spacer of sheet metal used for adjusting.

shoulder A shoulder is an enlarged portion of the body of a threaded fastener or shank of an unthreaded fastener.

single thread A single-start thread having lead equal to the pitch.

sliver A sliver is an irregular shaped piece of metal clinging loosely to the finished fastener.

slot depth The slot depth on a headed fastener is the distance measured parallel to the axis of the fastener from the highest part of the head to the intersection of the bottom of the slot with the head or bearing surface. The slot depth on a nut or headless fastener is the distance measured parallel to the fastener axis from the top surface to the extreme bottom of the slot.

slot eccentricity Slot eccentricity is the amount that a slot in a slotted head is eccentric with the body of the fastener.

slot width The slot width is the distance measured in a plane perpendicular to the axis of the fastener over the intersection of the sides of the slot with the head surface of a headed fastener or top surface of a nut.

slotted head A slotted head is a head having a slot centered across its top surface.

slotted nut A hexagon nut having opposed slots at the end opposite to the bearing face which are perpendicular to the axis, designed for insertion of a cotter to secure the nut in place when used with a drilled fastener.

slotting Slotting is the process of forming or cutting the slot on the head of a fastener during either the primary or secondary operation.

socket depth The socket depth is the distance measured parallel to the fastener axis from the intersection of the socket with the head surface to the extreme end of the socket. In socket head screws, the effective socket depth is most often specified as "Key Engagement" which is the distance from the intersection of the socket with the head surface to that depth to which the key or wrench will penetrate, measured in a like manner.

socket diameter The socket diameter is the diameter measured in a plane perpendicular to the axis of the fastener over the intersection of the outermost extremities of the socket with the head surface.

socket head. A recessed head in which the sides of the recess are parallel to the body axis, into which a wrench fits. The recess is commonly hexagonal or fluted in form. Also designated as "internal wrenching head."

socket width The socket width is the distance measured in a plane perpendicular to the axis of the fastener over the intersection of opposite sides of the socket with the head surface.

soft Soft describes the condition of a fastener which though made from a material which can be, and normally is, hardened by heat treatment, has been left in the as-fabricated temper.

special fastener A special fastener is a fastener which differs in any respect from recognized standards.

spherical point A spherical point is an oval point in which the point radius is equal to half the shank diameter.

spot-face To finish a round spot on a rough surface, usually around a drilled hole, to give a good seat to a screw or bolthead, cut, usually 1/16 inch deep.

square neck A square neck is a style of neck consisting of a square shoulder formed integral with the underside of the head.

stainless steel Stainless steel is a corrosion resistant type of alloy steel which contains a minimum of 12 per cent chromium.

standard fastener A standard fastener is a fastener which conforms in all respects to recognized standards.

strain hardening Strain hardening is the increase in hardness and hence strength, resulting from plastic deformation at a temperature below the recrystallization range. Sometimes called work hardening.

stress relieve Heating to a suitable temperature, holding long enough to reduce residual stresses and then cooling to minimize the development of new residual stresses.

stripper bolt Discarded term for a shoulder screw.

stud A stud is a cylindrical rod of moderate length, threaded on either one or both ends or throughout its entire length (See Mechanical Fastener Section for description of stud types.)

surface heat treatment Surface heat treatment is a process that improves the hardness or other mechanical property of the fastener in any surface area.

surface treatment Surface treatment is any treatment which changes the chemical, physical, or mechanical properties of a surface.

swell neck A swell neck is a tapered or variable neck.

T

tap (v) To cut threads in a hole with a rotating tool called a "tap," having threads on it and fluted to give cutting edges.

taper Taper, as used for fasteners, is the angle between one side and the axis of the fastener. Taper may refer to head, shank or some other feature of a fastener.

tapped hole A threaded hole in a part.

temper Temper is the state of a metal or alloy involving its structure and mechanical properties. Temper varies from the annealed temper (soft) to spring temper.

tempering Reheating a quench-hardened or normalized ferrous alloy to a temperature below the transformation range and then cooling at any rate desired.

tension fastener A tension fastener is a fastener whose primary function is to resist forces which tend to elongate it.

thickness of thread The distance between the flanks of the thread measured at a specified position and parallel to the axis.

thread A thread is a ridge of uniform section in the form of a helix on the external or internal surface of a cylinder. This is known as a straight or parallel thread to distinguish it from a taper thread which is formed on a cone or frustum of a cone.

screw thread (See: thread)

thread insert A thread insert is an internally threaded bushing designed to be molded in or inserted into soft or brittle materials to provide greater strength and minimize wear of threaded assembly. (See Mechanical Fastener section for descriptions of insert types.)

thread rolling A cold-forging process whereby screw threads are formed by displacement of metal as opposed to cutting. The process is usually used for producing external threads and is seldom practical for small quantities.

thread series Groups of diameter-pitch combinations distinguished from each other by the number of threads per inch applied to a specific diameter.

threads per inch The reciprocal of the lead in inches.

tolerance A tolerance is the total permissible variation of a size; the tolerance is the difference between the limits of size.

tolerance limit The variation, positive or negative, by which a size is permitted to depart from the design size.

total thread The total thread includes the complete or effective thread and the incomplete thread.

toughness Toughness is the ability of a material to absorb considerable energy without fracturing.

trim heads Similar to the 82° flat and oval head machine screws except that the size of head for a given size screw is one or two sizes smaller than the regular flat and oval sizes.

trimming Trimming is the term applied to the process of shaping or sizing by forcing a part through a die of desired size and shape.

truss head Low rounded top surface with a flat bearing surface.

tumbling Tumbling is the process of cleaning or abrading parts in a rotating container, either with or without cleaning or abrasive materials.

turn To machine on a lathe.

turnbuckle A turnbuckle is a loop or sleeve usually internally threaded with a left-hand thread at one end and a right-hand thread at the other end, intended for assembly with a threaded stud, eye, hook, or jaw at each end, used for applying tension to rods, wire rope, etc. Turnbuckles are some times made with a swivel feature at one end.

twelve point head A standard head form for externally wrenched screws designed for use in counter-bored holes.

U

undercut head For short lengths of flat and oval head machine screws, the heads are undercut to 70 per cent of normal side height to afford greater length of thread on the screws.

underhead fillet An underhead fillet is the fillet at the junction of the head and shank of a headed fastener.

under-size body The reduced body of a bolt or screw. the diameter of which may range from below the pitch diameter to the minimum major diameter of the thread. Such a body diameter is found on some bolts or screws having roller threads.

unfinished fastener An unfinished fastener is a fastener made to the same basic dimensions as a finished fastener but having relatively wider tolerances than a finished fastener and having all surfaces in their formed condition. (See Finished Fastener, Semi-Finished Fastener.)

unified thread standards The basic American standards for fastening screw threads as agreed upon by standards bodies of Canada, the United Kingdom, and the United States. They are a complete and integrated system of threads for fastening purposes. Their outstanding characteristic is general interchangeability of threads achieved through the standardization of thread form, diameter-pitch combinations, and limits of size.

upsetting Upsetting is the process of increasing the cross sectional area by displacement of material longitudinally and radially.

V

voids Voids are internal fissures in ferrous materials. They are sometimes called "chrome checks," "fish eyes," "shatter cracks" and "snow flakes."

W

washer A washer is a part usually thin, having a centrally located hole or partial slot. The washer performs various functions when assembled between the bearing surface of a fastener and the part being attached. Insulation, lubrication, spanning of large clearance holes, and improved stress distribution are a few design uses. (See Mechanical Fasteners Section for descriptions of washer types.)

washer face A washer face is a circular boss on the bearing surface of a bolt or nut.

whitworth thread A screw thread, also known as the British Standard Whitworth (B.S.W.), used principally in Great Britain.

work hardening See: Strain Hardening

work hardness Hardness developed in metal as a result of cold working.

wrenching head A wrenching head is a head having provision for driving or holding by means of a wrench. "External wrenching" designates the application of a wrench externally to the sides of the head. "Internal Wrenching" designates the application of a wrench internally to a socket, the sides of which are parallel to the fastener axis.

Y

yield point The stress necessary to produce an elongation under load of 0.50 per cent of the specimen's original length. Expressed as psi.

yield strength The stress at which a material exhibits a specified limiting set, commonly taken by the offset method as 0.20 per cent of the specimen's original length. Expressed as psi.

An International Fasteners Vocabulary

English	German	French	Spanish
Fastener Types			
eye bolt	Augenschraube	boulon basculant	tornillo con ojo
hexagon bolt	Sechskantschraube	vis a tête hexagonale (ou six pans)	tornillo de cabeza hexagonal
square head bolt	Vierkantschraube	vis à tête carrée	tornillo de cabeza cuadrada
U-bolt	Bügelschraube	étrier fileté	Horquilla roscada
hexagon socket head cap screw	Zylinderschraube mit Innensechskant	vis à tête cylindrique à six pans creux	tornillo de cabeza cilindrica con hexágono interior
knurled thumb screw	Rändelschraube	vis à tête molétee	tornillo de cabeza moleteade
tapping screws	Blechschrauben	vis à tôle	tornillos de rosca cortante
wing screw	Flügelschraube	vis à oreilles	tornillo con cabeza de mariposa
stud	Stiftschraube	goujon	espárrago liso
cap nut	Hutmutter niedrige Form	écrou borgne bas	tuerca ciega baja
hexagon castle nut	Kronenmutter	écrou à créneaux dégagés	tuerca hexagonal almenada alta, con collarete
hexagon lock nut	flache Sechskantmutter	écrou hexagonal bas	tuerca hexagonal baja
hexagon nut	Sechskantmutter	écrou hexagonal	tuerca hexagonal
hexagon slotted nut	Kronenmutter	écrou à créneaux	tuerca hexagonal almenada lisa
knurled nut	flache Rändelmutter	écrou moleté bas	tuerca moleteada baja
square nut	Vierkantmutter	écrou carré	tuerca cuadrada
wing nut	Flügelmutter	écrou à oreilles	tuerca de mariposa
helical spring washer	glatter Federring	rondelle Grower	arandela espiral
plain washer	Scheibe	rondelle plate sans chanfrein	arandela plana
tooth lock washer	federnde Zahnscheibe	rondelle a dents	arandela elástica dentada
Thread Forms			
inch thread	Zollgewinde	filetage en inches	rosca en pulgadas
metric thread	Metrisches Gewinde	filetage métrique	rosca métrica rosca gas
pipe thread	Rohrgewinde	filetage au pas du gaz	rosca gas
screw thread	Schraubengewinde	filetage	rosca cilindrica
tapping screw thread	Blechschrauben-gewinde	filetage de vis à tôle	rosca cortante

English	German	French	Spanish
Head Styles			
binding head	Linsenflachkopf	tête cylindrique bombée	cabeza cilindrica reducida y bombeada
countersunk (flat) head	Senkkopf	tête fraisée	cabeza avellanada
fillister head	Linsenzylinderkopf	tête cylindrique bombée	cabeza cilindrica y bombeada
hexagon head	Sechskantkopf	tête hexagonale	cabeza hexagonal exterior
mushroom (truss) head	Flachrundkopf	tête goutte de suif	cabeza gota de sebo
oval head	Linsensenkkopf	tête fraisée bombéer	cabeza avellanada y bombeada
pan head	Linsenkopf	tête cylindre avec déponille	cabeza bombeada plana
round head	Halbrundkopf	tête ronde	cabeza redonda
square head	Vierkantkopf	tête carrée	cabeza cuadrada
Fastener Dimensions			
angle of chamfer	Fasenwinkel am Kopf	angle du chanfrein	angulo de chaflan
effective pitch diameter	Flankendurchmesser	diamètre sur flancs de filet	diámetro del núcleo
grip length	Klemmlänge	longueur de serrage	longitud de apriete del perno
head angle	Senkwinkel	angle de fraisure	ángulo de la avellanado
head diameter	Kopfdurchmesser	diamètre de la tête	diametro de la cabeza
head height	Kopfhöhe	hauteur de tête	altura de la cabeza
nominal diameter	Nenndurchmesser	diamètre nominal	diámetro nominal
shank diameter	Schaftdurchmesser	diamètre de la partie lisse	diámetro de la espiga
shank length	Länge des gewindefreien Schaftteiles	longueur de la partie lisse	longitud de la parte no roscada
thread diameter	Gewindedurchmesser	diamètre du filetage	diámetro nominal de la rosca
thread length	Gewindelänge	longueur filetée	longitud roscada
width across flats	Schlüsselweite des Sechskantes	dimension sur plats	ancho entre caras
Points and Ends			
chamfered end	Kegelkuppe	à bout plat chanfreiné	extremo achaflanado
cone point	Spitze	à bout pointu	punta cónica
cup point	Ringschneide	extrémité à cuvette	extremo achaflanado con cono embutido
full dog point	Zapfen	extrémité à téton long	punta con largo tetón
half dog point	Kernansatz	extrémité à téton court	punta con corto tetón
plain end	ohne Kuppe	à bout plat	extremo plano
rounded end	Linsenkuppe	extrémité bombée	extremo bombeado
Driving Elements			
cross recess	Kreuzschlitz	empreinte cruciforme	ranura en cruz
hexagon socket	Innensechskant	six pans creux	hexagono interior
slot	Schlitz	fente	ranura recta
wing	Flügel	oreilles	mariposa

PRESSURE-SENSITIVE TAPES

for assembly operations

Pressure-sensitive tapes are essentially self-adhering materials. They consist of a backing, coated on one or both sides with an adhesive that forms a bond when light pressure is applied.

The common tape-backing materials are impregnated-treated paper, glass cloth, metal foils, and films of cellophane, cellulose acetate, vinyl, polyethylene, polyester, unplasticized polyvinylchloride, and TFE fluorocarbon resin. Nearly all adhesives consist of an elastomer and resin. The elastomer, which may be synthetic, crude, or reclaimed rubber, provides internal strength and conformability. The resin provides the desired adhesive properties.

The important physical properties of pressure sensitive tapes are:

1. Thickness: The total thickness of the tape. This dimension is usually expressed in thousandths of an inch or mils.
2. Tensile Strengths: The longitudinal force (lb) required to break a one inch wide tape strip.
3. Stretch Ratio: The amount a tape may be stretched over its original length before breaking. This property is expressed as a percentage of the original length.
4. Adhesion: The force (ounces) required to pull a one inch wide strip of tape from a steel surface. (Federal test method STD No. 147)

TAPE TYPES

Paper-Backed Tapes are among the most widely used industrial tapes because of their adaptability and comparatively low cost. They are made in a range of tensile strength and thicknesses, and are ideally suited for masking, holding, protecting, labeling, and sealing jobs. Some of these tapes are printable, and can be obtained in different colors.

Basic types of paper tapes are: crepe and flatback. In general, crepe backed tapes have moderate stretch and high conformability. Tape thicknesses range from 7.5 to 10 mils. Tensile strength is around 20 lb. Stretch ratio varies from 8 to 17% and adhesion is usually between 30 and 40 oz.

Flatback tapes have a wider range of variation. Thickness varies from 6 to 16 mils, tensile strength from 10 to 80 lb, stretch ratio from 2 to 12%, and adhesion from 2 oz for protective tapes to more than 50 oz for heavy duty sealing tapes.

Because flatback tapes have little stretch and high tensile strength, they are primarily suited for straight line masking work.

Typical paper tape applications include sealing access holes, holding resistors and capacitors in place during assembly, forming a vacuum seal in high-temperature bonding operations, and holding staples and other parts in belt fashion to allow automatic dispensing.

Cellophane Tapes were the first synthetic film tapes designed for industrial use. There are at least six types of cellophane tapes: transparent for sealing, holding, and splicing; colored for decorating and coding; high-tack transparent for heavy duty sealing; special high-tack for stain-resistant applications; printable for advertising and labeling; and low-tack for light duty sealing and other applications where clean removal of tape from a surface without causing surface delamination is required.

These tapes are generally 3 to 4 mils thick. Tensile strength is about 30 lb, stretch ratio is 15%, and adhesion averages approximately 40 oz.

Polyester Tapes have excellent chemical, mechanical, and electrical properties. Currently, there are as many as 15 different tape types. Some of these materials are specifically designed for decoration, solvent resistance, high and low temperature, long-aging, and splicing. Backings may be transparent, colored, metallized silver and gold, or special-purpose polyester materials. Many of these tapes provide a strong moisture-vapor barrier and are often used in place of chrome and gold leaf trim on appliances.

Physical properties of these tapes are: thickness, 2 to 2.5 mils; tensile strength, 25 lb; stretch ratio, 110%; and adhesion, 25 to 40 oz.

PLASTIC TAPES can be used for sealing parts, protecting surfaces from acid and caustic materials, masking, identifying, holding, and insulating. This group of tapes includes backings made of plasticized vinyls, polyethylenes, polypropylenes, and other highly stretchable and conformable plastic materials. Plastic tapes can be made weather resistant for outdoor use. Some tapes can be used as separators for dissimilar metals, such as magnesium and aluminum, or as moisture barriers around tubing joint sections.

A relatively new tape in the plastic group has an unplasticized polyvinylchloride backing. It is designed to retain its properties under adverse atmospheric conditions and is exceptionally clear. It is expected to replace cellophane tapes in many industrial applications because of its greater resistance to changes in humidity.

Plastic tapes are normally available in 3 to 10 mil thicknesses. Tensile strengths range from 9 to 25 lb, stretch ratios from 40 to 250%, and adhesion from 18 to 35 oz.

ACETATE FIBRE TAPES usually have a backing of cellulose acetate film with rope-fibre tissue paper laminated to the film to provide tear resistance. These tapes are considered suitable for functions such as protecting, color coding, and labeling. They can have either a synthetic adhesive for adhering to waxy surfaces, or a rubber based adhesive which may be colored. They may be printable or non-printable. Typical uses for these tapes are identifying and labeling.

Physical properties of these tapes are: thickness 3.5 to 5 mils; tensile strength, averages 35 lb; stretch ratio, 2 to 4%; adhesion, 25 to 35 oz.

ALUMINUM FOIL TAPES can be used for sealing overlap joints of metal sheets, weather-proofing, and similar applications. They are sometimes used to form a dust seal on the yoke of radio speaker assemblies, to reduce noise, and to damp vibrations. Foil tape has also been used as a reflective heat shield for protecting delicate components in space vehicles.

Aluminum foil tapes generally are about 5 mils thick. Tensile strength is about 30 lb; stretch ratio is 5%, and adhesion is 50 oz.

LEAD FOIL TAPE is especially suited for masking areas in selective hard chrome plating because of its ability to "thieve" the plating, resulting in no edge buildup. Also, the lead backing enables the tape to seal windows because of low moisture vapor transmission rate and to provide an excellent barrier against the radioactive rays occasionally encountered during x-ray and atomic energy operations.

Lead foil tapes are about 6.5 to 7 mils thick. They have a tensile strength of 20 lb, a stretch ratio of 15%, and an adhesion of 40 oz.

FILAMENT TAPES are considered the strongest pressure-sensitive tapes available. The backing in these tapes contains thousands of continuous filaments of either glass or rayon yarns. These tapes, which are designed to provide maximum resistance to shock and impact, are particularly suited for reinforcing and closing. They are widely used for bundling, palletizing, and unitizing. The tapes have tensile strengths from 240 to 500 lb, yet are designed not to crush or cut, and can be easily and safely handled. They can be readily removed with a knife but are virtually tearproof. Other physical properties are: thickness, 7 to 10 mils; stretch ratio, 3 to 15%; and adhesion, 40 oz.

DOUBLE-COATED TAPES have a high-tack adhesive on both sides and are available with backings of paper, cloth, cellophane, polyester, unplasticized polyvinylchloride, and foam. Usually, one side is protected with a special paper liner which facilitates tape roll unwinding.

These tapes are generally used for fastening or holding metal or plastic parts in place. Double-coated tapes are generally 4 to 5 mils thick. Tensile strength is 15 to 30 lb; stretch ratio is 6 to 15%; and adhesion is 45 to 65 oz.

ADHESIVE TRANSFER TAPES consist of an exceptionally thin adhesive line in roll form. They have a removable paper liner over one side of the adhesive. When the paper liner is removed, only the adhesive remains on the taped surface. The adhesive line thickness is 2 mils or less. These tapes are commonly used for adhering metal and plastic name plates to surfaces, and for metal-to-metal bonding. Tensile strength and stretch ratio is indeterminate since there is no tape backing. When adhesive is laminated to a backing, adhesion ranges as high as 60 oz.

FOAM TAPES are one of the new types of pressure-sensitive tapes. They provide high shear strengths and can hold heavy metal and plastic objects to rough surfaces. Both double-coated and single-coated styles are available. Double-coated foam tapes are generally used for mounting parts and cushioning vibration. Some of the single-coated foam tapes are designed specifically for such applications as insulating, padding, cushioning, gasketing, and sealing. Thicknesses of foam tapes can range from 30 to 250 mils. Tensile strength varies from 100 to 150 lbs per cubic in., stretch ratio is 125%, and adhesion is 48 oz per inch of width. These tapes are available in both high and low density, and in various materials from sponge rubber to urethane foam. The high density urethane with long-aging adhesive provides permanent holding, and, in general, will support up to ¼ lb/sq in. Low density foam tapes do not provide a rigid bond for permanency.

BONDING FILM TAPES are the newest addition to the pressure-sensitive tape group. They can provide an effective method of permanent adhesive bonding, and are usually thicker in caliper (2 to 6.5 mils) than adhesive transfer tapes to permit use with large and heavy parts. These tapes can be die-cut into almost any shape and size for holding, sealing, and joining specific types and configurations of metal and plastic parts. Bonding film tape also can be used for holding metal grilles on instrument panels and for bonding metal name plates to surfaces, friction materials to brake or clutch assemblies, and plastic trim to non-plastic parts. When cured, these tapes provide an extremely good bond.

Courtesy Industrial Tape Division, Minnesota Mining and Mfg. Company, St. Paul, Minnesota

Associations

These groups are at work to help you with your business, your products and your manufacturing methods. If any one can be of service to you, or if you desire membership, we urge you to contact them for further information. A summary of the association's purpose is included where information was available.

ADHESIVE & SEALANT COUNCIL, INC.
1410 Higgins Road, Park Ridge, Ill. 60068
Phone: (312) 825-8196
Contact: Robert M. Stansel, Executive Secretary
Aim: To promote, encourage and advance the uses of the products in the industry.

ADHESIVES MANUFACTURERS ASSOCIATION OF AMERICA
441 Lexington Ave., New York, N.Y. 10017
Phone: (212) 678-5370
Contact: Kenyon Loomis, Executive Secretary

AEROSPACE INDUSTRIES ASSOCIATION OF AMERICA, INC.
1725 De Sales St., N.W., Washington, D.C. 20036
Phone: (202) 347-2317
Contact: Karl G. Harr, Jr., President & General Manager
Aim: To serve as a vehicle for the interchange of information among its member companies, and to represent them in their relations with their customers, government or commercial, on matters of industry-wide interest.

ALUMINUM ASSOCIATION, THE
420 Lexington Ave., New York, N.Y. 10017
Phone: (212) 679-5580
Contact: S. L. Goldsmith, Jr., Executive Vice President
Aim: To increase the use of aluminum by providing users with detailed technical and marketing information about aluminum.

AMERICAN BUREAU OF METAL STATISTICS
50 Broadway, New York, N.Y. 10004
Phone: (212) 944-1870
Contact: H. Fasting, Director
Aim: Compile and release economic statistical reports on the non-ferrous minerals and metals. Compile and publish a Year Book on these minerals and metals.

AMERICAN ELECTROPLATERS' SOCIETY, INC.
56 Melmore Gardens, East Orange, N.J. 07017
Phone: (201) 678-4100
Contact: Rodney Leeds, General Manager
Aim: Publish a journal and hold meetings on electroplating, metal finishing and allied coating processes.

AMERICAN INSTITUTE OF INDUSTRIAL ENGINEERS, INC.
345 East 47th St., New York, N.Y. 10017
Phone: (212) 752-6800, Ext. 235
Contact: J. F. Jericho, Executive Director
Aims: To advance the profession of Industrial Engineering through the resources and creative talents of its members and to promote the dissemination of knowledge and information in order to improve the art and science of Industrial Engineering.

AMERICAN PRODUCTION & INVENTORY CONTROL SOCIETY
Suite 504 Watergate Bldg., 2600 Virginia Ave., N.W., Washington, D.C. 20037
Phone: (202) 333-1660
Contact: H. C. "Hal" Hubbard, Executive Director
Aim: To develop efficiency through the study and application of scientific principles and methods of production and inventory control management.

AMERICAN ENGINEERING ASSOCIATION
Lakeside, Michigan 49116
Contact: T. Edward Black
Aim: To promote the economic viability of the United States of America through engineering excellence. To obtain this stated goal, the American Engineering Association shall dedicate itself to advancing the engineering profession through promoting engineering education, state registration and professional ethics.

AMERICAN SOCIETY FOR ABRASIVE METHODS
20 N. Wacker Dr., Chicago, Ill. 60606
Phone: (312) 236-2436
Contact: Joseph L. Koach, Executive Secretary
Aim: For the advancement of scientific knowledge in the field of abrasives; to provide the means and methods of applying such knowledge in practice and education.

AMERICAN SOCIETY FOR METALS
Metals Park, Ohio 44073
Phone: (216) 338-5151
Contact: Allan Ray Putnam, Managing Director
Aim: To advance the knowledge of metals and related materials in engineering, design, processing and fabricating.

AMERICAN SOCIETY FOR QUALITY CONTROL
161 West Wisconsin Ave., Milwaukee, Wis. 53203
Phone: (414) 272-8575
Contact: Robert W. Shearman, Administrative Secretary
Aim: To create, promote and stimulate interest in the advancement of the science of quality control and its application to industry and business.

AMERICAN SOCIETY FOR TESTING & MATERIALS
1916 Race St., Philadelphia, Pa. 19103
Phone: (215) 569-4200
Contact: T. A. Marshall, Jr., Executive Secretary
Aim: The promotion of knowledge of the materials of engineering, and the standardization of specifications and the methods of testing.

AMERICAN SOCIETY OF MECHANICAL ENGINEERS, THE
345 East 47th St., New York, N.Y. 10017
Phone: (212) 752-6800
Contact: O. B. Schier, II, Executive Director and Secretary
Aim: To develop and disseminate technical information in order to provide a continuing education for mechanical engineers.

AMERICAN SOCIETY OF TOOL AND MANUFACTURING ENGINEERS
20501 Ford Rd., Dearborn, Mich. 48128
Phone: (313) 271-1500
Contact: Robert M. Johnson, President
Aim: To advance the manufacturing sciences through the continuing education of manufacturing engineers and managers by means of clinics, seminars, workshops, conferences, symposia and tool expositions.

AMERICAN SUPPLY & MACHINERY MANUFACTURERS' ASSOCIATION, INC.
2130 Keith Building, Cleveland, Ohio 44115
Phone: (216) 241-7333
Contact: F. A. Petersen, Business Manager
Aim: To build a closer relationship and better cooperation between manufacturers and industrial distributors. Also helps members improve their marketing methods and techniques.

AMERICAN WELDING SOCIETY
345 E. 47th St., New York, N.Y. 10017
Phone: (212) 752-6800
Contact: Fred L. Plummer, Executive Director
Aim: The advancement of the science, technical knowledge and art of welding and the extension of its application.

ASSOCIATION OF ELECTRONIC MANUFACTURERS, INC.
Suite 1710, 100 S. Wacker Dr., Chicago, Ill. 60606
Phone: (312) 346-4800
Contact: Kenneth C. Prince, Executive Vice President
Aim: To foster better understanding and business practices between the manufacturers of electronic parts, equipment and related items and the customers to whom they see their products for purposes of resale.

CONVEYOR EQUIPMENT MANUFACTURERS ASSOCIATION
1000 Vermont Ave., N.W., Suite 504, Washington, D.C. 20005
Phone: (202) 628-4634
Contact: Charles E. Perry, Executive Secretary
Aim: To promote the proper application of continuous materials handling techniques and the use of conveying equipment to the mutual benefit of members and customers.

ELECTRICAL APPARATUS SERVICE ASSOCIATION, INC.
7730 Carondelet Ave., St. Louis, Mo. 63105
Phone: (314) 726-4944
Contact: August A. Baechle, Executive Vice President
Aim: Co-operative self-improvement, including collecting and disseminating engineering data, information about improved methods and materials, and education for better management.

ELECTRONIC INDUSTRIES ASSOCIATION
2001 Eye St., N.W., Washington, D.C. 20006
Phone : (202) 659-2200
Contact: James D. Secrest, Executive Secretary
Aim: To serve as a trade association on a national scale for all U.S. electronic manufacturers.

GAS APPLIANCE MANUFACTURERS ASSOCIATION, INC.
60 East 42nd St., New York, N.Y. 10017
Phone: (212) 682-8743
Contact: Harold Massey, Managing Director

INDUSTRIAL AUDIO-VISUAL ASSOCIATION
P.O. Box 656, Downtown Station, Chicago, Ill. 60690
Phone: (312) 662-3460
Contact: Frederic J. Woldt, Executive Secretary and Treasurer
Aim: To study all means of audio-visual communication; and to promote better standards and equipment, and to establish a high concept of ethics in relation to its members with associated interests.

INDUSTRIAL FASTENERS INSTITUTE
1717 E. Ninth St., Cleveland, Ohio 44114
Phone: (216) 241-1482
Contact: Frank Masterson, President
Aim: To advance fastener technology and application engineering through planned programs of research and education.

INDUSTRIAL GAS CLEANING INSTITUTE, INC.
150 Purchase St., Box 448, Rye, N.Y. 10580
Phone: (914) 907-7044
Contact: Herbert R. Herington, Executive Secretary
Aim: To promote the interests of all manufacturers of industrial gas cleaning equipment.

INDUSTRIAL MANAGEMENT SOCIETY
20 N. Wacker Drive, Chicago, Ill. 60606
Phone: (312) 939-4957, 4958
Contact: Joseph L. Koach, Executive Manager
Aim: To advance the science of industrial management through meetings, publications and journals.

INSTITUTE OF ELECTRICAL & ELECTRONICS ENGINEERS, INC., THE
345 E. 47th St., New York, N.Y. 10017
Phone: (212) 752-6800
Contact: Donald G. Fink, General Manager
Aim: Its purposes are scientific, literary and educational, directed toward the advancement of the theory and practice of electrical engineering, electronics, radio, allied branches of engineering or the related arts and sciences.

INSTRUMENT SOCIETY OF AMERICA
530 Wm. Penn Place, Pittsburgh, Pa. 15219
Phone: (412) 281-3171
Contact: Herbert S. Kindler, Executive Director

INTERNATIONAL MATERIAL MANAGEMENT SOCIETY
Landmark Bldg., 1343 H St., N.W., Washington, D.C. 20005
Phone: (202) 638-3330
Contact: Sam P. Triffy, Executive Director
Aim: To hold educational seminars and displays of materials.

MANUFACTURERS STANDARDIZATION SOCIETY OF THE VALVE & FITTINGS INDUSTRY
420 Lexington Ave., New York, N.Y. 10017
Phone: (212) 532-6146
Contact: Robert V. Warrick, Executive Secretary

MATERIAL HANDLING EQUIPMENT DISTRIBUTORS ASSOCIATION
20 N. Wacker Dr., Chicago, Ill. 60606
Phone: (312) 236-2436
Contact: Joseph L. Koach, Executive Director

MATERIAL HANDLING INSTITUTE, INC., THE
250 Gateway Towers, Pittsburgh, Pa. 15222
Phone: (412) 471-1580
Contact: L. West Shea, Managing Director
Aim: To promote education in the use of material handling equipment; to compile, distribute and exchange helpful information, and to cooperate with industry and government in the promotion of standards.

METAL TREATING INSTITUTE
150 Purchase St., Box 448, Rye, N.Y. 10580
Phone: (914) 967-3360
Contact: Herbert R. Herington, Executive Secretary
Aim: To protect, promote, foster and advance the interests of persons, firms or corporations engaged in the business of performing and selling in the open market the service of heat treating metals and metal products.

NATIONAL ASSOCIATION OF MANUFACTURERS
277 Park Ave., New York, N.Y. 10017
Phone: (212) 826-2100
Contact: W. P. Gullander, President
Aim: To promote the industrial interest of the United States; foster domestic and foreign commerce of the U.S.

NATIONAL ASSOCIATION OF METAL FINISHERS
248 Lorraine Ave., Upper Montclair, N. J. 07043
Phone: (201) 746-8933
Contact: P. Peter Kovatis, Executive Director
Aim: To provide its members with educational seminars and offers Washington representation.

NATIONAL ELECTRICAL MANUFACTURERS ASSOCIATION
155 E. 44th St., New York, N.Y. 10017
Phone: (212) 682-1500
Contact: Joseph F. Miller, Executive Vice President
Aim: To assist member companies to operate more efficiently and profitably by providing a necessary and helpful exchange of useful information on statistical and other phases of the industry.

NATIONAL FLUID POWER ASSOCIATION
P.O. Box 49, Thiensville, Wis. 53092
Phone: (414) 242-4160
Contact: Theodore Pearce, Executive Vice President
Aim: Standardization, marketing and statistical services; educational projects, public information, management services and government liaison.

NATIONAL INDUSTRIAL DISTRIBUTORS' ASSOCIATION
1900 Arch St., Philadelphia, Pa. 19103
Phone: (215) 564-3484
Contact: Robert C. Fernley, Managing Director
Aim: To improve conditions in the industrial distribution industry. Provides educational seminars and maintains close liaison with suppliers of the industry.

NATIONAL INSTITUTE OF JIG & FIXTURE COMPONENT MANUFACTURERS
294 Fairmount Ave., Oakland, Calif. 94611
Phone: (415) 451-6299
Contact: George S. Eaton, Executive Secretary
Aim: To provide standardization of products and annual meetings.

NATIONAL MACHINE TOOL BUILDERS' ASSOCIATION
2130 Wisconsin Ave., N.W., Washington, D.C. 20007
Phone: (202) 337-8270
Contact: James A. Gray, Executive Vice President
Aim: To be an effective spokesman for the industry in promoting a better understanding of the Machine Tool Industry by the government, the public and all other interested groups; to provide educational, technical, economic and informational services to its members; to encourage higher technical and safety standards, and to promote technological advancements and improvements in the design, engineering, manufacturing and selling of machine tools.

NATIONAL SAFETY COUNCIL
425 N. Michigan Ave., Chicago, Ill. 60611
Phone: (312) 527-4800
Contact: Howard Pyle, President
Aim: Safety in use, operation and application.

NATIONAL SCREW MACHINE PRODUCTS ASSOCIATION
2860 E. 130th St., Cleveland, Ohio 44120
Phone: (216) 751-0909
Contact: Frank T. McGinnis, Executive Vice President
Aim: To collect and disseminate pertinent statistics for the industry; conducts seminars on technical and management subjects, and serves as industry liaison on government matters. Develops training materials.

NATIONAL STANDARDS ASSOCIATION, INC.
1321 Fourteenth St., N.W., Washington, D.C. 20005
Phone: (202) 332-3900
Contact: Clarence H. Bennett, President
Aim: Publishes and distributes National Aerospace Standards and provides related technical data and service.

NATIONAL TOOL, DIE & PRECISION MACHINING ASSOCIATION
1411 K St., N.W., Suite 600, Washington, D.C. 20005
Phone: (202) 347-9011
Contact: William E. Hardman, Executive Vice President
Aim: To represent its members and the industry in relationships with government; to assist the membership in developing, improving and expanding its markets; to publicize the services of the industry and the Association to all their various publics; to promote continuing education in application of management techniques; to develop a comprehensive structure for training, on all levels.

NUMERICAL CONTROL SOCIETY
44 Nassau St., Princeton, N.J. 08540
Phone: (609) 924-4781
Contact: Mary Ann De Vries, Executive Secretary
Aim: To promote the advancement and diffusion of knowledge and application of numerical control, the advancement of education in numerical control; the development, recommendation and dissemination of numerical control evaluation and justification analysis, the collection, storage and distribution of information of all phases of numerical control; the dissemination of information to further numerical control technology; the recommendation and dissemination of numerical control standards; the exchange of numerical control information with other non-profit organizations, and cooperation with educational institutions in fostering interest in numerical control.

POWER TOOL INSTITUTE
604 Davis St., P.O. Box 1406, Evanston, Ill. 60204
Phone: (312) 869-2020
Contact: J. A. Proven, Executive Manager

RESISTANCE WELDER MANUFACTURERS ASSOCIATION
1900 Arch St., Philadelphia, Pa. 19103
Phone: (215) 564-3484
Contact: R. Bruce Wall, Secretary-Treasurer
Aim: To sponsor research programs for improvement of the various resistance welding processes and establish proper techniques for specific applications; to develop equipment standards, and to promote high standard of ethics in the industry.

SOCIETY OF AUTOMOTIVE ENGINEERS, INC.
Two Pennsylvania Plaza, New York, N.Y. 10001
Phone: (212) 594-5700
Contact: Joseph Gilbert, Secretary and General Manager
Aim: To promote the arts, sciences, standards and engineering practices connected with the design, construction and utilization of self-propelled mechanisms, prime movers, components thereof and related equipment.

SOCIETY OF DIE CASTING ENGINEERS, INC., THE
14530 W. 8 Mile Rd., Detroit, Mich. 48237
Phone: (404) 436-9508
Contact: William A. Campbell, Executive Secretary
Aim: Technical and educational, devoted to the improvement and dissemination of knowledge in the arts and sciences of die casting, in the finishing of metals and in allied arts. To foster education programs, to encourage investigation of new products, and advance through education and research, the development and growth of the die casting industry.

SOCIETY OF AMERICAN VALUE ENGINEERS, INC.
Windy Hill Suite A-1, 1741 Roswell St., Smyrna, Ga. 30080
Phone¾ (404) 436-9508
Contact: Mrs. Joyce A. Marting, Executive Secretary
Aim: To promote the advancement of the knowledge of value engineering.

SOCIETY OF THE PLASTICS INDUSTRY, INC., THE
250 Park Ave., New York, N.Y. 10017
Phone: (212) 687-2675
Contact: Ralph L. Harding, Jr., Executive Vice President
Aim: To assemble and disseminate scientific, engineering and general informative data on plastics; to cooperate with the military and allied departments of the U.S. and Canadian governments in the furtherance of plastics projects, and to promote and advance the application and use of plastics through greater public acceptance and favorable recognition of plastic products.

SPRING MANUFACTURERS INSTITUTE
81 Main St., Bristol, Conn. 06010
Phone: (203) 583-2403
Contact: George E. Underwood, Secretary
Aim: To keep members informed on the newest development in the spring industry.

TOOL & DIE INSTITUTE
777 Busse Highway, Park Ridge, Ill. 60068
Phone: (312) 825-1120
Contact: Thomas E. De Pinto, Executive Vice President
Aim: To foster and advance the special tooling industry through educational programs, public relations, special studies and apprentice programs.

UNITED STATES OF AMERICA STANDARDS INSTITUTE
10 E. 40th St., New York, N.Y. 10016
Phone: (212) 685-3058
Contact: Donald L. Peyton, Managing Director & Secretary
Aim: To act as the national coordinating institution for voluntary standardization, through which interested organizations may cooperate in establishing, recognizing and improving voluntary standards of the United States of America; to promote knowledge, understanding and use of voluntary national standards; to simplify the development of engineering, commercial, consumer and safety standards; to encourage competent organizations to develop standards in accordance with institute procedures; to serve as the national clearing house for information on all U.S. and foreign standards.

WELDED STEEL TUBE INSTITUTE
522 Westgate Tower, Cleveland, Ohio 44116
Phone: (216) 333-4550
Contact: Lyle Jenkins, Executive Director
Aim: To make available to tubing users information common to fabricators of welded stainless and carbon steel tubing; to advance and encourage engineering knowledge and practices; to exchange ideas of precise and economical methods of production; to keep abreast of user requirements and engineering practices; to expand proven uses for welded steel tubing and to stimulate developments and acceptance of new applications.

WIRE ASSOCIATION, INC., THE
209 Montowese St., Branford, Conn. 06405
Phone: (203) 488-6321
Contact: J. Edward Donnellan, Executive Secretary
Aim: To provide a clearing house for information on manufacturing and processing of wire and wire products.

AN42-AN49 EYE BOLTS

Eye bolts are manufactured to Specification MIL-B-6812 and AN Standard drawings AN42 thru AN49. Material is non-corrosion-resistant steel, cadmium plated to specification QQ-P-416.

Example: AN44-10A has a shank diameter of 5/16"; 24 threads per inch; hole diameter ¼"; undrilled shank with length of 1". (AN44-10 is the same bolt, but having a drilled shank).

NOTE: *Available with or without cotter pin hole drilled in shank.*

HOW TO ORDER

Order by AN part number. Add the letter "A" after the dash number when ordering undrilled eye bolts.

AN NUMBER	SHANK DIAMETER	THREADS PER INCH	HOLE DIAMETER
AN42B.	No. 10	32	3/16
AN43B.	¼	28	3/16
AN44.	5/16	24	¼
AN45.	5/16	24	5/16
AN46.	⅜	24	⅜
AN47.	7/16	20	⅜

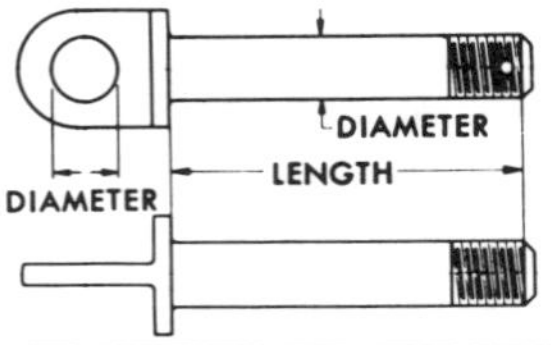

ORDERING TABLE

SHANK LENGTHS Cadmium-plated Steel Eye Bolts

−4 through −22

AN DASH NUMBER	−4	−5	−6	−7	−10	−11	−12	−13	−14	−15	−16	−17	−20	−21	−22
NOMINAL LENGTH (inches)	½	⅝	¾	⅞	1	1⅛	1¼	1⅜	1½	1⅝	1¾	1⅞	2	2⅛	2¼

−23 through −40

AN DASH NUMBER	−23	−24	−25	−26	−27	−30	−31	−32	−33	−34	−35	−36	−37	−40
NOMINAL LENGTH (inches)	2⅜	2½	2⅝	2¾	2⅞	3	3⅛	3¼	3⅜	3½	3⅝	3¾	3⅞	4

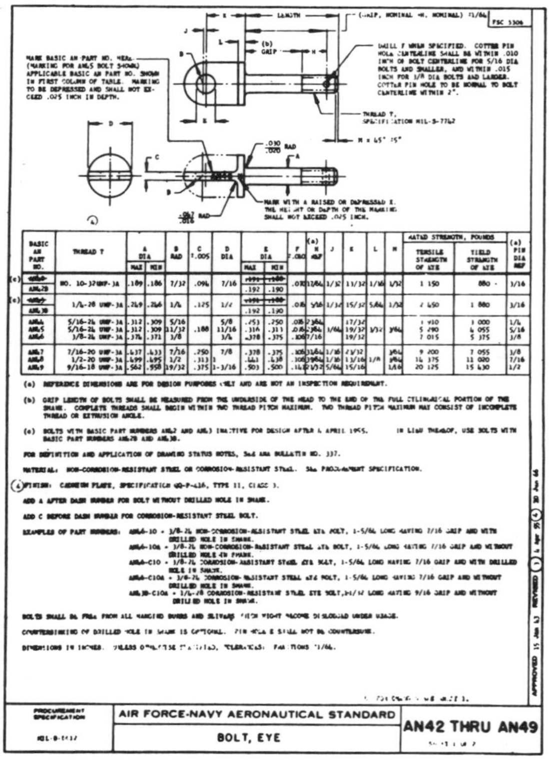

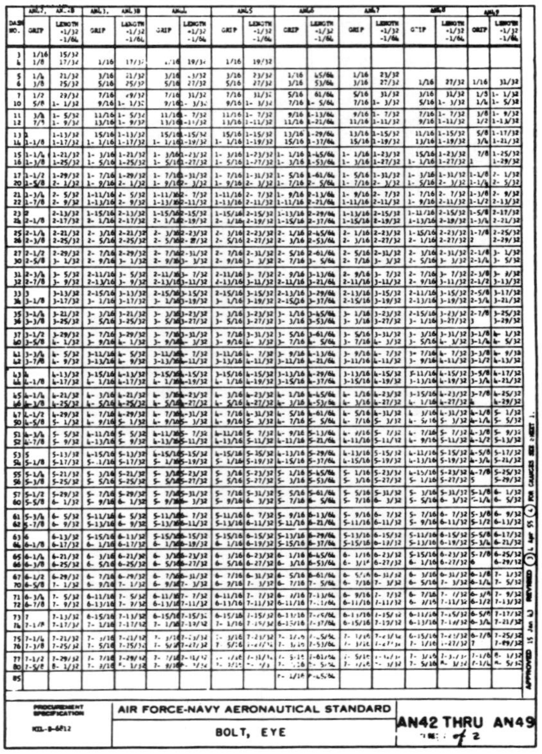

"6 DIGIT" AN BOLTS

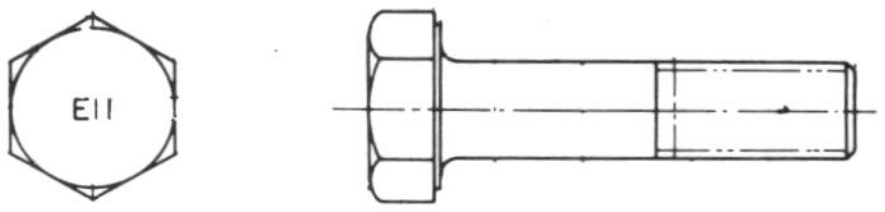

ORDERING TABLE FOR HEX HEAD (PLAIN) BOLTS ONLY

TABLE I

Order by part number. Example: AN101028 is a cadmium plated steel hex head bolt having a diameter of No. 10-32 and a shank length of 2".

SHANK LENGTH (INCHES)	DIAMETER AND THREAD								
	#10-32	1/4-28	5/16-24	3/8-24	7/16-20	1/2-20	9/16-18	5/8-18	3/4-16
3/8	AN101006								
7/16	AN101007								
1/2	AN101008	AN101106							
9/16	AN101009	AN101107							
5/8	AN101010	AN101108	AN101206						
11/16	AN101011	AN101109	AN101207						
3/4	AN101012	AN101110	AN101208	AN101306					
13/16	AN101013	AN101111	AN101209	AN101307	AN101406				
7/8	AN101014	AN101112	AN101210	AN101308	AN101407				
15/16	AN101015	AN101113	AN101211	AN101309	AN101408	AN101506			
1	AN101016	AN101114	AN101212	AN101310	AN101409	AN101507			
1 1/16	AN101017	AN101115	AN101213	AN101311	AN101410	AN101508	AN101606		
1 1/8	AN101018	AN101116	AN101214	AN101312	AN101411	AN101509	AN101607	AN101706	
1 3/16	AN101019	AN101117	AN101215	AN101313	AN101412	AN101510	AN101608	AN101707	
1 1/4	AN101020	AN101118	AN101216	AN101314	AN101413	AN101511	AN101609	AN101708	AN101806
1 5/16	AN101021	AN101119	AN101217	AN101315	AN101414	AN101512	AN101610	AN101709	AN101807
1 3/8	AN101022	AN101120	AN101218	AN101316	AN101415	AN101513	AN101611	AN101710	AN101808
1 7/16	AN101023	AN101121	AN101219	AN101317	AN101416	AN101514	AN101612	AN101711	AN101809
1 1/2	AN101024	AN101122	AN101220	AN101318	AN101417	AN101515	AN101613	AN101712	AN101810
1 9/16		AN101123	AN101221	AN101319	AN101418	AN101516	AN101614	AN101713	AN101811
1 5/8	AN101025	AN101124	AN101222	AN101320	AN101419	AN101517	AN101615	AN101714	AN101812
1 11/16		AN101125	AN101223	AN101321	AN101420	AN101518	AN101616	AN101715	AN101813
1 3/4	AN101026	AN101126	AN101224	AN101322	AN101421	AN101519	AN101617	AN101716	AN101814
1 13/16							AN101618	AN101717	AN101815
1 7/8	AN101027	AN101127	AN101225	AN101323	AN101422	AN101520	AN101619	AN101718	AN101816
1 15/16									AN101817
2	AN101028	AN101128	AN101226	AN101324	AN101423	AN101521	AN101620	AN101719	AN101818
2 1/8	AN101029	AN101129	AN101227	AN101325	AN101424	AN101522	AN101621	AN101720	AN101819
2 1/4	AN101030	AN101130	AN101228	AN101326	AN101425	AN101523	AN101622	AN101721	AN101820
2 3/8	AN101031	AN101131	AN101229	AN101327	AN101426	AN101524	AN101623	AN101722	AN101821
2 1/2	AN101032	AN101132	AN101230	AN101328	AN101427	AN101525	AN101624	AN101723	AN101822
2 5/8	AN101033	AN101133	AN101231	AN101329	AN101428	AN101526	AN101625	AN101724	AN101823
2 3/4	AN101034	AN101134	AN101232	AN101330	AN101429	AN101527	AN101626	AN101725	AN101824
2 7/8	AN101035	AN101135	AN101233	AN101331	AN101430	AN101528	AN101627	AN101726	AN101825
3	AN101036	AN101136	AN101234	AN101332	AN101431	AN101529	AN101628	AN101727	AN101826
3 1/8	AN101037	AN101137	AN101235	AN101333	AN101432	AN101530	AN101629	AN101728	AN101827
3 1/4	AN101038	AN101138	AN101236	AN101334	AN101433	AN101531	AN101630	AN101729	AN101828
3 3/8	AN101039	AN101139	AN101237	AN101335	AN101434	AN101532	AN101631	AN101730	AN101829
3 1/2	AN101040	AN101140	AN101238	AN101336	AN101435	AN101533	AN101632	AN101731	AN101830
3 5/8	AN101041	AN101141	AN101239	AN101337	AN101436	AN101534	AN101633	AN101732	AN101831
3 3/4	AN101042	AN101142	AN101240	AN101338	AN101437	AN101535	AN101634	AN101733	AN101832
3 7/8	AN101043	AN101143	AN101241	AN101339	AN101438	AN101536	AN101635	AN101734	AN101833
4		AN101144	AN101242	AN101340	AN101439	AN101537	AN101636	AN101735	AN101834
4 1/8		AN101145	AN101243	AN101341	AN101440	AN101538	AN101637	AN101736	AN101835
4 1/4		AN101146	AN101244	AN101342	AN101441	AN101539	AN101638	AN101737	AN101836
4 3/8		AN101147	AN101245	AN101343	AN101442	AN101540	AN101639	AN101738	AN101837
4 1/2		AN101148	AN101246	AN101344	AN101443	AN101541	AN101640	AN101739	AN101838
4 5/8		AN101149	AN101247	AN101345	AN101444	AN101542	AN101641	AN101740	AN101839
4 3/4		AN101150	AN101248	AN101346	AN101445	AN101543	AN101642	AN101741	AN101840
4 7/8		AN101151	AN101249	AN101347	AN101446	AN101544	AN101643	AN101742	AN101841
5		AN101152	AN101250	AN101348	AN101447	AN101545	AN101644	AN101743	AN101842
5 1/8		AN101153	AN101251	AN101349	AN101448	AN101546	AN101645	AN101744	AN101843
5 1/4		AN101154	AN101252	AN101350	AN101449	AN101547	AN101646	AN101745	AN101844
5 3/8		AN101155	AN101253	AN101351	AN101450	AN101548	AN101647	AN101746	AN101845
5 1/2		AN101156	AN101254	AN101352	AN101451	AN101549	AN101648	AN101747	AN101846
5 5/8		AN101157	AN101255	AN101353	AN101452	AN101550	AN101649	AN101748	AN101847
5 3/4		AN101158	AN101256	AN101354	AN101453	AN101551	AN101650	AN101749	AN101848
5 7/8		AN101159	AN101257	AN101355	AN101454	AN101552	AN101651	AN101750	AN101849
6		AN101160	AN101258	AN101356	AN101455	AN101553	AN101652	AN101751	AN101850

TO ORDER "6 DIGIT" BOLTS OTHER THAN PLAIN HEX HEAD use Table II. The first 4 digits indicate the type and diameter. The last 2 digits (--) indicate the shank length. Pick up the shank length from Table I. Example: You want the part number for a cres. hex head bolt with a ¼-28 diameter and a (drilled) shank and a length of 4". Table II tells you that the first 4 digits are "1056"; Table I tells you that the final 2 digits are "44". The complete part number, then is AN105644.

FOR "6 DIGIT" NUTS WHICH ARE USED WITH "6 DIGIT" BOLTS SEE TABLE III. Example: AN121576 is a stainless steel castle nut having a No. 10-32 diameter.

TABLE II

Hex Head (drilled shank)	AN1019(--)	AN1020(--)	AN1021(--)	AN1022(--)	AN1023(--)	AN1024(--)	AN1025(--)	AN1026(--)	AN1027(--)
Hex Head (drilled head; 1 hole)	AN1028(--)	AN1029(--)	*	*	*	*	*	*	*
Hex Head (drilled head; 6 holes)	AN1037(--)	AN1038(--)	*	*	*	*	*	*	*
Cres Hex Head (undrilled)	AN1046(--)	AN1047(--)	AN1048(--)	AN1049(--)	AN1050(--)	AN1051(--)	AN1052(--)	AN1053(--)	AN1054(--)
Cres Hex Head (drilled shank)	AN1055(--)	AN1056(--)	AN1057(--)	AN1058(--)	AN1059(--)	AN1060(--)	AN1061(--)	AN1062(--)	AN1063(--)
Cres Hex Head (drilled head; 1 hole)	AN1064(--)	AN1065(--)	*	*	*	*	*	*	*
Cres Hex Head (drilled head; 6 holes)	AN1073(--)	AN1074(--)	*	*	*	*	*	*	*

TABLE III

PLAIN NUT:									
Cadmium Plated	AN121501	AN121502	AN121503	AN121504	AN121512	AN121506	AN121507	AN121508	AN121509
Corr. Resis. Steel	AN121526	AN121527	AN121528	AN121529	AN121537	AN121531	AN121532	AN121533	AN121534
CASTLE NUT:									
Cadmium Plated	AN121551	AN121552	AN121553	AN121554	AN121562	AN121556	AN121557	AN121558	AN121559
Corr. Resis. Steel	AN121576	AN121577	AN121578	AN121579	AN121587	AN121581	AN121582	AN121583	AN121584

* Above Hex Head Ordering table cannot be used to determine part number. Refer to AN Standard Drawing.

AN310 AN315 AN320 — AIRFRAME NUTS

All nuts on this page are manufactured to the latest revision of specification MIL-N-6034. Steel nuts are cadmium plated and dichromate dipped per specification QQ-P-416, Type II, Class 3. Aluminum alloy nuts are anodized per specification QQ-A-696. Corrosion-resisting steel nuts are passivated as specified by Government standards.

HOW TO ORDER
Order by AN part number. Designate desired material by adding the letter "D" before the dash number for aluminum alloy; the letter "C" before the dash number for corrosion-resistant steel.
EXAMPLE: AN315D4R is a plain nut; ¼" diameter; 28 right hand threads per inch; material alum. alloy.

NOTE: *Only AN315 and AN316 nuts are available with right- or left hand thread. Add "R" or "L" after the dash number.*

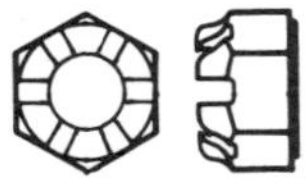
AN310—CASTLE

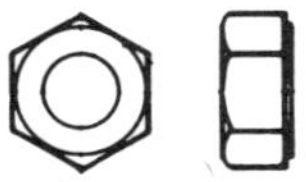
AN315—PLAIN

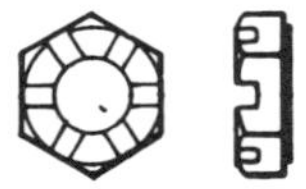
AN320—SHEAR

DASH NUMBER	DIAMETER & THREAD	COTTER PINS USED WITH AN310 & AN320
—2	No. 8-36	MS24665-132
—3	No. 10-32	MS24665-132
—4	¼-28	MS24665-132
—5	5/16-24	MS24665-132
—6	⅜-24	MS24665-283
—7	7/16-20	MS24665-283
—8	½-20	MS24665-283
—9	9/16-18	MS24665-353
—10	⅝-18	MS24665-353
—12	¾-16	MS24665-355
—14	⅞-14	MS24665-355
—16	1-14	MS24665-357
—20	1¼-12	MS24665-360

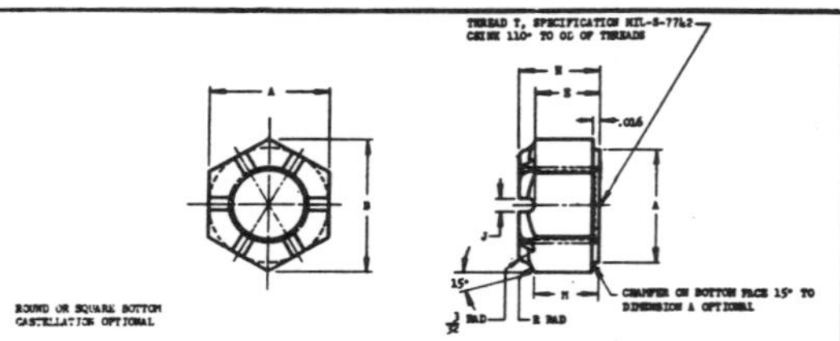

AN PART NO.	(b) THREAD T	ULTIMATE TENSILE STRENGTH MINIMUM, POUNDS		(a) A	B APPROX	E	K	J +1/32 -0	M	R
		STEEL	AL ALLOY							
AN310-3	NO. 10-32NF-3	2 210	1 100	.375 +.002/-.010	7/16	7/64	1/4	5/64	.110	3/32
AN310-4	1/4 -28NF-3	4 080	2 030	.438 +.002/-.010	1/2	1/8	9/32	5/64	.125	3/32
AN310-5	5/16-24NF-3	6 500	3 220	.500 +.002/-.010	37/64	11/64	21/64	5/64	.172	3/32
AN310-6	3/8 -24NF-3	10 100	5 020	.563 +.002/-.010	21/32	7/32	13/32	1/8	.218	3/32
AN310-7	7/16-20NF-3	13 600	6 750	.625 +.002/-.011	23/32	17/64	29/64	1/8	.265	3/32
AN310-8	1/2 -20NF-3	18 500	9 180	.750 +.002/-.012	7/8	23/64	9/16	1/8	.359	1/8
AN310-9	9/16-18NF-3	23 600	11 700	.875 +.002/-.012	1- 1/64	25/64	39/64	5/32	.390	5/32
AN310-10	5/8 -18NF-3	30 100	14 900	1.000 +.002/-.014	1- 5/32	15/32	23/32	5/32	.468	5/32
AN310-12	3/4 -16NF-3	44 000	21 800	1.125 +.002/-.016	1-19/64	9/16	13/16	5/32	.562	3/16
AN310-14	7/8 -14NF-3	60 000	29 800	1.313 +.002/-.017	1-33/64	21/32	29/32	5/32	.656	3/16
AN310-16	1 -14NF-3	80 700	40 000	1.500 +.002/-.019	1-47/64	3/4	1	5/32	.750	3/16
AN310-18	1-1/8 -12NF-3	101 800	50 500	1.688 +.002/-.021	1-61/64	13/16	1- 5/32	5/32	.844	1/4
AN310-20	1-1/4 -12NF-	130 200	64 400	1.875 +.002/-.023	2-11/64	7/8	1- 1/4	5/32	.938	1/4

(a) FOR AL-ALLOY NUTS LARGER THAN -5 SIZE, TOLERANCES ON A MAY CONFORM TO APPLICABLE MATERIAL SPECIFICATIONS FOR BAR AND ROD.

MATERIAL: STEEL, AL ALLOY AND CORROSION-RESISTANT STEEL. SEE PROCUREMENT SPECIFICATION.

FINISH: SEE PROCUREMENT SPECIFICATION.

ADD D BEFORE DASH NUMBER FOR AL-ALLOY NUTS.

ADD C BEFORE DASH NUMBER FOR CORROSION-RESISTANT STEEL NUTS.

EXAMPLES OF PART NO.: AN310-5 = STEEL NUT, 5/16-24NF-3.
AN310D5 = AL-ALLOY NUT, 5/16-24NF-3.
AN310C5 = CORROSION-RESISTANT STEEL NUT, 5/16-24NF-3.

REMOVE ALL BURRS.

DIMENSIONS IN INCHES. UNLESS OTHERWISE SPECIFIED, TOLERANCES: FRACTIONS ±1/64, DECIMALS ±.010, ANGLES ±1°.

(4) (b) UNF-3B THREADS ARE ACCEPTABLE.

SUPERSEDES FORMER USAF AND NAVY STANDARD ISSUES OF AN310

PROCUREMENT SPECIFICATION	AIR FORCE-NAVY AERONAUTICAL STANDARD	AN310
MIL-N-6034	NUT, CASTELLATED, AIRFRAME	

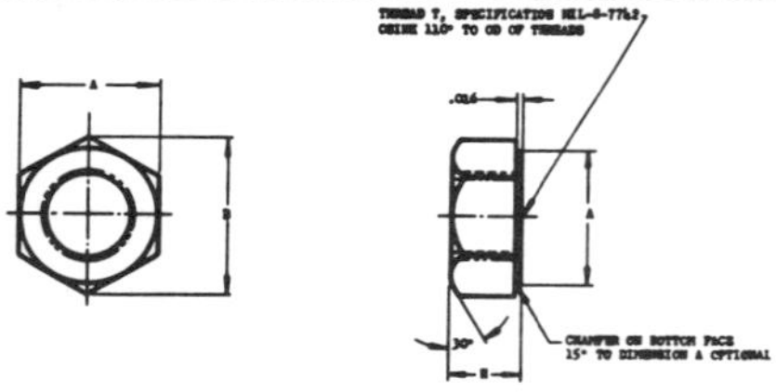

AN PART NO.	(b) THREAD T	ULTIMATE TENSILE STRENGTH MINIMUM POUNDS		(a) A	B APPROX	E
		STEEL	AL ALLOY			
AN315-640	NO. 6-40NF-3	-	-	.313 +.002/-.010	23/64	7/64
AN315-3	NO. 10-32NF-3	2 210	1 100	.375 +.002/-.010	7/16	9/64
AN315-4	1/4 -28NF-3	4 080	2 030	.438 +.002/-.010	1/2	3/16
AN315-5	5/16-24NF-3	6 500	3 220	.500 +.002/-.010	37/64	15/64
AN315-6	3/8 -24NF-3	10 100	5 020	.563 +.002/-.010	21/32	9/32
AN315-7	7/16-20NF-3	13 600	6 750	.625 +.002/-.011	23/32	21/64
AN315-8	1/2 -20NF-3	18 500	9 180	.750 +.002/-.012	7/8	3/8
AN315-9	9/16-18NF-3	23 600	11 700	.875 +.002/-.012	1- 1/64	27/64
AN315-10	5/8 -18NF-3	30 100	14 900	1.000 +.002/-.014	1- 5/32	15/32
AN315-12	3/4 -16NF-3	44 000	21 800	1.125 +.002/-.016	1-19/64	5/8
AN315-14	7/8 -14NF-3	60 000	29 800	1.313 +.002/-.017	1-33/64	21/32
AN315-16	1 -14NF-3	80 700	40 000	1.500 +.002/-.019	1-47/64	3/4
AN315-18	1-1/8 -12NF-3	101 800	50 500	1.688 +.002/-.021	1-61/64	13/16
AN315-20	1-1/4 -12NF-3	130 200	64 400	1.875 +.002/-.023	2-11/64	7/8

(a) FOR ALUMINUM-ALLOY NUTS LARGER THAN -5 SIZE, TOLERANCES ON A MAY CONFORM TO APPLICABLE MATERIAL SPECIFICATIONS FOR BAR AND ROD.

MATERIAL: STEEL, ALUMINUM ALLOY, AND CORROSION-RESISTANT STEEL. SEE PROCUREMENT SPECIFICATION.

FINISH: SEE PROCUREMENT SPECIFICATION.

ADD C BEFORE DASH NUMBER FOR CORROSION-RESISTANT STEEL NUTS.

ADD D BEFORE DASH NUMBER FOR ALUMINUM-ALLOY NUTS.

ADD R AFTER DASH NUMBER FOR RIGHT-HAND THREAD.

ADD L AFTER DASH NUMBER FOR LEFT-HAND THREAD.

EXAMPLES OF PART NO.: AN315-7L = STEEL NUT, 7/16-20NF-3, LEFT-HAND THREAD.
AN315D7R = ALUMINUM-ALLOY NUT, 7/16-20NF-3, RIGHT-HAND THREAD.
AN315C7R = CORROSION-RESISTANT STEEL NUT, 7/16-20NF-3, RIGHT-HAND THREAD.

REMOVE ALL BURRS.

DIMENSIONS IN INCHES. UNLESS OTHERWISE SPECIFIED, TOLERANCES: FRACTIONS ±1/64, DECIMALS ±.010, ANGLES ±1°.

(4) (b) UNF-3B THREADS ARE ACCEPTABLE.

SUPERSEDES FORMER USAF AND NAVY STANDARD ISSUES OF AN315

PROCUREMENT SPECIFICATION	AIR FORCE-NAVY AERONAUTICAL STANDARD	AN315
MIL-N-6034	NUT, PLAIN, AIRFRAME	

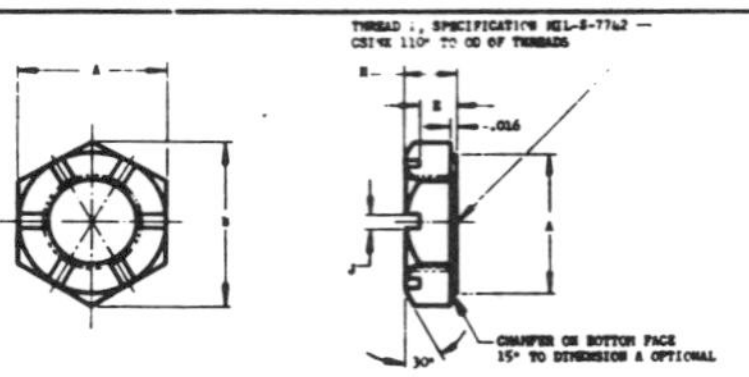

AN PART NO.	(b) THREAD T	ULTIMATE TENSILE STRENGTH MINIMUM POUNDS		(a) A	B APPROX	E	H	J +1/32 -0
		STEEL	AL ALLOY					
AN320-1	NO. 6-40NF-3	-	-	.313 +.002/-.010	23/64	5/64	5/32	5/64
AN320-2	NO. 8-36NF-3	-	-	.344 +.002/-.010	25/64	5/64	5/32	5/64
AN320-3	NO. 10-32NF-3	1 105	550	.375 +.002/-.010	7/16	3/32	3/16	5/64
AN320-4	1/4 -28NF-3	2 040	1 015	.438 +.002/-.010	1/2	3/32	3/16	5/64
AN320-5	5/16-24NF-3	3 250	1 610	.500 +.002/-.010	37/64	3/32	3/16	5/64
AN320-6	3/8 -24NF-3	5 050	2 510	.563 +.002/-.010	21/32	7/64	7/32	1/8
AN320-7	7/16-20NF-3	6 800	3 375	.625 +.002/-.011	23/32	7/64	7/32	1/8
AN320-8	1/2 -20NF-3	9 250	4 590	.750 +.002/-.012	7/8	9/64	1/4	1/8
AN320-9	9/16-18NF-3	11 840	5 850	.875 +.002/-.012	1- 1/64	3/16	5/16	5/32
AN320-10	5/8 -18NF-3	15 050	7 450	1.000 +.002/-.014	1- 5/32	3/16	5/16	5/32
AN320-12	3/4 -16NF-3	22 000	10 900	1.125 +.002/-.016	1-19/64	1/4	3/8	5/32
AN320-14	7/8 -14NF-3	30 000	14 900	1.313 +.002/-.017	1-33/64	5/16	7/16	5/32
AN320-16	1 -14NF-3	40 350	20 000	1.500 +.002/-.019	1-47/64	3/8	1/2	5/32
AN320-18	1-1/8 -12NF-3	50 900	25 250	1.688 +.002/-.021	1-61/64	13/32	9/16	5/32
AN320-20	1-1/4 -12NF-3	65 100	32 200	1.875 +.002/-.023	2-11/64	15/32	5/8	5/32

(a) FOR ALUMINUM-ALLOY NUTS LARGER THAN -5 SIZE, TOLERANCES ON A MAY CONFORM TO APPLICABLE MATERIAL SPECIFICATIONS FOR BAR AND ROD.

MATERIAL: STEEL, ALUMINUM ALLOY, AND CORROSION-RESISTANT STEEL. SEE PROCUREMENT SPECIFICATION.

FINISH: SEE PROCUREMENT SPECIFICATION.

ADD D BEFORE DASH NUMBER FOR ALUMINUM-ALLOY NUTS.

ADD C BEFORE DASH NUMBER FOR CORROSION-RESISTANT STEEL NUTS.

EXAMPLES OF PART NO.: AN320-7 = NUT, STEEL, 7/16-20NF-3.
AN320D7 = NUT, ALUMINUM ALLOY, 7/16-20NF-3.
AN320C7 = NUT, CORROSION-RESISTANT STEEL, 7/16-20NF-3.

REMOVE ALL BURRS.

DIMENSIONS IN INCHES. UNLESS OTHERWISE SPECIFIED, TOLERANCES: FRACTIONS ±1/64, DECIMALS ±.010, ANGLES ±1°.

(5) (b) UNF-3B THREADS ARE ACCEPTABLE.

SUPERSEDES FORMER USAF AND NAVY STANDARD ISSUES OF AN320

PROCUREMENT SPECIFICATION	AIR FORCE-NAVY AERONAUTICAL STANDARD	AN320
MIL-N-6034	NUT, CASTLE SHEAR	

ENGINE NUTS

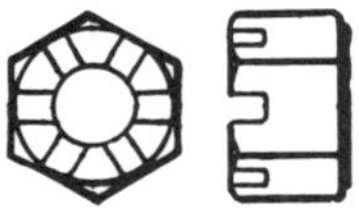

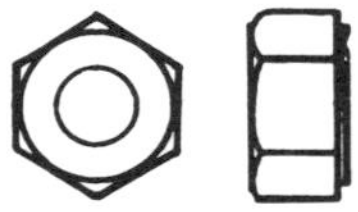

AN355— SLOTTED

AN360 — PLAIN

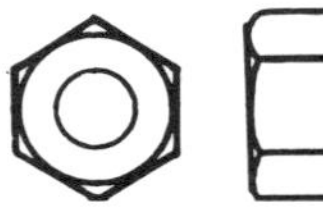

AN121501 - AN121550 PLAIN

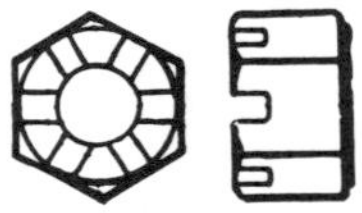

AN121551 - AN121600 CASTLE

AN355 slotted engine nuts are of cadmium plated steel, heat treated 125,000 PSI. Manufactured to AN Standard drawings.

AN360 plain engine nuts are of cadmium plated steel, heat treated 125,000 PSI. Manufactured to AN Standard drawings and to Specification AN-N-2.

AN121501 thru **AN121525** plain nuts are of AMS6322 **steel; cadmium plated** to AMS2400; surface roughness AS107; Rockwell hardness, C19-26.

AN121526 thru **AN121550** plain nuts are of **corrosion resistant steel** per AMS-5628; surface roughness, AS107; finished nuts are degreased, then cleaned in a nitric acid solution.

AN121551 thru **AN121575** castle nuts are of AMS6322 **steel; cadmium plated** to AMS2400; surface roughness, AS107; Rockwell hardness, C19-26.

AN121576 thru **AN121600** castle nuts are of **corrosion resistant steel** per AMS-5628; surface roughness AS107; finished nuts are degreased then cleaned in a nitric acid solution.

ORDERING TABLE AN355 & AN360 NUTS

DASH NUMBER	DIAMETER & THREAD
−3	No. 10-32
−4	¼-28
−5	5/16-24
−6	⅜-24
−7	7/16-20
−8	½-20
−9	9/16-18
−10	⅝-18
−12	¾-16

ORDERING TABLE FOR "6 DIGIT" AN NUTS

DIA. & THREAD	PLAIN NUT		CASTLE NUT	
	CAD. PLATED	STAINLESS	CAD. PLATED	STAINLESS
10-32	AN121501	AN121526	AN121551	AN121576
1/4-28	AN121502	AN121527	AN121552	AN121577
5/16-24	AN121503	AN121528	AN121553	AN121578
3/8-24	AN121504	AN121529	AN121554	AN121579
7/16-20	AN121512	AN121537	AN121562	AN121587
1/2-20	AN121506	AN121531	AN121556	AN121581
9/16-18	AN121507	AN121532	AN121557	AN121582
5/8-18	AN121508	AN121533	AN121558	AN121583
3/4-16	AN121509	AN121534	AN121559	AN121584
7/8-14	AN121510	AN121535	AN121560	AN121585
1-12	AN121511	AN121536	AN121561	AN121586

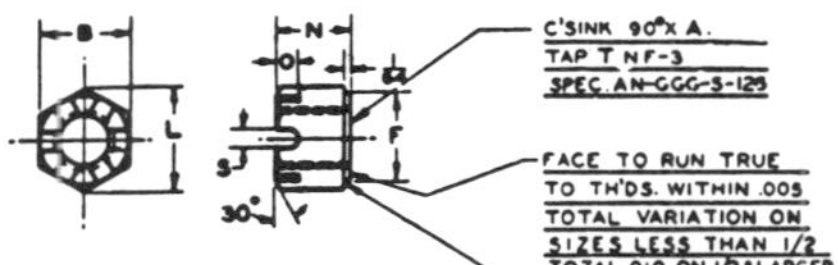

ROUND OR SQUARE BOTTOM CASTELLATION OPTIONAL.

DASH NOS.	TAP T			A	B	F	N	O	S	L
3	#10-32			.190	$.375^{+.002}_{-.010}$	3/8	1/4	3/32	5/64	7/16
4	1/4-28			1/4	$.4375^{+.002}_{-.010}$	7/16	9/32	3/32	5/64	1/2
5	5/16-24			5/16	$.500^{+.002}_{-.010}$	1/2	21/64	3/32	5/64	37/64
6	3/8-24			3/8	$.5625^{+.0025}_{-.0100}$	9/16	13/32	1/8	1/8	21/32
7	7/16-20			7/16	$.625^{+.0025}_{-.0100}$	5/8	29/64	1/8	1/8	23/32
8	1/2-20			1/2	$.750^{+.0025}_{-.0100}$	3/4	9/16	3/16	1/8	7/8
9	9/16-18			9/16	$.875^{+.0025}_{-.0100}$	7/8	39/64	3/16	5/32	1 1/64
10	5/8-18			5/8	$1.000^{+.0025}_{-.0100}$	1	23/32	1/4	5/32	1 5/3[illegible]
12	3/4-16			3/4	$1.125^{+.003}_{-.010}$	1 1/8	13/16	1/4	5/32	1 19/64

EXAMPLE OF PART NUMBER - AN355-4

PACK 50 IN A CARTON IN ACCORDANCE WITH SPEC. 40560.

THESE NUTS ARE DESIGNED SPECIFICALLY FOR ENGINES AND SHALL NOT BE USED ON AIRCRAFT.

SPECIFICATION 29-26 EXCEPT MATERIAL
MATERIAL - STEEL, SPEC. AN-QQ-S-689
OPTIONAL MATERIAL AN-QQ-S-690 OR AN-QQ-S-687
HEAT TREAT - 125,000 LBS. PER SQ. IN.
LIMITS ON DIMENSIONS ±.010 UNLESS OTHERWISE SPECIFIED

REVISED 7-1-42 8-25-41

APPROVED	AIR CORPS STOCK CLASSIFICATION 04A	ARMY & NAVY STD.
3-6-33	NUT - ENGINE - SLOTTED	AN355

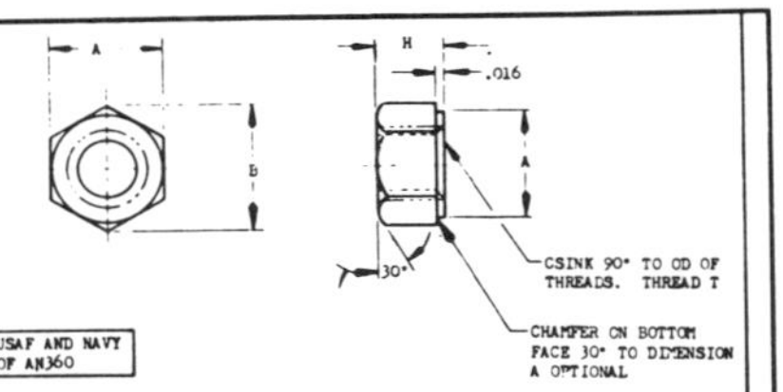

SUPERSEDES FORMER USAF AND NAVY STANDARD ISSUES OF AN360

AN PART NO.	THREAD T NF-3	A DIA	B APPROX	H
AN360-3	NO. 10-32	$.375^{+.000}_{-.010}$	7/16	5/32
AN360-4	1/4 -28	$.438^{+.000}_{-.010}$	1/2	7/32
AN360-5	5/16-24	$.500^{+.000}_{-.011}$	37/64	17/64
AN360-6	3/8 -24	$.563^{+.000}_{-.012}$	21/32	21/64
AN360-7	7/16-20	$.625^{+.000}_{-.013}$	23/32	3/8
AN360-8	1/2 -20	$.750^{+.000}_{-.014}$	7/8	7/16
AN360-9	9/16-18	$.875^{+.000}_{-.014}$	1- 1/64	31/64
AN360-10	5/8 -18	$.938^{+.000}_{-.016}$	1- 5/64	35/64
AN360-12	3/4 -16	$1.063^{+.000}_{-.018}$	1-15/64	21/32
AN360-14	7/8 -14	$1.250^{+.000}_{-.019}$	1- 7/16	49/64
AN360-16	1 -14	$1.438^{+.000}_{-.021}$	1-21/32	7/8
AN360-18	1-1/8 -12	$1.625^{+.000}_{-.023}$	1- 7/8	63/64
AN360-20	1-1/4 -12	$1.813^{+.000}_{-.025}$	2- 3/32	1- 3/32
AN360-22	1-3/8 -12	$2.000^{+.000}_{-.027}$	2- 5/16	1-13/64
AN360-24	1-1/2 -12	$2.188^{+.000}_{-.029}$	2-17/32	1- 5/16

THESE NUTS ARE DESIGNED SPECIFICALLY FOR ENGINES AND SHALL NOT BE USED ON AIRCRAFT.
MATERIAL: STEEL, 125,000 PSI TS MIN. SEE PROCUREMENT SPECIFICATION.
FINISH: SEE PROCUREMENT SPECIFICATION.
THREADS IN ACCORDANCE WITH AN-S-126.
DIMENSIONS IN INCHES. UNLESS OTHERWISE SPECIFIED, TOLERANCES: FRACTIONS ±1/64, DECIMALS ±.010, ANGLES ±1°.

④ CANCELED 7 JULY 1954. SEE DRAWING AN121501 THRU AN121525.

FOR DEFINITION AND APPLICATION OF DRAWING STATUS NOTES, SEE ANA BULLETIN NO. 337.

APPROVED 22 Jul 43 REVISED ① 15 Nov 43 ② 7 Dec 43 ③ 10 Mar 50 ④ 7 Jul 54

PROCUREMENT SPECIFICATION	AIR FORCE-NAVY AERONAUTICAL STANDARD	AN360
AN-N-2	NUT - PLAIN ENGINE	

The famous red nylon elastic collar for all standard applications . . . to 250°F.

The inside diameter of the red nylon insert is deliberately smaller than the major diameter of the matching bolt. The nut spins freely on the bolt until the bolt threads enter the locking insert where they impress, but do not cut mating threads in the nylon. This compression forces a metal to metal contact between the top flanks of the nut threads and the bottom flanks of the bolt threads. This friction hold plus the compression hold of the nylon insert guarantee that the nut will stay "put" anywhere on the bolt. Severe vibration will not loosen its grip and it will maintain a pre-set torque-tension relationship. The resilient action of the nylon insert also absorbs shock and impact loads while the "elastic memory" of nylon permits it to closely resume its original shape when the bolt is removed. This "memory" plus excellent wear resistance means that on standard bolts, nylon insert Elastic Stop® nuts can be re-used through more than 50 on-off application cycles. They fully meet the vibration and re-use requirements of specification MIL-N-25027

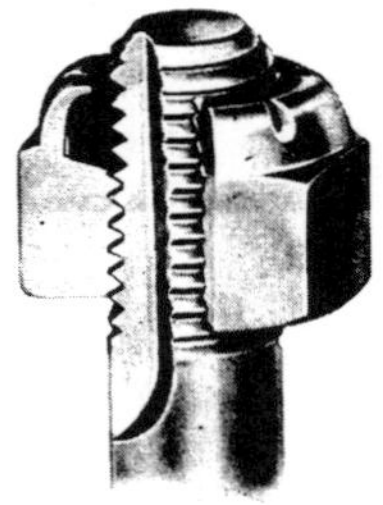

N type

NYLON LOCKING DEVICE

The all-metal elliptical closure for lightweight, close-clearance applications . . . to 1400°F.

The reduced height, lightweight, thin walled series of Elastic Stop® nuts use an offset oval locking device. The upper section of the nut cylinder is elliptically offset to provide a controlled degree of deflection in the threads which creates a consistent, dependable locking torque when engaged by the bolt. These parts fully meet the vibration and re-use requirements of specification MIL-N-25027

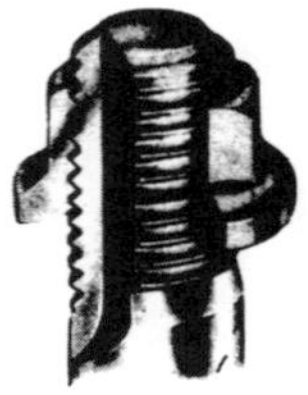

LH type

ELLIPTICAL LOCKING DEVICE

The all-metal beam type locking device for high performance applications . . . to 1400°F.

These all-metal Elastic Stop® nuts utilize a tapered beam type locking device. The long beam sections are deflected inwardly to produce a controlled locking torque at the crown of the nut. The tapered beamform provides superior load distribution along the entire threaded area of the nut assuring high performance at very elevated temperatures.

ESNA's "Z" nuts have become the accepted standard of the aircraft-missile engine industry. They fully meet the vibration and re-use requirements of specification MIL-N-25027

Z type

"BEAM" LOCKING DEVICE

HOW TO USE ELASTIC STOP NUTS

Use full length bolts: The first thread or two on a bolt may be undersize, therefore, at least one and one-half threads should extend beyond the nut collar for maximum locking effect.

Use bolts having clean, full threads: Burrs or rough edges may tear the locking insert. Stripped or worn threads do not provide a satisfactory holding surface. Whenever possible use new undrilled bolts.

Temperature consideration: Use at temperatures below 250°F. High temperatures reduce locking effectiveness of the collar. Esna high temperature nuts are available for temperatures above 250°F.

Do not tap threads in collar: Apply metal threads of nut first to insure proper self-locking action.

ESNA — GANG CHANNEL NUTS

STRAIGHT STAINLESS CHANNEL, COUNTER-BORED NUT
LHTG51, LHTG51-2860
8-32 thru ¼-28 to 450° F., to 900° F.
MS21064, MS21065, NAS688, NAS689, NAS690, NAS691, NAS692, NAS1034, NAS1035, NAS1036, NAS1037

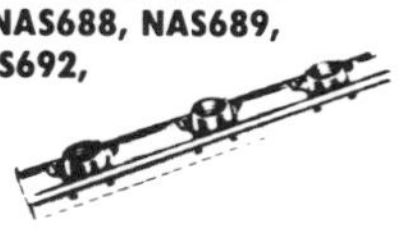

RADIUS CHANNEL
NRG1
8-32 thru 5/16-24
to 250° F.

STRAIGHT CHANNEL
NG1
8-32 thru ⅜-24
to 250° F.

STRAIGHT HIGH STRENGTH ALUMINUM CHANNEL, COUNTERBORED NUT
LHTG51(J)
8-32 thru 5/16-24 to 450° F.
MS21063, NAS688, 689, 690
NAS691, 692, NAS1034
NAS1035, 1036, 1037

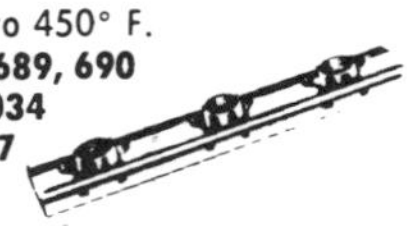

RADIUS CHANNEL, HIGH STRENGTH ALUMINUM, 100° COUNTERSUNK NUT
NRG5
8-32 and 10-32
to 250° F.

STRAIGHT CHANNEL, HIGH STRENGTH ALUMINUM (Dyed Blue)
NGJ1
8-32 thru ¼-28
to 250° F.

STRAIGHT HIGH STRENGTH ALUMINUM CHANNEL, 100° COUNTERSUNK NUT
LHTG55(J)
8-32 and 10-32 to 450° F.
MS21066, NAS693, NAS694, NAS695, NAS1039, NAS1040, NAS1041

RADIUS HIGH STRENGTH ALUMINUM CHANNEL, COUNTERBORED NUT
RG51
8-32 thru 5/16-24
to 450° F.

STRAIGHT CHANNEL, LIGHTWEIGHT, NARROW
NG41
8-32 thru ¼-28
to 250° F.
MS21079, NAS1035

STRAIGHT HIGH STRENGTH ALUMINUM CHANNEL, DEEP COUNTERBORED NUT
G15
10-32 and ¼-28
to 450° F.

RADIUS STAINLESS CHANNEL, STAINLESS NUT
RG38-3980
¼-28
1200° F. and above

STRAIGHT CHANNEL, 100° COUNTERSUNK NUT
NG5
8-32 thru ¼-28
to 250° F.
NAS1039, NAS1040, NAS1041

STRAIGHT STAINLESS CHANNEL, STAINLESS NUT
LHG3493
10-32 thru 5/16-24
to 1200° F.

STRAIGHT CHANNEL, HIGH STRENGTH ALUMINUM, 100° COUNTERSUNK NUT
NGJ5 (Dyed Blue)
8-32 and 10-32
to 250° F.
NAS1039, NAS1040

STRAIGHT CHANNEL, BASE SEAL, REMOVABLE NUT
G2500
10-32 thru 5/16-24
to 250° F.

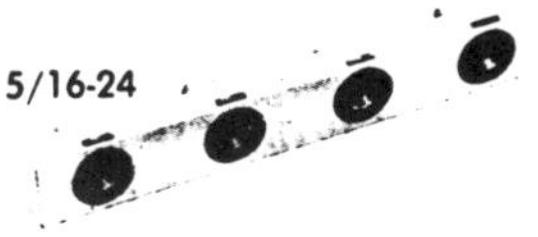

TINNERMAN NUTS

(NAS444) ANCHOR TYPE

Riveted or welded in position for blind location assembly. Double-lug, offset or corner types are available in a wide range of screw sizes.

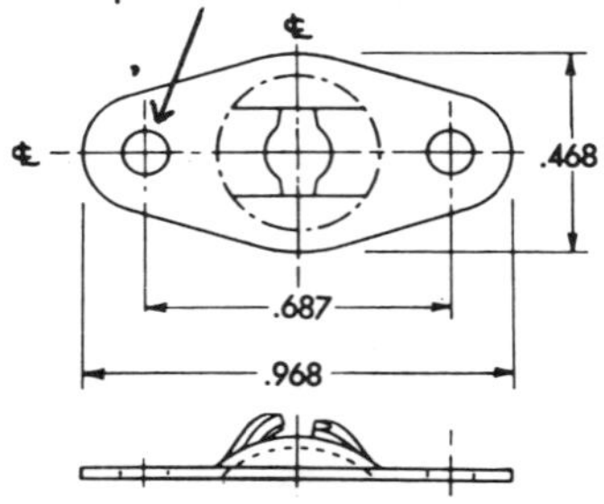

COUNTERSUNK FOR FLAT HEAD SCREWS

This countersunk variation, for use with sheets dimpled for 100 degree flat head screws, is charted only on NAS sheet below. The chart adjacent to the NAS sheet covers the plain anchor only.

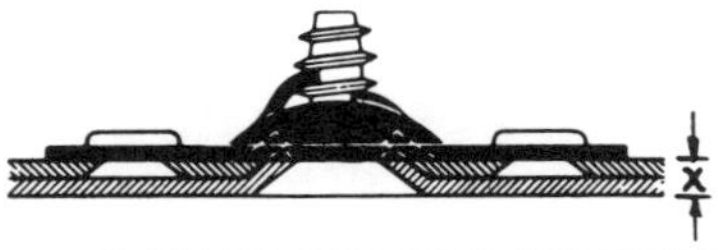

SCREW SIZE	SCREW LENGTH
6B or Z	"X" plus 5/16"
8B or Z	"X" plus 3/8"
10B or Z	

PLAIN FOR ROUND HEAD SCREWS

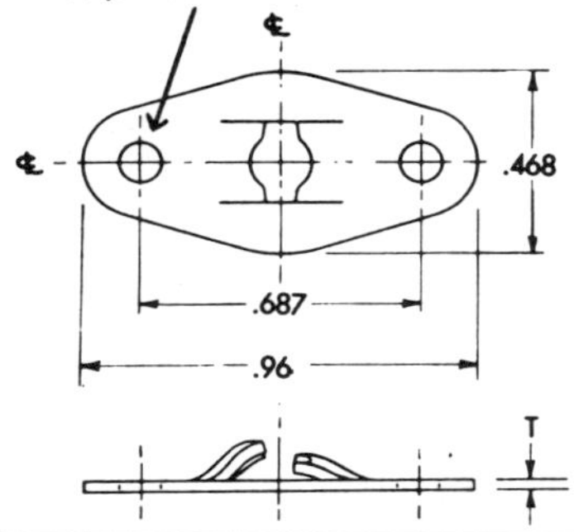

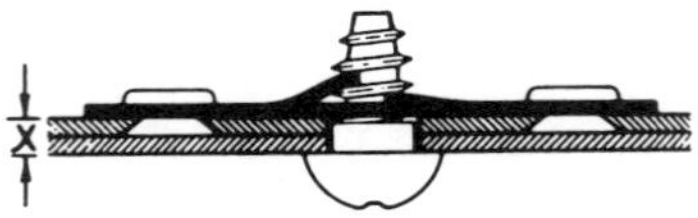

DESIGN VARIATIONS

A

Use with Sheets Dimpled for 100° Flat Head Rivets

B

Welding Nibs Stainless Steel Only

SCREW SIZE	SCREW LENGTH
4B or Z	"X" plus 1/4"
6B or Z	
8B or Z	"X" plus 5/16"
10B or Z	

NATIONAL AIRCRAFT STANDARDS COMMITTEE

AIRCRAFT INDUSTRIES ASSOCIATION OF AMERICA, INC., 610 SHOREHAM BUILDING, WASHINGTON 5, D. C.

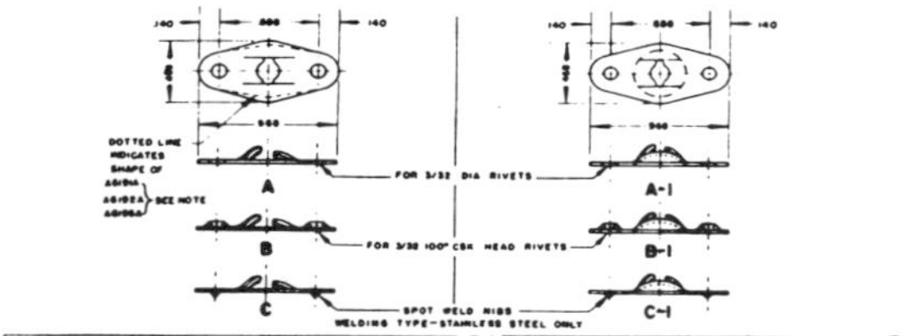

TINNERMAN NO.	NAS NO.	TYPE	SCREW NO.	MATERIAL THICKNESS ①
A6191-6Z	NAS-444-1	A	AN530-6	.022 (.025 SS)
A6195-6Z	NAS-444-2	B		
A6192-6Z	NAS-444-3	C		
A6191-8Z	NAS-444-4	A	AN530-8	.025 (.028 SS)
A6195-8Z	NAS-444-5	B		
A6192-8Z	NAS-444-6	C		
A6191-10Z	NAS-444-7	A	AN530-10	.028 (.031 SS)
A6195-10Z	NAS-444-8	B		
A6192-10Z	NAS-444-9	C		

TINNERMAN NO.	NAS NO.	TYPE	MATERIAL THICKNESS ①
A6203-6Z	NAS-444-10	A-1	.022 (.025 SS)
A6162-6Z	NAS-444-11	B-1	
A6164-6Z	NAS-444-12	C-1	
A6203-8Z	NAS-444-13	A-1	.025 (.028 SS)
A6162-8Z	NAS-444-14	B-1	
A6164-8Z	NAS-444-15	C-1	
A6203-10Z	NAS-444-16	A-1	.028 (.031 SS)
A6162-10Z	NAS-444-17	B-1	
A6164-10Z	NAS-444-18	C-1	

NOTE: ADD SUFFIX "A" IMMEDIATELY AFTER FIRST THREE PART NUMBERS FOR NUT .38" WIDE. EXAMPLE: A6191A-6Z NAS-444-1-A

PROCUREMENT: TINNERMAN PRODUCTS, INC., CLEVELAND, OHIO. SIMILAR ITEMS OF COMPARABLE FUNCTION MUST BE EQUIVALENT PRIOR TO USE ON AIRCRAFT IN LIEU OF ITEMS SHOWN HEREON. ①

ENGINEERING INFORMATION:
PROCUREMENT SPECIFICATION: AAF25533.
SCREW LENGTH: (TYPE A, B, C; 6Z, X PLUS 1/4"; 8Z & 10Z, X PLUS 5/16". (TYPE A-1, B-1, C-1; 6Z, X PLUS 5/16"; 8Z & 10Z, X PLUS 3/8".
THE DOUBLE LUG ANCHOR TYPE NUT IS TO BE USED WHEREVER POSSIBLE IN APPLICATIONS THAT ARE NOT STRUCTURAL OR SUBJECT TO EXCESSIVE VIBRATION. FOR ADDITIONAL SIZES AND SHAPES CONSULT THE STANDARDS UNIT.
MATERIAL: SAE 1060 STEEL OR STAINLESS STEEL.
SUFFIXES: ADD SUFFIX "SS" FOR STAINLESS STEEL (EXAMPLE: A6191SS-6Z; NAS-444-1-SS). ADD SUFFIX "C" FOR RIVET HOLES FOR 1/8" DIA RIVETS (EXAMPLE: A6191-6Z-C; NAS-444-1-C).
COMMERCIAL PART CODE: A6191-6Z = DOUBLE LUG ANCHOR TYPE SPEED NUT .968 LONG, .468 WIDE.
NAS PART CODE: NAS-444-1 = DOUBLE LUG ANCHOR TYPE SPEED NUT .968 LONG, .468 WIDE.
FINISH: SAE 1060 STEEL = METALLIC SPRAY (ZINC). AN-M-6 ① STAINLESS STEEL = NONE.

② INACTIVE FOR NEW DESIGN

APPROVAL DATE JAN 1948 REVISION ① DEC. 1949. (2) 5-15-51

EXAMPLE OF CALL-OUT:

A6191-6Z	NUT - DOUBLE LUG ANCHOR TYPE NAS-444-1			TINNERMAN PRODUCTS, INC., CLEVELAND, OHIO (OR EQUIVALENT)	
PART NUMBER (OPTIONAL)	NAME OF PART NAS PART NUMBER	ZONE CODE	SIZE - GAGE SEC. NO.	MATERIAL	H. T.

TITLE	CLASSIFICATION
NUT - DOUBLE LUG ANCHOR TYPE	COMMERCIAL PART
	NAS 444

TYPICAL PART NUMBER: A6191-4Z-1

SCREW SIZE	DESIGN VARIATION	PART NUMBER	H Rivet Diameter	T Material Thickness
4B or Z		A6191-4Z-	3/32	.022
6B or Z		A6191-6Z-		.025
8B or Z		A6191-8Z-		.028
10B or Z		A6191-10Z-		.031
6B or Z		A8579-6Z-	1/8	.025
8B or Z		A8579-8Z-		.028
10B or Z		A8579-10Z-		.031
4B or Z	A	A6195-4Z-	3/32	.022
6B or Z		A6195-6Z-		.025
SB or Z		A6195-8Z-		.028
10B or Z		A6195-10Z-		.031
4B or Z		A8583-4Z-	1/8	.020
6B or Z		A8583-6Z-		.025
8B or Z		A8583-8Z-		.028
10B or Z		A8583-10Z-		.031
4B or Z	B	A6192SS-4Z-		.022
6B or Z		A6192SS-6Z-		.025
8B or Z		A6192SS-8Z-		.028
10B or Z		A6192SS-10Z-		.031

CORNER ANCHOR TYPE TINNERMAN NUTS

FOR ROUND HEAD SCREWS

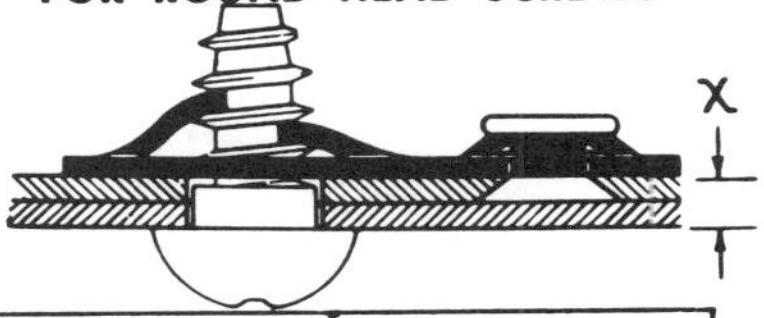

SCREW SIZE	SCREW LENGTH
4B or Z 6B or Z	"X" plus 1/4"
8B or Z 10B or Z	"X" plus 5/16"

DESIGN VARIATIONS

A

Use with Sheets Dimpled for 100° Flat Head Rivets

B

Welding Nibs Stainless Steel Only

FOR FLAT HEAD SCREWS

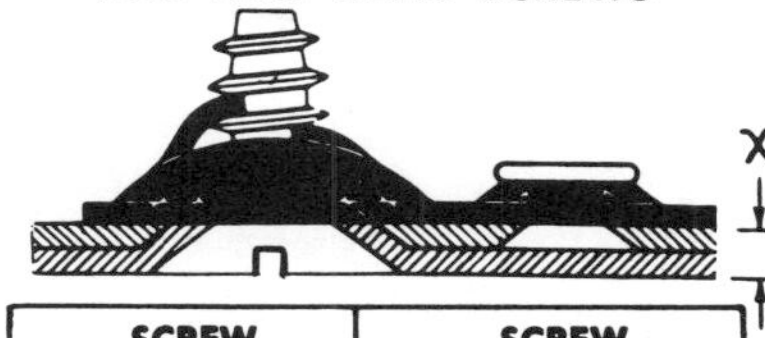

SCREW SIZE	SCREW LENGTH
4B or Z 6B or Z	"X" plus 5/16"
8B or Z	"X" plus 3/8"

ORDERING AND WEIGHT TABLE Corner Anchor Type Speed Nuts

PLAIN

"H" Two clearance holes for required rivet diameters

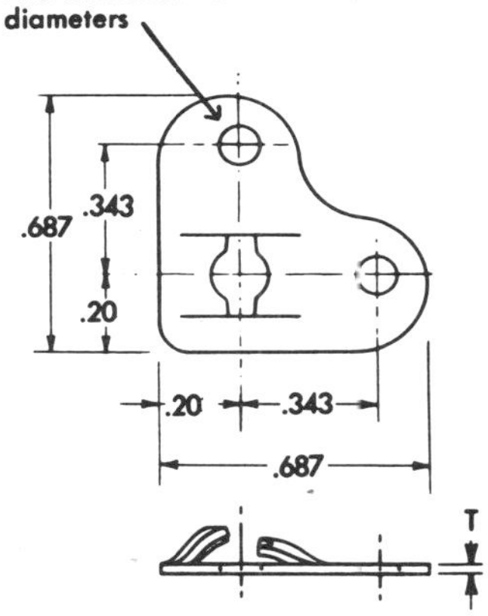

STYLE	SCREW SIZE	DESIGN VARIATION	PART NUMBER*	H RIVET DIAMETER	T MATERIAL THICKNESS	WEIGHT LBS. PER M
PLAIN	4B or Z		A6200-4Z-1	3/32	.020	1.80
	6B or Z		A6200-6Z-1		.025	2.00
	8B or Z		A6200-8Z-1		.028	2.30
	10B or Z		A6200-10Z-1		.031	2.60
	8B or Z		A8588-8Z-1	1/8	.028	2.30
	10B or Z		A8588-10Z-1		.031	2.60
	6B or Z	A	A6201-6Z-1	3/32	.025	2.00
	8B or Z		A6201-8Z-1		.028	2.30
	10B or Z		A6201-10Z-1		.031	2.60
	8B or Z		A8589-8Z-1	1/8	.028	2.30
	4B or Z	B	A6202SS-4Z-1		.020	1.80
	6B or Z		A6202SS-6Z-1		.025	2.00
	8B or Z		A6202SS-8Z-1		.028	2.30
	10B or Z		A6202SS-10Z-1		.031	2.60

COUNTERSUNK

"H" Two clearance holes for required rivet diameters

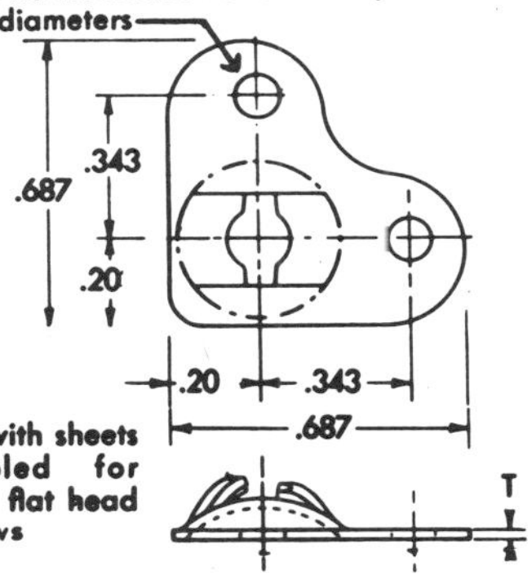

Use with sheets dimpled for 100° flat head Screws

STYLE	SCREW SIZE	DESIGN VARIATION	PART NUMBER*	H RIVET DIAMETER	T MATERIAL THICKNESS	WEIGHT LBS. PER M
COUNTERSUNK	4B or Z		A6197-4Z-1	3/32	.020	1.80
	6B or Z		A6197-6Z-1		.025	2.00
	8B or Z		A6197-8Z-1		.028	2.30
	8B or Z		A8586-8Z-1	1/8	.028	2.30
	4B or Z	A	A6198-4Z-1	3/32	.020	1.80
	6B or Z		A6198-6Z-1		.025	2.00
	8B or Z		A6198-8Z-1		.028	2.30
	4B or Z	B	A6199SS-4Z-1		.020	1.80
	6B or Z		A6199SS-6Z-1		.025	2.00
	8B or Z		A6199SS-8Z-1		.028	2.30

* -1 at end of part number specifies olive drab phenolic paint finish. For zinc metal finish replace with —3.

TINNERMAN NUTS (NAS 395) "U" TYPE

Performs same function as "J" Type. Used where full bearing on lower leg of the SPEED NUT is required.

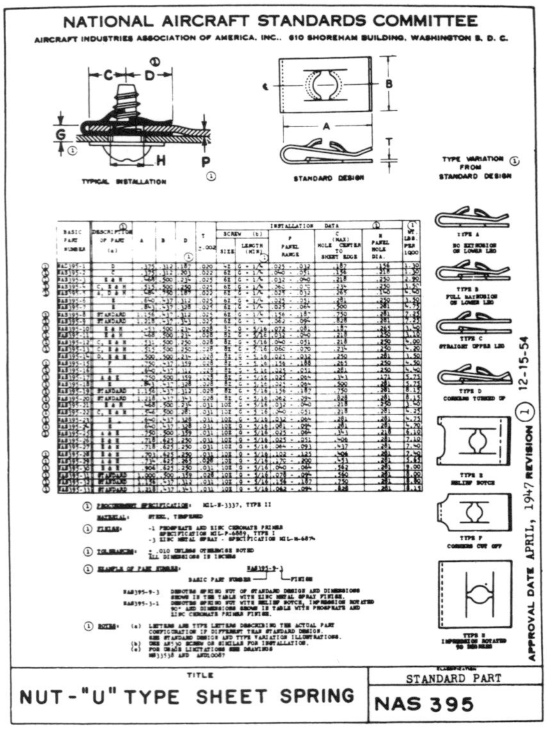

SCREW SIZE	SCREW LENGTH
4B or Z 6B or Z	"X" plus ¼"
8B or Z 10B or Z	"X" plus 5/16"

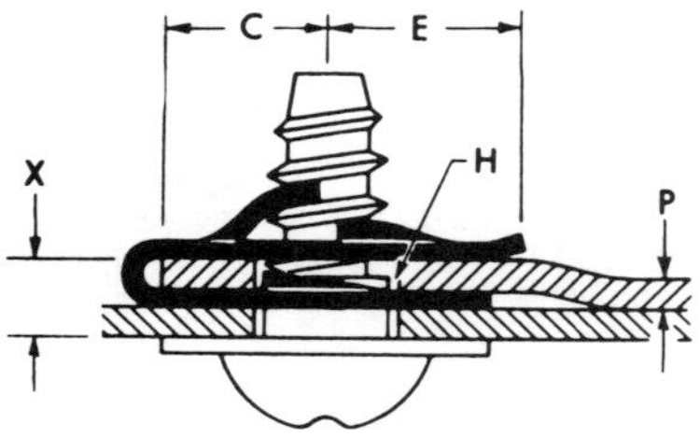

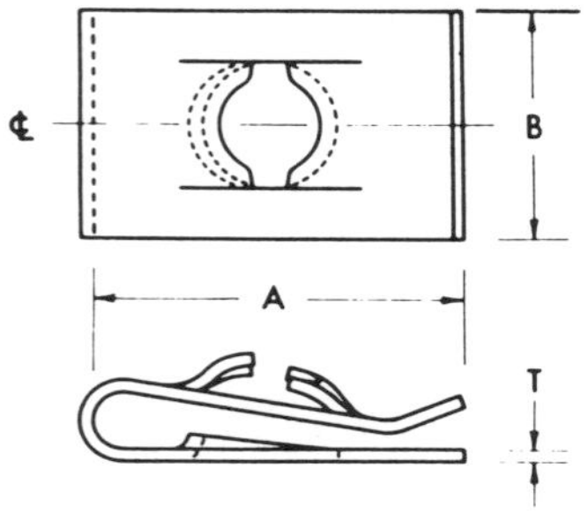

Type U speed nuts are featured in a number of design variations as well as standard. Speed Nut part numbers are preceded (see Ordering and Weight Table) by a column titled "Design Variation." If there are no letters in that column the Speed Nut is standard as shown above. If there are letters in the column they indicate that the nut is available only as listed, with that number of deviations from standard combined in the one nut. Example:

Design Variation	Part Number
ADH	A7344-6Z-1

The example shows that the nut has:

- A No extrusion on lower leg
- D Corner turned up
- H Speed Nut impression turned 90 degrees

ORDERING AND WEIGHT TABLE Type U Speed Nuts

(NOTE: This table is continued on following page)

SCREW SIZE	P PANEL RANGE	DESIGN VARI-ATION	PART NUMBER*	NAS PART NUMBER	A Length	B Width	E Center of Im-pression to End	C Max. Center of Hole to Edge	H Panel Hole Dia.	T Mat'l Thick-ness	WT. LBS. PER M
4B or Z	.020-.030	CD	A9028-4Z-1	....	.37	.31	.17	.187	.156	.020	1.30
	.025-.032		A6187-4Z-1	395-1	.38		.18			.020	1.30
	.036-.047		A9029SS-4Z-1	....	.37		.17			.022	1.30
6B or Z	.025-.032	ACD	A188-6Z-1	395-5	.48	.75	.187	.265	.140	.022	4.40
	.025-.051	E	A1784-6Z-1	395-6	.61	.437	.30	.281	.250	.025	3.50
	.025-.064	E	A1785-6Z-1	395-7	.84	.437	.312	.500	.281	.025	4.75
	.032-.040	EH	A6052-6Z-1	395-3	.45	.50	.218	.218	.250	.025	2.90
	.032-.050	AFH	A6607-6Z-1		.34	.50	.14	.171	.156	.025	2.50
	.032-.050	BE	A1932-6Z-1	396-1	.58	.50	.27	.265	.343	.025	4.10
	.035-.045	ADH	A7344-6Z-1	....	.468	.75	.187	.265	.140	.022	4.40
	.040-.051	C	A880-6Z-1	395-2	.36	.31	.18	:156	.218	.022	1.30
	.051-.064	EH	A6452-6Z-1	....	.46	.50	.23	.187	.265	.025	3.40
	.055-.065	AFH	A9222-6Z-1	....	.34	.50	.14	.164	.156	.025	2.20

* -1 at end of part number specifies olive drab phenolic paint finish. For zinc metal finish replace with —3.

MS SCREWS

MS27039

Available in aluminum alloy, anodized finish, per MIL-A-8625, Type I or II; Manganese bronze, cadmium plated per QQ-P-416, Type II, Class 3, or plain finish per code B; alloy steel, cadmium plated per QQ-P-416, Type II, Class 3; or corrosion and heat resistant steel, passivated finish. All meet procurement specification MIL-S-7839.

Examples of part numbers:

MS27039DD0809 = 8-32UNC-3A, anodized aluminum alloy screw, .594 inch length.

MS27039BP1-13 = 10-32UNF-3A, cadium plated manganese bronze screw, .844 inch length.

MS27039B4-10 = ¼-28UNF-3A, plain finished manganese bronze screw, .656 inch length.

MS27039-5-14 = 5/16-24UNF-3A, cadium plated alloy steel screw, .922 inch length.

MS27039C6-10 = ⅜-24UNF-3A, corrosion and heat resis--tant steel screw, .719 inch length

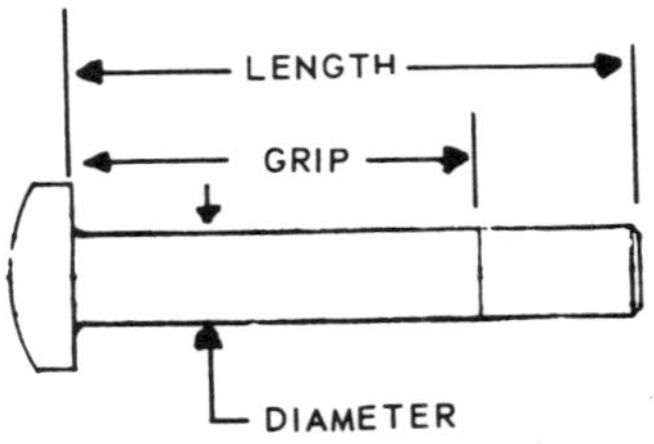

NAS SCREWS

NAS 220 - NAS227 BRAZIER HEAD

Available in aluminum alloy, alloy steel, corrosion resistant steel, bare manganese bronze and cadmium plated manganese bronze. Order by NAS part number, per Standard Drawings at right. Example: NAS220DD-7 is an aluminum alloy screw, 8-32 thread, 15/32" long.

NAS517 FLAT HEAD (100 DEGREE)

Available in cadmium plated alloy steel only. Heat treated to 160,000 p.s.i. Plated per QQ-P-416; recess in accordance with MS9006. First dash number indicates diameter; second dash number indicates length and grip. Order by NAS part number, per Standard Drawings at bottom of page 63. Example: NAS517-3-8 has a 10-32 thread, 1/2" grip length, 29/32" total length.

NAS600 - NAS606 PAN HEAD

Available in low alloy steel, cadmium plated to QQ-P-416, Type II (regular or dull black) and to Type I (regular only). Order by NAS part number per Standard Drawing below, right. Example: NAS601-8B has a 6-32 thread, 1/2" shank length, plated dull black.

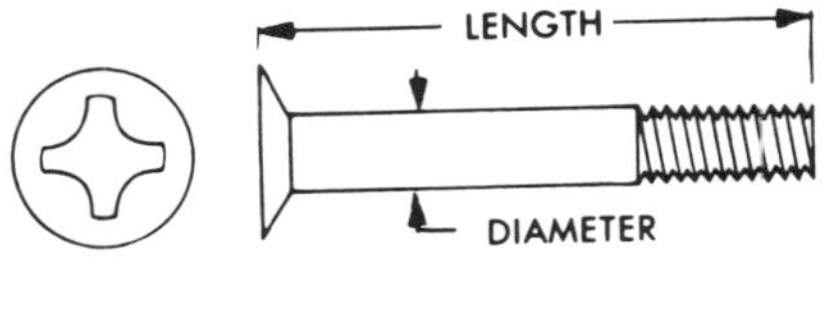

MACHINE SCREWS

ORDERING TABLE

MACHINE SCREWS—AN500, AN501, AN502, AN503, AN505, AN510, AN515, AN520, and AN525*.

The following table shows the 1st and 2nd dash numbers used as a suffix after the basic AN Number to indicate diameter and length of all machine screws except AN507, AN509 and AN526.

DIAMETER COURSE THREAD FINE THREAD	No. 2 56 64	No. 3 48 56	No. 4 40 48	No. 5 40 44	No. 6 32 40	No. 8* 32 36	No. 10* 24 32	¼* 20 28	5/16 18 24	3/8 16 24
LENGTH (inches)	**DASH NUMBERS**									
1/8	−2-2	−3-2	....	....	....	....	....	....	....	....
3/16	−2-3	−3-3	−4-3	−5-3	−6-3	....	....	....	....	....
1/4	−2-4	−3-4	−4-4	−5-4	−6-4	−8-4	−10-4	....	....	....
5/16	−2-5	−3-5	−4-5	−5-5	−6-5	−8-5	−10-5	−416-5	....	....
3/8	−2-6	−3-6	−4-6	−5-6	−6-6	−8-6	−10-6	−416-6	−516-6	−616-6
7/16	−2-7	−3-7	−4-7	−5-7	−6-7	−8-7	−10-7	−416-7	−516-7	−616-7
1/2	−2-8	−3-8	−4-8	−5-8	−6-8	−8-8	−10-8	−416-8	−516-8	−616-8
5/8	−2-10	−3-10	−4-10	−5-10	−6-10	−8-10	−10-10	−416-10	−516-10	−616-10
3/4	−2-12	−3-12	−4-12	−5-12	−6-12	−8-12	−10-12	−416-12	−516-12	−616-12
7/8	−2-14	−3-14	−4-14	−5-14	−6-14	−8-14	−10-14	−416-14	−516-14	−616-14
1	−2-16	−3-16	−4-16	−5-16	−6-16	−8-16	−10-16	−416-16	−516-16	−616-16
1 1/8	−2-18	−3-18	−4-18	−5-18	−6-18	−8-18	−10-18	−416-18	−516-18	−616-18
1 1/4	−2-20	−3-20	−4-20	−5-20	−6-20	−8-20	−10-20	−416-20	−516-20	−616-20
1 1/2	−2-24	−3-24	−4-24	−5-24	−6-24	−8-24	−10-24	−416-24	−516-24	−616-24
1 3/4	−2-28	−3-28	−4-28	−5-28	−6-28	−8-28	−10-28	−416-28	−516-28	−616-28
2	−2-32	−3-32	−4-32	−5-32	−6-32	−8-32	−10-32	−416-32	−516-32	−616-32
2 1/4	−2-36	−3-36	−4-36	−5-36	−6-36	−8-36	−10-36	−416-36	−516-36	−616-36
2 1/2	−2-40	−3-40	−4-40	−5-40	−6-40	−8-40	−10-40	−416-40	−516-40	−616-40
2 3/4	−2-44	−3-44	−4-44	−5-44	−6-44	−8-44	−10-44	−416-44	−516-44	−616-44
3	−2-48	−3-48	−4-48	−5-48	−6-48	−8-48	−10-48	−416-48	−516-48	−616-48

*AN525 screws available only in No. 8-32, No. 8-36, No. 10-32 and ¼"-28 diameters and threads. No. 8 diameter can be supplied in both coarse and fine threads. No. 10 and ¼" diameters available only in fine thread. No. 8 diameter in coarse thread should be designated by substituting -832 for the first dash number rather than -8 as shown in the table. When ordering AN525 fine thread series, use dash numbers as they appear in table.

HOW TO ORDER

Use AN numbers in individual tables showing choice of machine screw types. Add to the AN number the first and second dash numbers which appear in the Ordering Table above.

For example: AN 515B-4-7 would be the correct designation for a round head machine screw, material brass, thread coarse, head slotted, diameter No. 4 and length 7/16".

HEAD MARKINGS FOR AN MACHINE SCREWS

AN502, AN503 and AN509 Cadmium-plated Steel	AN505, AN507, AN510, AN515, AN520, AN526 Corrosion-resisting Steel	AN509 Bronze
XX	— —	= =
Only one "X" need be visible	Only one "—" need be visible	Only one "=" need be visible

NOTE: All machine screws other than those AN series (and specific materials) shown above do *not* have head markings.

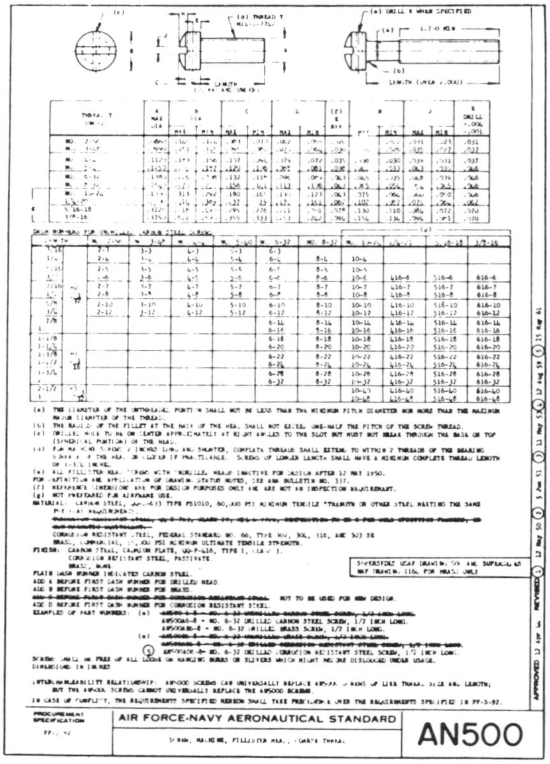

MACHINE SCREWS

AN507
100° FLAT HEAD
MACHINE SCREWS

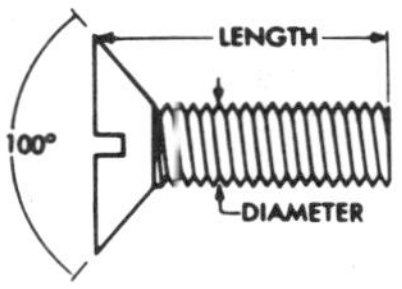

AN507 machine screws are manufactured in several varieties, as indicated in the table below. For key to 1st and 2nd dash numbers, indicating diameter and length, see Ordering Table below.

MATERIAL	FINISH	COARSE THREAD		FINE THREAD	
		SLOTTED HEAD	RECESSED HEAD	SLOTTED HEAD	RECESSED HEAD
STEEL	CADMIUM PLATE	AN507	AN507R	AN507	AN507R
BRASS	BLACK OXIDE	AN507B	AN507BR	AN507B	AN507BR
BRASS	NONE	AN507UB	AN507UBR	AN507UB	AN507UBR
COR. RES. STEEL	PASSIVATED	AN507C	AN507CR	AN507C	AN507CR
ALUMINUM ALLOY	ANODIZED	AN507DD	AN507DDR	AN507DD	AN507DDR

AN526
BUTTON HEAD
MACHINE SCREWS

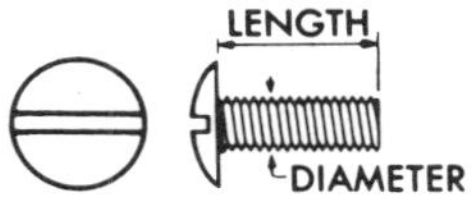

Button head machine screws are manufactured in the many varieties shown in the table to the right. For key to 1st and 2nd dash numbers used as a suffix after the basic AN number to indicate diameter and length, see Ordering Table below.

MATERIAL	FINISH	COARSE THREAD		FINE THREAD	
		SLOTTED HEAD	RECESSED HEAD	SLOTTED HEAD	RECESSED HEAD
CARBON STEEL	CADMIUM PLATE	AN526	AN526R	AN526	AN526R
COR. RES. STEEL	PASSIVATED	AN526C	....	AN526C	....
ALUMINUM ALLOY	ANODIZED	AN526DD	....	AN526DD	....

ORDERING TABLE AN507 and AN526 MACHINE SCREWS

DIAMETER	NO. 6		NO. 8		NO. 10		1/4	
THREADS PER INCH	COARSE 32	FINE 40	COARSE 32	FINE 36	COARSE 24	FINE 32	COARSE 20	FINE 28
LENGTH	DASH NUMBERS							
1/4	−632-4	−640-4	−832-4	−836-4	−1024-4	−1032-4	....	....
5/16	−632-5	−640-5	−832-5	−836-5	−1024-5	−1032-5	....	....
3/8	−632-6	−640-6	−832-6	−836-6	−1024-6	−1032-6	−420-6	−428-6
7/16	−632-7	−640-7	−832-7	−836-7	−1024-7	−1032-7	−420-7	−428-7
1/2	−632-8	−640-8	−832-8	−836-8	−1024-8	−1032-8	−420-8	−428-8
9/16	−632-9	−640-9	−832-9	−836-9	−1024-9	−1032-9	−420-9	−428-9
5/8	−632-10	−640-10	−832-10	−836-10	−1024-10	−1032-10	−420-10	−428-10
3/4	−632-12	−640-12	−832-12	−836-12	−1024-12	−1032-12	−420-12	−428-12
7/8	−632-14	−640-14	−832-14	−836-14	−1024-14	−1032-14	−420-14	−428-14
1	−632-16	−640-16	−832-16	−836-16	−1024-16	−1032-16	−420-16	−428-16
1 1/8	−632-18	−640-18	−832-18	−836-18	−1024-18	−1032-18	−420-18	−428-18
1 1/4	−632-20	−640-20	−832-20	−836-20	−1024-20	−1032-20	−420-20	−428-20
1 3/8	−632-22	−640-22	−832-22	−836-22	−1024-22	−1032-22	−420-22	−428-22
1 1/2	−632-24	−640-24	−832-24	−836-24	−1024-24	−1032-24	−420-24	−428-24
1 5/8	−632-26	−640-26	−832-26	−836-26	−1024-26	−1032-26	−420-26	−428-26
1 3/4	−632-28	−640-28	−832-28	−836-28	−1024-28	−1032-28	−420-28	−428-28
1 7/8	−632-30	−640-30	−832-30	−836-30	−1024-30	−1032-30	−420-30	−428-30
2	−632-32	−640-32	−832-32	−836-32	−1024-32	−1032-32	−420-32	−428-32
2 1/4	−632-36	−640-36	−832-36	−836-36	−1024-36	−1032-36	−420-36	−428-36
2 1/2	−632-40	−640-40	−832-40	−836-40	−1024-40	−1032-40	−420-40	−428-40
2 3/4	−632-44	−640-44	−832-44	−836-44	−1024-44	−1032-44	−420-44	−428-44
3	−632-48	−640-48	−832-48	−836-48	−1024-48	−1032-48	−420-48	−428-48

MACHINE SCREWS

AN509
100° FLAT HEAD
MACHINE SCREWS
(Structural)

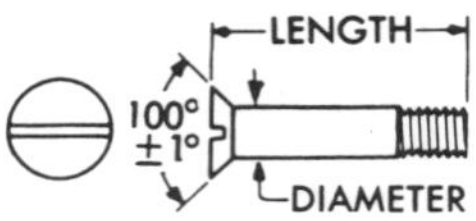

These screws are manufactured in several varieties as indicated. The Ordering Table below shows the 1st and 2nd dash numbers used as a suffix after the basic AN Number to indicate diameter and length.

MATERIAL	FINISH	SLOTTED HEAD	RECESSED HEAD
CARBON STEEL	CADMIUM PLATE	AN509	AN509R
ALUMINUM ALLOY	ANODIZED	AN509DD	AN509DDR
BRONZE	CADMIUM PLATE	AN509P	AN509PR
BRONZE	NONE	AN509Z	AN509ZR

ORDERING TABLE — AN509* MACHINE SCREWS

DIAMETER	No. 8	No. 10	1/4	5/16
THREAD	32	32	28	24
LENGTH	DASH NUMBERS			
9/32 . . .	8-4	10-4	....	....
11/32 . . .	8-5	10-5	416-5	516-5
13/32 . . .	8-6	10-6	416-6	516-6
15/32 . . .	8-7	10-7	416-7	516-7
17/32 . . .	8-8	10-8	416-8	516-8
19/32 . . .	8-9	10-9	416-9	516-9
21/32 . . .	8-10	10-10	416-10	516-10
23/32 . . .	8-11	10-11	416-11	516-11
25/32 . . .	8-12	10-12	416-12	516-12
27/32 . . .	8-13	10-13	416-13	516-13
29/32 . . .	8-14	10-14	416-14	516-14
31/32 . . .	8-15	10-15	416-15	516-15
1 1/32 . . .	8-16	10-16	416-16	516-16
1 3/32 . . .	8-17	10-17	416-17	516-17
1 5/32 . . .	8-18	10-18	416-18	516-18
1 7/32 . . .	8-19	10-19	416-19	516-19
1 9/32 . . .	8-20	10-20	416-20	516-20
1 11/32 . . .	8-21	10-21	416-21	516-21
1 13/32 . . .	8-22	10-22	416-22	516-22
1 15/32 . . .	8-23	10-23	416-23	516-23
1 17/32 . . .	8-24	10-24	416-24	516-24
1 19/32 . . .	8-25	10-25	416-25	516-25
1 21/32 . . .	8-26	10-26	416-26	516-26
1 23/32 . . .	8-27	10-27	416-27	516-27
1 25/32 . . .	8-28	10-28	416-28	516-28
1 27/32 . . .	8-29	10-29	416-29	516-29
1 29/32 . . .	8-30	10-30	416-30	516-30
1 31/32 . . .	8-31	10-31	416-31	516-31
2 1/32 . . .	8-32	10-32	416-32	516-32
2 3/32 . . .	8-33	10-33	416-33	516-33
2 5/32 . . .	8-34	10-34	416-34	516-34
2 7/32 . . .	8-35	10-35	416-35	516-35
2 9/32 . . .	8-36	10-36	416-36	516-36
2 11/32 . . .	8-37	10-37	416-37	516-37
2 13/32 . . .	8-38	10-38	416-38	516-38
2 15/32 . . .	....	10-39	416-39	516-39
2 17/32 . . .	....	10-40	416-40	516-40
2 19/32 . . .	....	10-41	416-41	516-41
2 21/32 . . .	....	10-42	416-42	516-42
2 23/32 . . .	....	10-43	416-43	516-43
2 25/32 . . .	....	10-44	416-44	516-44
2 27/32 . . .	....	10-45	416-45	516-45
2 29/32 . . .	....	10-46	416-46	516-46
2 31/32 . . .	....	10-47	416-47	516-47
3 1/32 . . .	....	10-48	416-48	516-48

*Larger sizes in various types and materials are available on special order.

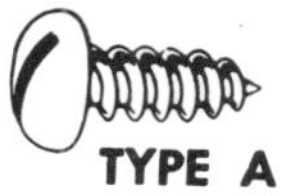

TYPE A

TYPE A SCREWS

A coarse-threaded screw with gimlet point recommended for use on light-gage sheet metal, plywoods and asbestos compositions. Available with slotted or Phillips recessed heads in sizes listed in the Ordering Table below. When ordering please include type of screw, head style, diameter number, and length.

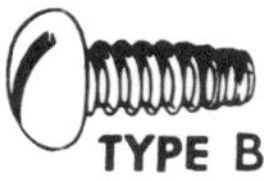

TYPE B

TYPE B SCREWS AN530-AN531

Blunt-pointed screws with coarse threads pitched slightly finer than Type A. AN530 round head, and AN531 flat head screws are used in sheet metal from 28 to 6 gage; for non-ferrous castings, plastics, plywoods, etc. Cadmium-plated to specification QQ-P-416. Available with slotted or Phillips recessed heads.

When ordering include type of screw, head style, diameter number, and length. Use AN numbers when they apply — example; AN 530-6-8 (Type B round head, size No. 6-½"). For AN screws with recessed heads add "R" between first and second dash numbers —AN530-6R8.

HEAD STYLES

R Round Head | P Pan Head | T Truss Head | F Countersunk Flat Head | O Countersunk Oval Head

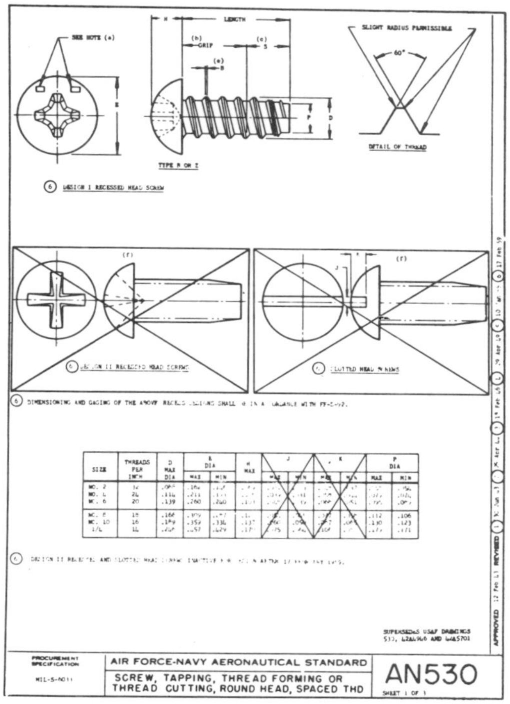

SIZE	LENGTH (inches)	HEAD STYLES AVAILABLE: TYPE "A" FLAT (F) ROUND (R) OVAL (O) PAN (P) SLTD.	TYPE "A" TRUSS (T) REC.	TYPE "B" AN 530 RD. HD. DASH NO. SLTD.	AN 530 RD. HD. DASH NO. REC.	AN 531 FLT. HD. DASH NO. SLTD.	AN 531 FLT. HD. DASH NO. REC.	OVAL (O) PAN (P) TRUSS (T) SLTD.	OVAL (O) PAN (P) TRUSS (T) REC.
2	1/8	-----	-----	2-2	---	---	---	- P -	---
2	3/16	-----	-----	2-3	---	2-3	---	- P -	- P -
2	1/4	-----	-----	2-4	---	2-4	---	P	- P -
2	5/16	-----	-----	2-5	---	---	---	- P -	- P -
2	3/8	-----	-----	2-6	---	---	---	- P -	- P -
4	3/16	-----	-----	4-3	---	---	---	- PT	---
4	¼	FROPT	FR-PT	4-4	4R4	4-4		OPT	PT
4	5/16	-----	-----	4-5	4R5	---	---	- PT	- P
4	3/8	FROPT	FROPT	4-6	4R6	4-6	4R6	OPT	OPT
4	½	FROPT	FROPT	4-8	4R8		4R8	OP-	- P -
6	3/16	-----	-----	6-3	---	---	---	- PT	---
6	¼	FROPT	FR-PT	6-4	6R4	6-4	---	OPT	OPT
6	5/16	-----	-----	6-5	6R5			- PT	OPT
6	3/8	FROPT	FROPT	6-6	6R6	6-6	6R6	PT	OPT
6	½	FROPT	FROPT	6-8	6R8	6-8	6R8	P-	OPT
6	5/8	FROP-	FROP-	6-10	6R10	---	6R10	OP-	OP-
6	¾	FROP-	FROPT	6-12	---	---	6R12	- P -	OP-
8	¼	-R-P-	-----	8-4	---	---	---	-PT	---
8	5/16	-----	-----	8-5	---	---	---	-PT	---
8	3/8	FROPT	FROPT	8-6	8R6	8-6	8R6	OPT	PT
8	½	FROPT	FROPT	8-8	8R8	8-8	8R8	OPT	OPT
8	5/8	FROPT	FROPT	8-10	8R10	8-10	8R10	P	OPT
8	¾	FROPT	FROPT	8-12	8R12	8-12	8R12	OP-	OP-
8	7/8	ROP-	FROP-	8-14	8R14	8-14	8R14	OP-	OP-
8	1	FROP-	FROP-	8-16	8R16	8-16	8R16	P-	OP-
10	¼	-----	-----	10-4	---	---	---	-PT	---
10	5/16	-----	-----	10-5	---	---	---	-PT	---
10	3/8	FROPT	FROPT	10-6	10R6	---	---	-PT	-PT
10	½	FROPT	FROPT	10-8	10R8	10-8	10R8	OPT	OPT
10	5/8	FROPT	FROPT	10-10	10R10	10-10	10R10	OPT	P
10	¾	FROPT	FROPT	10-12	10R12	10-12	10R12	OPT	OP-
10	7/8	ROP-	FROP-	10-14	10R14	10-14	10R14	OP-	OP-
10	1	FROPT	FROP-	10-16	10R16	10-16	10R16	OP-	P-

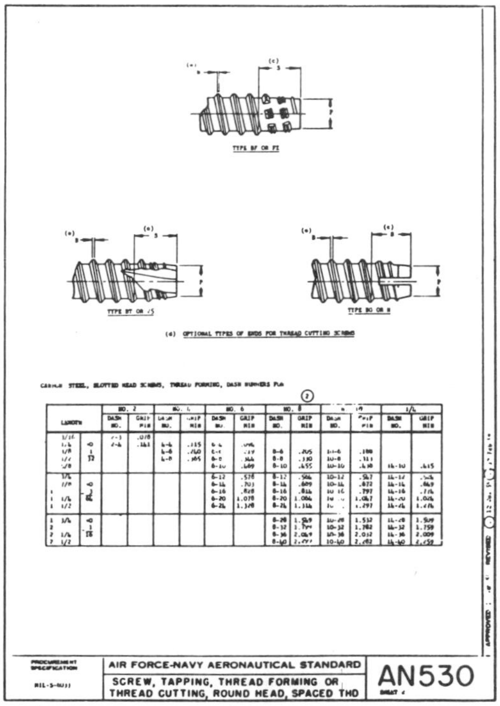

LOCK WASHERS

AN 935 MS35338

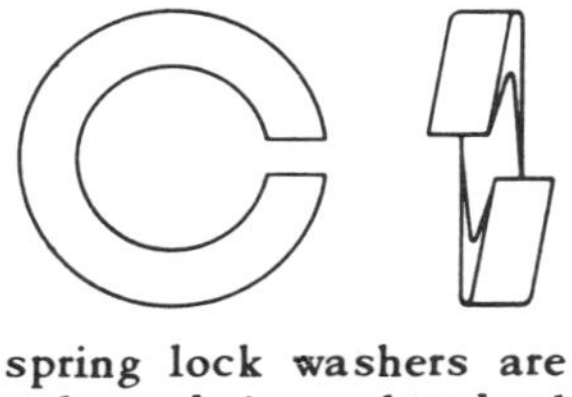

AN935-L spring lock washers are available in cadmium plated carbon steel, cadmium plated phosphor bronze and corrosion resistant steel. Plating is per QQ-P-416. Available in lightweight type only. For medium weight series see MS35338 below.

Examples of part numbers:

AN935-10L = steel lock washer, light series, for #10 bolt.
AN935-B10L = Bronze lock washer, light series, for #10 bolt.

MS35338 split, helical, lock washers made of carbon steel per specification QQ-S-633. Cadium plated per QQ-P-416, Type 2, Class C. MS35338 washers supersede, in part, AN935 medium washers.

Examples of part numbers:

MS35338-40 (size #4, medium series)

SCREW OR BOLT SIZE	AN935 DASH NUMBER	MS35338 DASH NUMBER
#4	-4L	-40
#6	-6L	-41
#8	-8L	-42
#10	-10L	-43
1/4"	-416L	-44
5/16"	-516L	-45
3/8"	-616L	-46
7/16"	-716L	-47
1/2"	-816L	-48
9/16"	-916L	-49
5/8"	-1016L	-50

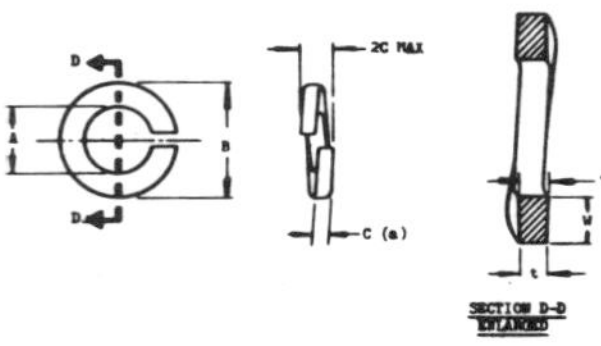

DASH NUMBERS MEDIUM SERIES	DASH NUMBERS LIGHT SERIES	BOLT SIZE	A -.000	MEDIUM SERIES B MAX	MEDIUM SERIES C -.000	MEDIUM SERIES W MIN	LIGHT SERIES B MAX	LIGHT SERIES C -.000	LIGHT SERIES W MIN
2	2L	NO. 2(.086)	.088 +.009	.175	.020 +.006	.035	.165	.015 +.006	.030
3	3L	NO. 3(.099)	.102 +.009	.198	.025 +.006	.040	.188	.020 +.006	.035
4	4L	NO. 4(.112)	.115 +.009	.212	.025 +.006	.040	.202	.020 +.006	.035
5	5L	NO. 5(.125)	.128 +.009	.239	.031 +.006	.047	.225	.025 +.006	.040
6	6L	NO. 6(.138)	.141 +.010	.253	.031 +.006	.047	.239	.025 +.006	.040
8	8L	NO. 8(.164)	.168 +.010	.296	.040 +.006	.055	.280	.031 +.006	.047
10	10L	NO.10(.190)	.194 +.011	.337	.047 +.006	.062	.323	.040 +.006	.055
12	12L	NO.12(.216)	.221 +.011	.380	.056 +.006	.070	.364	.047 +.006	.062
416	416L	1/4	.255 +.012	.493	.062 +.010	.109	.489	.047 +.010	.107
516	516L	5/16	.319 +.014	.591	.078 +.010	.125	.575	.056 +.010	.117
616	616L	3/8	.382 +.016	.688	.094 +.010	.141	.678	.070 +.010	.136
716	716L	7/16	.446 +.018	.784	.109 +.010	.156	.780	.085 +.010	.154
816	816L	1/2	.509 +.020	.879	.125 +.010	.171	.877	.099 +.010	.170
916	916L	9/16	.573 +.022	.979	.141 +.010	.188	.975	.113 +.010	.186
1016	1016L	5/8	.636 +.024	1.086	.156 +.010	.203	1.082	.126 +.010	.201
1216	1216L	3/4	.763 +.028	1.279	.188 +.010	.234	1.277	.153 +.010	.233

(a) MEAN THICKNESS $C = \frac{T+t}{2}$

INCREASE IN THICKNESS FROM OUTER PERIPHERY TO INNER PERIPHERY SHALL BE: (T-t) MIN = .032W; (T-t) MAX = .064W.

MATERIALS: CARBON STEEL: CLASS A MATERIAL IN PROCUREMENT SPECIFICATION; PHOSPHOR BRONZE: CLASS E MATERIAL IN PROCUREMENT SPECIFICATION OR SPECIFICATION QQ-B-746, COMPOSITION A, CONDITION HARD; SILICON BRONZE: CLASS F MATERIAL IN PROCUREMENT SPECIFICATION OR SPECIFICATION QQ-C-591, CLASS A, CONDITION HARD.

(4) FINISH: STEEL: CADMIUM PLATE, SPECIFICATION QQ-P-416, TYPE II, CLASS 3.
PHOSPHOR AND SILICON BRONZE: CADMIUM PLATE, SPECIFICATION QQ-P-416, TYPE II, CLASS 3, AND DYE DEEP YELLOW IN ACCORDANCE WITH FEDERAL STANDARD NO. 595, COLOR NO. 23538.

TABULATED DASH NUMBERS DESIGNATE CARBON STEEL LOCK WASHERS.
ADD B BEFORE DASH NUMBER FOR PHOSPHOR BRONZE OR SILICON BRONZE LOCK WASHERS.
EXAMPLES OF PART NUMBERS:
AN935-10 = STEEL LOCK WASHER, MEDIUM SERIES, FOR NO. 10 BOLT.
AN935B10 = BRONZE LOCK WASHER, MEDIUM SERIES, FOR NO. 10 BOLT.
AN935-10L = STEEL LOCK WASHER, LIGHT SERIES, FOR NO. 10 BOLT.
AN935B10L = BRONZE LOCK WASHER, LIGHT SERIES, FOR NO. 10 BOLT.

REMOVE ALL BURRS AND SHARP EDGES.

DIMENSIONS IN INCHES.

IN CASE OF CONFLICT, THE REQUIREMENTS SPECIFIED HEREON SHALL TAKE PRECEDENCE OVER THE REQUIREMENTS SPECIFIED IN SPECIFICATION FF-W-84.

USE IN CONFORMANCE WITH AND10476, WASHER-LIMITATIONS ON USAGE OF LOCK.

SUPERSEDES FORMER AIR FORCE AND NAVY STANDARD ISSUES OF AN935.

APPROVED 26 Aug 47 REVISED (1) 22 May 44 (2) 28 Oct 49 (1) 3 Jun 53 (4) 12 Aug 59

PROCUREMENT SPECIFICATION	AIR FORCE-NAVY AERONAUTICAL STANDARD	AN935
FF-W-84	WASHER - LOCK, SPRING	

FED. SUP CLASS 5310

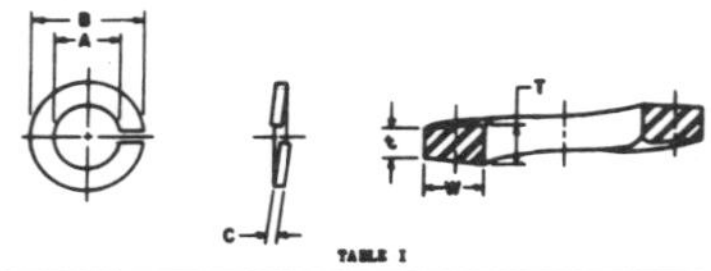

TABLE I

NOMINAL SIZE		A INSIDE DIAMETER MIN	A INSIDE DIAMETER MAX	B OUTSIDE DIA MAX	WASHER SECTION W WIDTH MIN	WASHER SECTION C THICK $\frac{T+t}{2}$ MIN	STEEL, CARBON [illegible] (5)	STEEL, CARBON [illegible] (5)	STEEL, CARBON CAD	STEEL, CARBON PHOS (5)	CRES [illegible] (5)	CRES 300 SERIES PASSIVATED	CRES 420 PASSIVATED	PH BRZ OR TIN BRS CADMIUM (3)	NICKEL COPPER ALLOY PLAIN
							DASH NUMBERS				DASH NUMBERS			DASH NUMBERS	
#2	.086	.088	.094	.172	.035	.020	[illegible]	[illegible]	39	58	[illegible]	134	153	96	115
#4	.112	.115	.121	.209	.040	.025	[illegible]	[illegible]	40	59	[illegible]	135	154	97	116
#6	.138	.141	.148	.250	.047	.031	[illegible]	[illegible]	41	60	[illegible]	136	155	98	117
#8	.164	.168	.175	.293	.055	.040	[illegible]	[illegible]	42	61	[illegible]	137	156	99	118
#10	.190	.194	.202	.334	.062	.047	[illegible]	[illegible]	43	62	[illegible]	138	157	100	119
1/4	.250	.255	.263	.489	.109	.062	[illegible]	[illegible]	44	63	[illegible]	139	158	101	120
5/16	.312	.318	.328	.586	.125	.078	[illegible]	[illegible]	45	64	[illegible]	140	159	102	121
3/8	.375	.382	.393	.683	.141	.094	[illegible]	[illegible]	46	65	[illegible]	141	160	103	122
7/16	.438	.446	.459	.779	.156	.109	[illegible]	[illegible]	47	66	[illegible]	142	161	104	123
1/2	.500	.509	.523	.873	.171	.125	[illegible]	[illegible]	48	67	[illegible]	143	162	105	124
9/16	.562	.572	.587	.971	.188	.141	[illegible]	[illegible]	49	68	[illegible]	144	163	106	125
5/8	.625	.636	.653	1.079	.203	.156	[illegible]	[illegible]	50	69	[illegible]	145	164	107	126
3/4	.750	.763	.783	1.271	.234	.188	[illegible]	[illegible]	51	70	[illegible]	146	165	108	127
7/8	.875	.890	.912	1.464	.266	.219	[illegible]	[illegible]	52	71	[illegible]	147	166	109	128
1	1.000	1.017	1.042	1.661	.297	.250	[illegible]	[illegible]	53	72	[illegible]	148	167	110	129
1-1/8	1.125	1.144	1.172	1.853	.328	.281	[illegible]	[illegible]	54	[illegible]	[illegible]	149	168	111	130
1-1/4	1.250	1.271	1.302	2.045	.359	.312	[illegible]	[illegible]	55	[illegible]	[illegible]	150	169	112	131
1-3/8	1.375	1.398	1.432	2.239	.391	.344	[illegible]	[illegible]	56	[illegible]	[illegible]	151	170	113	132
1-1/2	1.500	1.525	1.561	2.430	.422	.375	[illegible]	[illegible]	57	[illegible]	[illegible]	152	171	114	133

MATERIAL: Steel, Carbon, C1060 to C1080 Rockwell "C" 45-53, Federal Standard No. 66.
Steel, Corrosion Resisting, 300 Series, Rockwell "C" 35-43, and Type 420, Rockwell "C" 45-53, Federal Standard No. 66.
Phosphor Bronze, Composition Hard, Specification QQ-B-750.
Tin Brass, Copper Development Association Alloy Number 425.
Nickel Copper Alloy, Class A, Specification QQ-N-286.

PROTECTIVE COATING: Cadmium Plate, Type II, Class 3, Specification QQ-P-416.
Passivate, Specification QQ-P-35.
Phosphate Coating, Type Z, Class 2, Specification MIL-P-16232.

MAGNETIC PERMEABILITY: Corrosion resisting steel (300 series only) washers have a magnetic permeability of 2.0 maximum (Air = 1.0) for a field strength of H = 200 oersteds using a magnetic indicator per MIL-I-17214.

DIMENSIONS: All dimensions are in inches unless otherwise specified.

PART NUMBER: The MS part number consists of the MS number, plus the dash number. Example: MS35338-39.

NOTES:
(1) For design feature purposes, this standard takes precedence over procurement documents referenced herein.
(2) Referenced documents shall be of the issue in effect on date of invitations for bid.
(3) Phosphor bronze or tin brass washers which are cadmium plated, shall be dyed deep yellow in accordance with Color No. 23538, Federal Standard No. 595.
(4) These dimensions, in agreement with ASA B27.1-1965, are applicable to the active washers only.
(5) CANCELLED, IN PART, FOR NEW DESIGN AND ENGINEERING. For supersessions see Table II, sheet 2.

PROPOSED MARCH 4, 1964 REVISED (A) SEP 10, 1964 (B) 18 APR 66 (C) 21 MAR 1966

FED. SUP CLASS 5310

TABLE II - INTERCHANGEABILITY

After 21 MAR 1966, (1) washers given on MS16214, MS35337, NAS1061, and this standard in part, are cancelled, but their existing stocks should be used for maintenance purposes until depleted and (2) washers given on AN935 are inactivated for new design. USE ONLY THE FOLLOWING SUPERSEDING WASHERS FOR NEW DESIGN AND REPLACEMENT.

CROSS REFERENCE OF PART NUMBERS

CANCELLED INACTIVATED	SUPERSESSION	CANCELLED INACTIVATED	SUPERSESSION	CANCELLED INACTIVATED	SUPERSESSION	CANCELLED INACTIVATED	SUPERSESSION
MS16214-1	MS35338-135	MS35337-37	MS35338-56	MS35337-108	MS35338-108	AN935-2	MS35338-39
2	136	38	57	109	109	-3	•
3	137	39	39	110	110	-4	40
4	138	40	40	111	111	-5	•
5	139	41	41	112	112	-6	41
6	140	42	42	113	113	-8	42
7	141	43	43	114	114	-10	43
8	142	44	44			-12	•
9	143	45	45	MS35338-1	MS35338-39	-416	44
10	145	46	46	2	40	-516	45
11	146	47	47	3	41	-616	46
12	147	48	48	4	42	-716	47
13	148	49	49	5	43	-816	48
14	149	50	50	6	44	-916	49
15	150	51	51	7	45	-1016	50
16	151	52	52	8	46	-1216	51
17	152	53	53	9	47	-2L	39
18	97	54	54	10	48	-3L	•
19	98	55	55	11	49	-4L	40
20	99	56	56	12	50	-5L	•
21	100	57	57	13	51	-6L	41
22	101	58	58	14	52	-8L	42
23	102	59	59	15	53	-10L	43
24	103	60	60	16	54	-12L	•
25	104	61	61	17	55	-416L	44
26	105	62	62	18	56	-516L	45
27	107	63	63	19	57	-616L	46
28	108	64	64	20	39	-716L	47
29	109	65	65	21	40	-816L	48
30	110	66	66	22	41	-916L	49
31	111	67	67	23	42	-1016L	50
32	112	68	68	24	43	-1216L	51
33	113	69	69	25	44	B2	96
34	114	70	70	26	45	B3	•
		71	71	27	46	B4	97
MS35337-1	MS35338-39	72	72	28	47	B5	•
2	40	73	54	29	48	B6	98
3	41	74	55	30	49	B8	99
4	42	75	56	31	50	B10	100
5	43	76	57	32	51	B12	•
6	44	77	134 OR 153	33	52	B416	101
7	45	78	135 OR 154	34	53	B516	102
8	46	79	136 OR 155	35	54	B616	103
9	47	80	137 OR 156	36	55	B716	104
10	48	81	138 OR 157	37	56	B816	105
11	49	82	139 OR 158	38	57	B916	106
12	50	83	140 OR 159	73	54	B1016	107
13	51	84	141 OR 160	74	55	B1216	108
14	52	85	142 OR 161	75	56	B2L	96
15	53	86	143 OR 162	76	57	B3L	•
16	54	87	144 OR 163	77	134 OR 153	B4L	97
17	55	88	145 OR 164	78	135 OR 154	B5L	•
18	56	89	146 OR 165	79	136 OR 155	B6L	98
19	57	90	147 OR 166	80	137 OR 156	B8L	99
20	39	91	148 OR 167	81	138 OR 157	B10L	100
21	40	92	149 OR 168	82	139 OR 158	B12L	•
22	41	93	150 OR 169	83	140 OR 159	B416L	101
23	42	94	151 OR 170	84	141 OR 160	B516L	102
24	43	95	152 OR 171	85	142 OR 161	B616L	103
25	44	96	96	86	143 OR 162	B716L	104
26	45	97	97	87	144 OR 163	B816L	105
27	46	98	98	88	145 OR 164	B916L	106
28	47	99	99	89	146 OR 165	B1016L	107
29	48	100	100	90	147 OR 166	B1216L	108
30	49	101	101	91	148 OR 167		
31	50	102	102	92	149 OR 168	NAS1061-6	MS35338-136
32	51	103	103	93	150 OR 169	10	138
33	52	104	104	94	151 OR 170	416	139
34	53	105	105	95	152 OR 171	516	140
35	54	106	106			616	141
36	55	107	107				

PROPOSED MARCH 4, 1964 REVISED (C) FOR CHANGES SEE SHEET 1

MS35333, MS35335 LOCK WASHERS

Used for many aircraft and engine applications. Provide efficient locking by means of the spring tension and biting action of tapered-twisted teeth. Made of 1050 spring steel with .0002 cadmium plating and dichromate dip. MS35333 and MS35335 washers supersede AN936A and AN936B washers respectively.
HOW TO ORDER: To the basic MS35333 or MS35335 number add the appropriate dash number to indicate size. Examples:

MS35333-36 (Type A, Size No. 4)
MS35335-29 (Type B, Size No. 4)

TYPE A
Internal Teeth
(MS35333)

TYPE B
External Teeth
(MS35335)

SCREW OR BOLT SIZE	TYPE A MS35333 DASH NUMBER STEEL	TYPE A MS35333 DASH NUMBER BRONZE	TYPE A OLD AN936 DASH NO.	TYPE B MS35335 DASH NUMBER STEEL	TYPE B MS35335 DASH NUMBER BRONZE	TYPE B OLD AN936 NUMBER
No. 4	-36	-104	A4	-29	-85	B4
No. 6	-37	-105	A6	-30	-86	B6
No. 8	-38	-106	A8	-31	-87	B8
No. 10	-39	-107	A10	-32	-88	B10
1/4"	-40	-108	A416	-33	-89	B416
5/16"	-41	-109	A516	-34	-90	B516
3/8"	-42	-110	A616	-35	-91	B616
7/16"	-43	-111	A716	-36	-92	B716
1/2"	-44	-113	A816	-37	-93	B816
9/16"	-45	-114	A916	-38	-94	B916
5/8"	-46	-115	A1016	-39	-95	B1016
3/4"	-47	-116	A1216	-40	-96	B1216
7/8"	-48	-117	A1416	-41	-97	B1416
1"	-49	-118	A1616	-42	-98	B1616

FED. SUP CLASS 5310

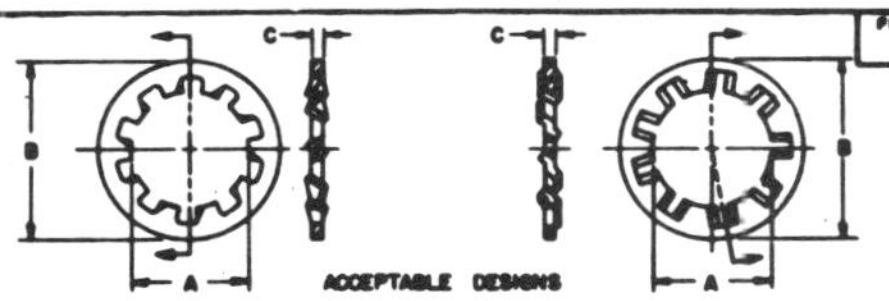

ACCEPTABLE DESIGNS

TABLE I

NOMINAL SIZE		A INSIDE DIAMETER MAX.	A INSIDE DIAMETER MIN.	B OUTSIDE DIAMETER MAX.	B OUTSIDE DIAMETER MIN.	C THICKNESS MAX.	C THICKNESS MIN.	STEEL CARBON CADMIUM PLATE DASH NO.	STEEL CORROSION RESISTING PASSIVATED DASH NO.	TIN-BRASS OR PHOSPHOR BRONZE OPTIONAL UNCOATED DASH NO.
#2	.086	.095	.089	.200	.175	.015	.010	35	69	103
#4	.112	.123	.115	.270	.255	.019	.015	36	70	104
#6	.138	.150	.141	.295	.275	.021	.017	37	71	105
#8	.164	.176	.168	.340	.325	.023	.018	38	72	106
#10	.190	.204	.195	.381	.365	.025	.020	39	73	107
1/4		.267	.256	.478	.460	.028	.023	40	74	108
5/16		.332	.320	.610	.594	.034	.028	41	75	109
3/8		.398	.384	.692	.670	.040	.032	42	76	110
7/16		.464	.448	.789	.740	.040	.032	43	77	111
7/16		.480	.472	.607	.593	.020	.016	121 (NOTE 4)	--	112 (NOTE 4)
1/2		.530	.512	.900	.867	.045	.037	44	78	113
9/16		.596	.576	.985	.957	.045	.037	45	79	114
5/8		.663	.640	1.071	1.045	.050	.042	46	80	115
3/4		.795	.769	1.245	1.220	.055	.047	47	81	116
7/8		.927	.894	1.410	1.364	.060	.052	48	82	117
1		1.060	1.019	1.637	1.590	.067	.059	49	83	118
1-1/8		1.192	1.144	1.830	1.799	.067	.059	50	84	119
1-1/4		1.325	1.275	1.975	1.921	.067	.059	51	85	120

MATERIAL: Steel, carbon; sheet, in accordance with procurement specification.
Corrosion resisting steel, in accordance with procurement specification.
Tin-brass, Copper and Brass Research Association Alloy Number 425.
PHOSPHOR BRONZE, COMPOSITION A, QQ-B-750.

PROTECTIVE COATING AND TREATMENT:
Carbon steel, cadmium plate in accordance with procurement specification.
Corrosion resisting steel, passivate in accordance with Specification QQ-P-35.

DIMENSIONS: All dimensions are in inches unless otherwise specified.

PART NUMBER: The MS part number consists of the MS number, plus the dash number. Example: MS35333-35.

NOTES: (1) The number of teeth shall be at the option of the manufacturer.
(2) For design feature purpose, this standard takes precedence over procurement documents referenced herein.
(3) Referenced documents shall be of the issue in effect on date of invitations for bid.
(4) This washer has the alternate teeth twisted in a direction opposite to that of the other teeth.

ED 4 MAR 54 REVISED (A) 18 APRIL 56 (B) 5 DEC 1963

FED. SUP CLASS 5310

INTERCHANGEABILITY TABLE

The following washers covered by dash numbers given in MS35333 Revision A in part and AN936 in part are cancelled after 5 December 1963. Use the dash numbers given in the preceding sheet. The cancelled washers cannot always replace the new washers and should be used until existing stocks are depleted. Use only the new washers for design and replacement. Replacement shall be in accordance with TABLE II. Interchangeability information regarding part numbers on the superseded drawings is also contained in TABLE II.

TABLE II

CANCELLED AN936 DASH NO.	CANCELLED MS35333 DASH NO.	REVISED MS35333 DASH NO.	CANCELLED AN936 DASH NO.	CANCELLED MS35333 DASH NO.	REVISED MS35333 DASH NO.
A2	1	35		26	43
	18	35		60	43
	52	35	A716S	--	121
A3	--	No Replacement	A616	10	44
A4	2	36		27	44
	19	36		61	44
	53	36	A916	11	45
A6	3	37		28	45
	20	37		62	45
	54	37	A1016	12	46
A8	4	38		29	46
	21	38		63	46
	55	39	A1216	13	47
A10	5	39		30	47
	22	39		64	47
	56	39	A1416	14	48
A416	6	40		31	48
	23	40		65	48
	57	40	A1616	15	49
A416H	--	No Replacement		32	49
A516	7	41		66	49
	24	41		16	50
	58	41		33	50
A616	8	42		67	50
	25	42		17	51
	59	42		34	51
A716	9	43		68	51

CANCELLED AN936 DASH NO.	CANCELLED MS35333 DASH NO.	REVISED MS35333 DASH NO.	CANCELLED AN936 DASH NO.	CANCELLED MS35333 DASH NO.	REVISED MS35333 DASH NO.
	86	103	A716RS		112
A4B	87	104		95	113
A6B	88	105		96	114
A8B	89	106		97	115
A10B	90	107		98	116
A416B	91	108		99	117
A516B	92	109		100	118
A616B	93	110		101	119
A716B	94	111		102	120

VED 4 MAR 54 REVISED (A) 18 APRIL 56 (B) FOR CHANGES SEE SHEET 1

AN960

FLAT WASHERS

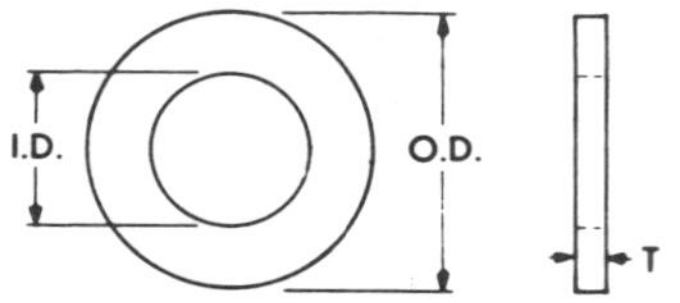

AN960 Flat Washers are manufactured in a variety of metals, sizes, and weights—weights vary because of differing washer thicknesses between the regular and light series. AN960 standard drawing requirements are adhered to, as well as all applicable government specifications.

ORDERING NOTE: Specify material by AN number; use AN dash number for size; and add "L" if light series is desired.

Examples: AN960B3 (Brass)
AN960C3L (Corrosion-resisting steel, light)

AN960 CADMIUM-PLATED STEEL: Made of carbon steel in accordance with AN-S-11 requirements, and cadmium-plated to QQ-P-416 specifications. Offered in regular and light series.

AN960B BRASS: Commercial brass washers conforming to Federal Specification QQ-B-611. Regular series only.

AN960C CORROSION-RESISTING STEEL: Meet MIL-S-5059 or MIL-S-6721 specifications for corrosion-resisting steel. Regular or light series.

AN960D ALUMINUM ALLOY: Produced from aluminum alloy – plain or anodized – in adherence to specification QQ-A-362. Anodizing conforms to AN-QQ-A-696 specifications. Available in regular and light series.

ORDERING TABLE

SCREW OR BOLT SIZE	I.D. (INCH)	O.D. (INCH)	REGULAR SERIES						LIGHT SERIES				
			THICK-NESS-T (INCH)	DASH NUMBERS					THICK-NESS-T (INCH)	DASH NUMBERS			
				CARBON STEEL	BRASS	COR. RES. STEEL	ALUMINUM ALLOY ANODIZED	ALUMINUM ALLOY PLAIN		CARBON STEEL	COR. RES. STEEL	ALUMINUM ALLOY ANODIZED	ALUMINUM ALLOY PLAIN
No. 2	.099	.250	.032	-2	-B2	-C2	-PD2	-D2	.016	-2L	-C2L	-PD2L	-D2L
No. 3	.109	.250	.032	-3	-B3	-C3	-PD3	-D3	.016	-3L	-C3L	-PD3L	-D3L
No. 4	.125	.312	.032	-4	-B4	-C4	-PD4	-D4	.016	-4L	-C4L	-PD4L	-D4L
No. 6	.149	.375	.032	-6	-B6	-C6	-PD6	-D6	.016	-6L	-C6L	-PD6L	-D6L
No. 8	.174	.375	.032	-8	-B8	-C8	-PD8	-D8	.016	-8L	-C8L	-PD8L	-D8L
No. 10	.203	.438	.064	-10	-B10	-C10	-PD10	-D10	.032*	-10L	-C10L	-PD10L	-D10L
1/4"	.265	.500	.064	-416	-B416	-C416	-PD416	-D416	.032*	-416L	-C416L	-PD416L	-D416L
5/16"	.326	.562	.064	-516	-B516	-C516	-PD516	-D516	.032*	-516L	-C516L	-PD516L	-D516L
3/8"	.390	.625	.064	-616	-B616	-C616	-PD616	-D616	.032*	-616L	-C616L	-PD616L	-D616L
7/16"	.453	.750	.064	-716	-B716	-C716	-PD716	-D716	.032*	-716L	-C716L	-PD716L	-D716L
1/2"	.515	.875	.064	-816	-B816	-C816	-PD816	-D816	.032*	-816L	-C816L	-PD816L	-D816L
9/16"	.578	1.062	.064	-916	-B916	-C916	-PD916	-D916	.032*	-916L	-C916L	-PD916L	-D916L
5/8"	.640	1.188	.091	-1016	-B1016	-C1016	-PD1016	-D1016	.032*	-1016L	-C1016L	-PD1016L	-D1016L
3/4"	.765	1.312	.091	-1216	-B1216	-C1216	-PD1216	-D1216	.032*	-1216L	-C1216L	-PD1216L	-D1216L
7/8"	.890	1.500	.091	-1416	-B1416	-C1416	-PD1416	-D1416	0.32*	-1416L	-C1416L	-PD1416L	-D1416L
1"	1.015	1.750	.091	-1616	-B1616	-C1616	-PD1616	-D1616	.032*	-1616L	-C1616L	-PD1616L	-D1616L

*Thickness for carbon and corrosion-resisting steel only. Aluminum alloy washers = .016"

COTTER PINS

MS24665

Manufactured in cadmium plated carbon steel, passivated corrosion resisting steel, brass and nickel copper alloy. The Ordering Table shows the dash numbers for cadmium plated carbon steel and corrosion resisting steel only, and up to 3" lengths only. Information on greater cotter pin lengths and on brass and nickel copper alloy will be supplied on request. Order by MS part number.

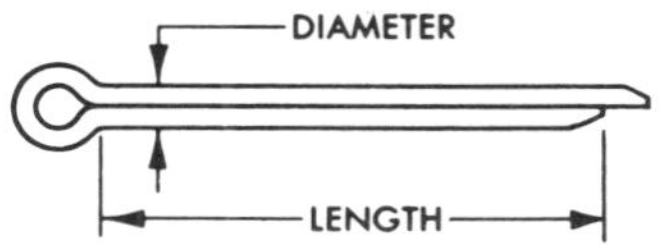

Examples: MS24665-7 (cadmium plated carbon steel; dia. 1/32"; length 3/4") MS24665-24 (corrosion resisting steel; dia. 1/32"; length 3/4")

LENGTH (INCHES)	DIAMETER (INCHES)															
	1/32		3/64		1/16		5/64		3/32		1/8		5/32		3/16	
	CARB.	CORR.	CARB.	CORR.	CARB.	CORR.	CARB.	CORR.	CARB.	CORR.	CARB.	CORR.	CARB.	CORR.	CARB.	CORR.
1/4	-1	-18	-65	-82												
5/16				-83		-148										
3/8	-3	-20	-67	-84	-130	-149										
7/16				-85		-150										
1/2	-5	-22	-69	-86	-132	-151	-208	-227	-281	-298	-349	-366	-417	-435		
5/8				-87		-152		-228		-299						
3/4	-7	-24	-71	-88	-134	-153	-210	-229	-283	-300	-351	-368	-418	-436	-490	-508
7/8				-89		-154		-230		-301		-369				
1	-9	-26	-73	-90	-136	-155	-212	-231	-285	-302	-353	-370	-419	-437	-491	-509
1 1/8								-232		-303		-371		-438		
1 1/4	-10	-27	-74	-91	-138	-157	-214	-233	-287	-304	-355	-372	-421	-439	-493	-511
1 3/8								-234		-305		-373		-440		
1 1/2	-11	-28	-75	-92	-140	-159	-216	-235	-289	-306	-357	-374	-423	-441	-495	-513
1 5/8								-236		-307		-375		-442		-514
1 3/4	-12	-29	-76	-93	-142	-161	-218	-237	-291	-308	-359	-376	-425	-443	-497	-515
2	-13	-30	-77	-94	-143	-162	-219	-238	-292	-309	-360	-377	-426	-444	-498	-516
2 1/4									-293	-310	-361	-378	-427	-445	-499	-517
2 1/2	-14	-31	-78	-95	-144	-163	-220	-239	-294	-311	-362	-379	-428	-446	-500	-518
2 3/4													-429	-447	-501	-519
3			-79	-96	-145	-164	-221	-240	-295	-312	-363	-380	-430	-448	-502	-520

COWLING LOCK PINS

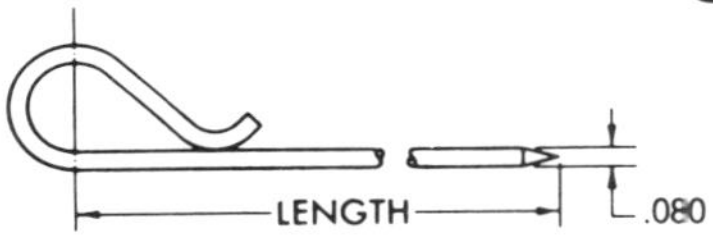

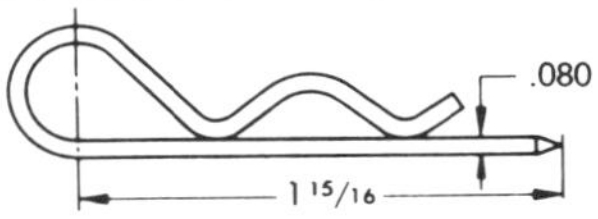

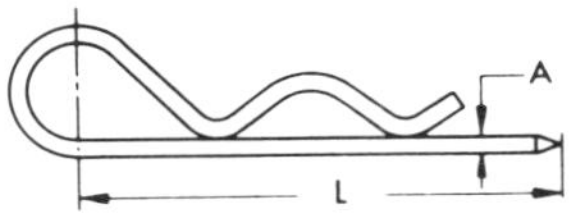

AN415-1

CORROSION-RESISTING STEEL

Used with AN226 cowl fasteners where pin passes through two or more studs. Made of corrosion-resisting steel wire in accordance with specification AN-W-23, condition B. Wire diameter is .080 inch. Available in any length in increments of one inch.

When ordering, specify length by adding a second dash number to indicate length in inches. Example: AN415-1-5 (Effective length — 5 inches)

AN415-2

CORROSION-RESISTING STEEL

Used with AN226 cowl fastener where the pin passes through one stud only. These cowling lock pins are manufactured of corrosion-resisting steel wire in conformance with Air Force-Navy specification AN-W-23, condition B. Wire diameter is .080 inch. Length is 1 15/16 inches.

AN415-2 pins are available in one size, therefore only the AN number need be used when ordering.

COMMERCIAL

CADMIUM-PLATED STEEL

Similar to AN415-2 lock pins, but made of cadmium-plated steel spring wire, these pins are available in two sizes as shown in the table below.

STOCK NUMBER	WIRE SIZE A (INCH)	LENGTH L (INCH)
481-200	.062	1¼
481-201	.080	1¼

COWLING SAFETY PINS

AN416

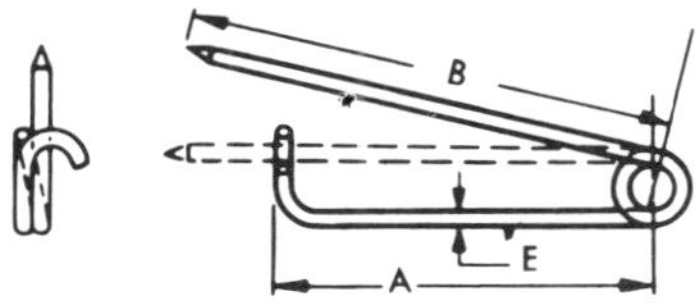

COMMERCIAL

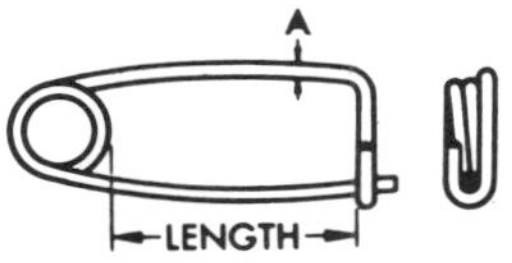

Positive locking pins that can be readily used without aid of tools. Made of spring steel in conformance with AN416 standard drawing, and cadmium-plated as required by QQ-P-416 specifications. Available in two sizes as shown in the Ordering Table below.

DASH NUMBER	DIMENSIONS (INCH)		
	A	B	E
AN416-1	1 1/16	1⅜	.051
AN416-2	¾	15/16	.041

Spring steel, cadmium-plated pins made to the same specifications as AN416 cowling safety pins. No tools required for installation or removal. Can be supplied in two sizes as shown in the following table.

STOCK NUMBER	WIRE SIZE A (INCH)	LENGTH (INCH)
481-202	.048	1 7/16
481-203	.062	1⅝

CLEVIS PINS

MS 20392

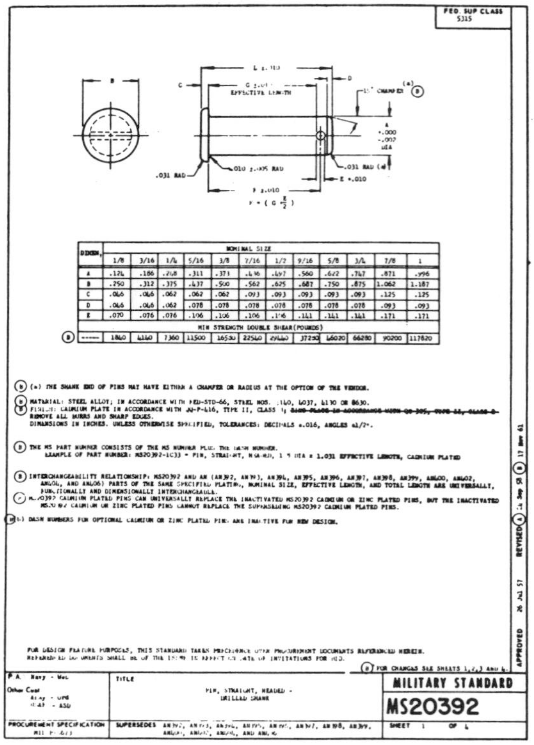

DIMEN.	NOMINAL SIZE											
	1/8	3/16	1/4	5/16	3/8	7/16	1/2	9/16	5/8	3/4	7/8	1
A	.124	.186	.248	.311	.373	.436	.497	.560	.622	.747	.871	.996
B	.250	.312	.375	.437	.500	.562	.625	.687	.750	.875	1.062	1.187
C	.046	.046	.062	.062	.062	.093	.093	.093	.093	.093	.125	.125
D	.046	.046	.062	.078	.078	.078	.078	.078	.078	.078	.093	.093
E	.070	.076	.076	.106	.106	.106	.106	.141	.141	.141	.171	.171
MIN STRENGTH DOUBLE SHEAR (POUNDS)												
(B) -----	1840	4140	7360	11500	16530	22540	29440	37250	46020	66280	90200	117820

(B) (a) THE SHANK END OF PINS MAY HAVE EITHER A CHAMFER OR RADIUS AT THE OPTION OF THE VENDOR.

(B) MATERIAL: STEEL ALLOY; IN ACCORDANCE WITH FED-STD-66, STEEL NOS. [illegible]140, 4037, 4130 OR 8630.
(B) FINISH: CADMIUM PLATE IN ACCORDANCE WITH QQ-P-416, TYPE II, CLASS 3; [illegible]
REMOVE ALL BURRS AND SHARP EDGES.
DIMENSIONS IN INCHES. UNLESS OTHERWISE SPECIFIED, TOLERANCES: DECIMALS ±.016, ANGLES ±1/2°.

(B) THE MS PART NUMBER CONSISTS OF THE MS NUMBER PLUS THE DASH NUMBER.
EXAMPLE OF PART NUMBER: MS20392-1C33 = PIN, STRAIGHT, HEADED, 1/8 DIA x 1.031 EFFECTIVE LENGTH, CADMIUM PLATED

(B) INTERCHANGEABILITY RELATIONSHIP: MS20392 AND AN (AN392, AN393), AN394, AN395, AN396, AN397, AN398, AN399, AN400, AN402, AN404, AND AN406) PARTS OF THE SAME SPECIFIED PLATING, NOMINAL SIZE, EFFECTIVE LENGTH, AND TOTAL LENGTH ARE UNIVERSALLY, FUNCTIONALLY AND DIMENSIONALLY INTERCHANGEABLE.
(C) MS20392 CADMIUM PLATED PINS CAN UNIVERSALLY REPLACE THE INACTIVATED MS20392 CADMIUM OR ZINC PLATED PINS, BUT THE INACTIVATED MS20392 CADMIUM OR ZINC PLATED PINS CANNOT REPLACE THE SUPERSEDING MS20392 CADMIUM PLATED PINS.

(B) (b) DASH NUMBERS FOR OPTIONAL CADMIUM OR ZINC PLATED PINS ARE INACTIVE FOR NEW DESIGN.

FOR DESIGN FEATURE PURPOSES, THIS STANDARD TAKES PRECEDENCE OVER PROCUREMENT DOCUMENTS REFERENCED HEREIN.
REFERENCED DOCUMENTS SHALL BE OF THE ISSUE IN EFFECT ON DATE OF INVITATIONS FOR BID.

(B) FOR CHANGES SEE SHEETS 1, 2, 3 AND 4.

P.A. Navy - Wep
Other Cust Army - Ord, Air Force - ASD

TITLE: PIN, STRAIGHT, HEADED - DRILLED SHANK

MILITARY STANDARD MS20392

PROCUREMENT SPECIFICATION MIL-P-5673

SUPERSEDES AN392, AN393, AN394, AN395, AN396, AN397, AN398, AN399, AN400, AN402, AN404, AND AN406

SHEET 1 OF 4

APPROVED 26 Jul 57 REVISED (A) 24 Sep 58 (B) 17 Nov 61

Made of steel per specification MIL-P-5673; cadmium plated per QQ-P-416. Drilled hole for cotter pin. Order by MS number, per table below.

Clevis Pins larger than those shown in the table are also available. Price and delivery on request.

Example of part number: MS20392-1C7 (1/8" dia; 7/32" length).

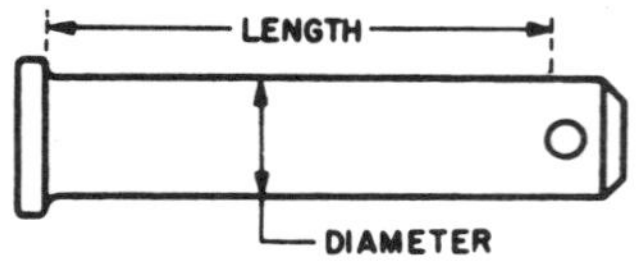

MS20392 DASH NUMBER	OLD 'AN' PART NUMBER	DIA. & LENGTH (inches)
-1C7	AN392-7	1/8 x 7/32
-1C9	-9	1/8 x 9/32
-1C11	-11	1/8 x 11/32
-1C13	-13	1/8 x 13/32
-1C15	-15	1/8 x 15/32
-1C17	-17	1/8 x 17/32
-1C19	-19	1/8 x 19/32
-1C21	-21	1/8 x 21/32
-1C23	-23	1/8 x 23/32
-1C25	-25	1/8 x 25/32
-1C27	-27	1/8 x 27/32
-1C29	-29	1/8 x 29/32
-1C31	-31	1/8 x 31/32
-1C33	-33	1/8 x 1 1/32
-1C35	-35	1/8 x 1 3/32
-1C37	-37	1/8 x 1 5/32
-1C39	-39	1/8 x 1 7/32
-1C41	-41	1/8 x 1 9/32
-2C7	AN393-7	3/16 x 7/32
-2C9	-9	3/16 x 9/32
-2C11	-11	3/16 x 11/32
-2C13	-13	3/16 x 13/32
-2C15	-15	3/16 x 15/32
-2C17	-17	3/16 x 17/32
-2C19	-19	3/16 x 19/32
-2C21	-21	3/16 x 21/32
-2C23	-23	3/16 x 23/32
-2C25	-25	3/16 x 25/32
-2C27	-27	3/16 x 27/32
-2C29	-29	3/16 x 29/32
-2C31	-31	3/16 x 31/32
-2C33	-33	3/16 x 1 1/32
-2C35	-35	3/16 x 1 3/32
-2C37	-37	3/16 x 1 5/32
-2C39	-39	3/16 x 1 7/32
-2C41	-41	3/16 x 1 9/32
-2C43	-43	3/16 x 1 11/32
-2C45	-45	3/16 x 1 13/32
-3C11	AN394-11	1/4 x 11/32
-3C13	-13	1/4 x 13/32
-3C15	-15	1/4 x 15/32
-3C17	-17	1/4 x 17/32
-3C19	-19	1/4 x 19/32
-3C21	-21	1/4 x 21/32
-3C23	-23	1/4 x 23/32
-3C25	-25	1/4 x 25/32
-3C27	-27	1/4 x 27/32
-3C29	-29	1/4 x 29/32
-3C31	-31	1/4 x 31/32
-3C33	-33	1/4 x 1 1/32
-3C35	-35	1/4 x 1 3/32
-3C37	-37	1/4 x 1 5/32
-3C39	-39	1/4 x 1 7/32
-3C41	-41	1/4 x 1 9/32
-3C43	-43	1/4 x 1 11/32
-3C45	-45	1/4 x 1 13/32
-3C47	-47	1/4 x 1 15/32
-3C49	-49	1/4 x 1 17/32
-3C51	-51	1/4 x 1 19/32
-3C53	-53	1/4 x 1 21/32
-3C55	-55	1/4 x 1 23/32
-4C17	AN395-17	5/16 x 17/32
-4C37	-37	5/16 x 1 5/32
-4C45	-45	5/16 x 1 13/32

COTTER PIN AND WASHER SIZES
To Use With Clevis Pins

MS NUMBER	USE AN24665 COTTER PIN		USE FLAT WASHER NUMBER
	CORROSION RESISTING STEEL	CARBON STEEL	
MS20392-1	-151	-132	AN960-6
MS20392-2	-151	-132	AN960-10
MS20392-3	-151	-132	AN960-416
MS20392-4	-300	-283	AN960-516
MS20392-5	-300	-283	AN960-616
MS20392-6	-300	-283	AN960-716
MS20392-7	-302	-285	AN960-816

SOLID RIVETS

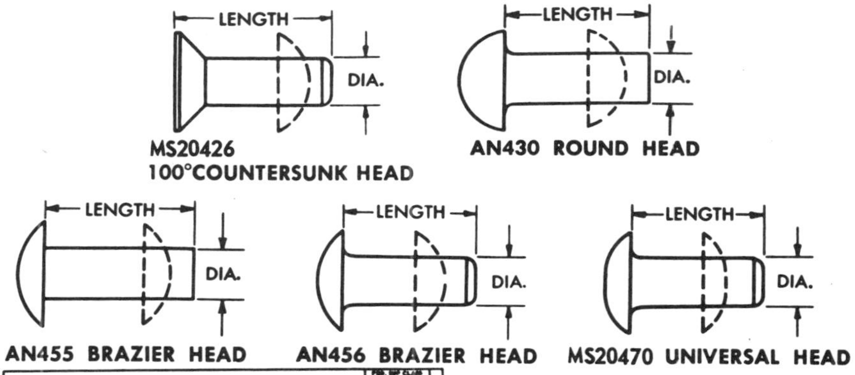

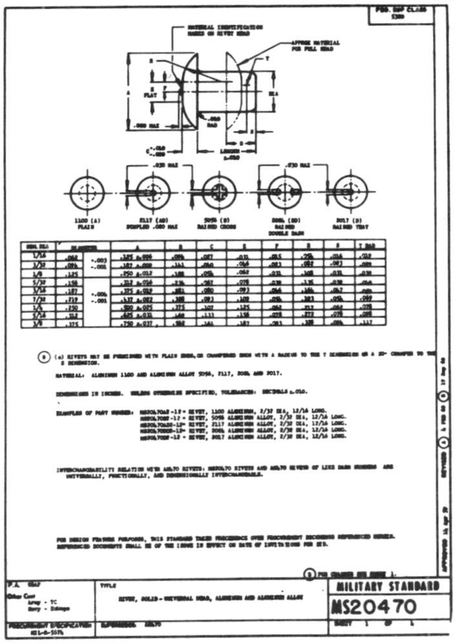

Of the five rivet head types shown above, the MS20426 and the MS20470 types are the most widely used today. Manufactured to Specification MIL-R-5674, these two types are available for immediate shipment in two materials: 'hard' 2117 aluminum alloy (AD) and 'soft' 1100 pure aluminum (A). 'A' rivets have a minimum tensile strength of 16,000 psi; 'AD' rivets are heat treated, have a shear strength of 26,000 psi and a tensile strength of 38,000 psi. MS20426 and MS20470 rivets of 5056 (B), 2024 (DD) and 2017 (D) aluminum alloy are not normally stock items but can be supplied on special order, as can rivets conforming to AN430, AN445 and AN456 specifications.

Examples of part numbers:
MS20426A3-12 (3/32" dia., 3/4" long, 1100 alum.)
MS20470AD3-12 (3/32" dia., 3/4" long, 2117 alum. alloy)

NOTE:
MS20426 and MS20470 rivets of like dash numbers are interchangeable with AN426 and AN470 rivets respectively.

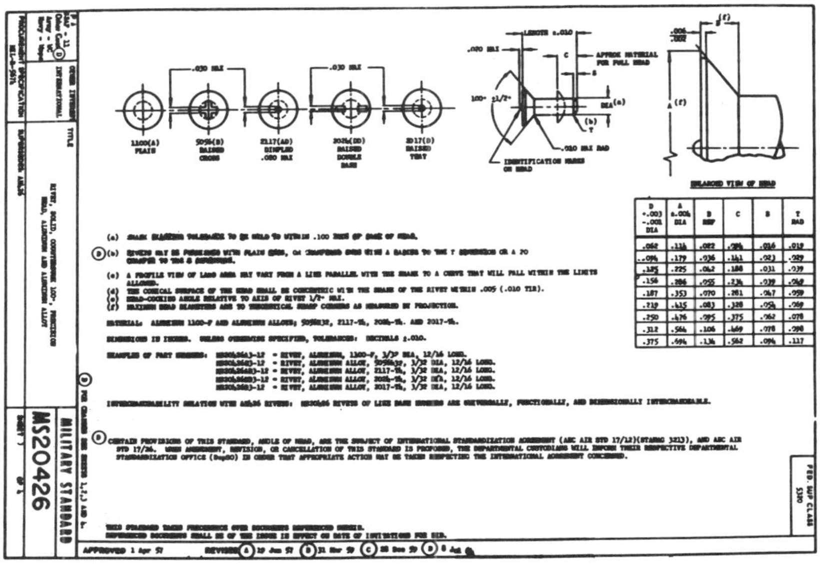

ORDERING AND WEIGHT TABLE

DASH NUMBER	SIZE (INCH)	QUANTITY PER POUND MS20470	QUANTITY PER POUND MS20426
2-2	1/16 x 1/8	....	23044
2-3	1/16 x 3/16	10456	16014
2-4	1/16 x 1/4	8701	12270
2-5	1/16 x 5/16	7450	9946
2-6	1/16 x 3/8	6514	8361
2-7	1/16 x 7/16	5787	7212
2-8	1/16 x 1/2	5205	6341
3-3	3/32 x 3/16	4107	6406
3-4	3/32 x 1/4	3485	5003
3-5	3/32 x 5/16	3027	4103
3-6	3/32 x 3/8	2675	3478
3-7	3/32 x 7/16	2397	3018
3-8	3/32 x 1/2	2171	2666
3-9	3/32 x 9/16	1983	2387
3-10	3/32 x 5/8	1826	2161
3-12	3/32 x 3/4	1576	1817
3-14	3/32 x 7/8	1386	1567
3-16	3/32 x 1	1237	1378
4-3	1/8 x 3/16	2003	....
4-4	1/8 x 1/4	1734	2810
4-5	1/8 x 5/16	1529	2308
4-6	1/8 x 3/8	1367	1958
4-7	1/8 x 7/16	1236	1700
4-8	1/8 x 1/2	1128	1502
4-9	1/8 x 9/16	1037	1346
4-10	1/8 x 5/8	960	1219
4-12	1/8 x 3/4	836	1025
4-14	1/8 x 7/8	740	885
4-16	1/8 x 1	664	755
5-4	5/32 x 1/4	994	1731
5-5	5/32 x 5/16	888	1440
5-6	5/32 x 3/8	802	1226
5-7	5/32 x 7/16	731	1079
5-8	5/32 x 1/2	672	946
5-9	5/32 x 9/16	621	849
5-10	5/32 x 5/8	578	770
5-12	5/32 x 3/4	507	649
5-14	5/32 x 7/8	452	561
5-16	5/32 x 1	407	494
6-4	3/16 x 1/4	628	......
6-5	3/16 x 5/16	566	......
6-6	3/16 x 3/8	515	846
6-7	3/16 x 7/16	472	738
6-8	3/16 x 1/2	436	654
6-9	3/16 x 9/16	406	588
6-10	3/16 x 5/8	379	534
6-12	3/16 x 3/4	334	451
6-14	3/16 x 7/8	299	390
6-16	3/16 x 1	271	343

PULL-THROUGH HOLLOW CHERRY RIVETS

Designed for installations not requiring the high shear strength of self-plugging or solid rivets. A conical head at the base of the stem forms a blind head against the back of the sheet being riveted, then partially collapses as the stem continues through to expand the rivet shank. The stem is finally ejected, leaving a hollow rivet which has been fully expanded to fill the hole.

Military Standard Pages MS20604 and MS20605 provide a means of specifying type of stem desired. The letter "K" in the MS number specifies a knob stem; the letter "W" in the MS number specifies a serrated stem; the absence of both "K" and "W" in the MS number, but replaced by a dash (—), indicates either a knob stem or serrated stem may be used. NOTE: *Other styles and types of Cherry Rivets are also available for which there are no corresponding military numbers.*

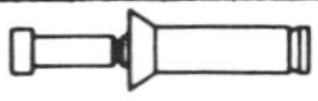
CR 116

CR 9116

CR 117

CR 9117

2117 aluminum alloy with steel stem and alumilite No. 205 finish, except in 3/32" diameter where 5056 aluminum alloy is used with an aluminum stem.

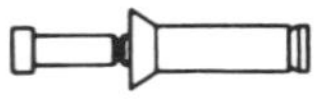
CR 126

CR 9126

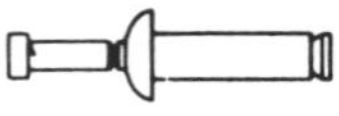
CR 127

CR 9127

5056 aluminum alloy with steel stem and alumilite finish, dyed orange. Used especially with magnesium.

CONVERSION TABLE
MILITARY NUMBERS TO CHERRY RIVET NUMBERS

UNIVERSAL HEAD						100° FLUSH HEAD					
2117 AL. ALLOY		5056 AL. ALLOY		MONEL		2117 AL. ALLOY		5056 AL. ALLOY		MONEL	
3/32 RIVET DIAMETER, NOMINAL						3/32 RIVET DIAMETER, NOMINAL					
CR No.	MS No.	CR No.	MS No.	CR No.	MS No.	CR No.	MS No.	CR No.	MS No.	CR No.	MS No.
117-3-4	20604AD3K2	127-3-4	20604B3K2			116-3-4	20605AD3K2	126-3-4	20605B3K2		
9117-3-2	20604AD3W2	9127-3-2	20604B3W2			9116-3-2	20605AD3W2	9126-3-2	20605B3W2		
117-3-6	AD3K3	127-3-6	B3K3			116-3-6	AD3K3	126-3-6	B3K3		
9117-3-3	AD3W3	9127-3-3	B3W3			9116-3-3	AD3W3	9126-3-3	B3W3		
117-3-8	AD3K4	127-3-8	B3K4			116-3-8	AD3K4	126-3-8	B3K4		
9117-3-4	AD3W4	9127-3-4	B3W4			9116-3-4	AD3W4	9126-3-4	B3W4		
1/8 RIVET DIAMETER, NOMINAL						1/8 RIVET DIAMETER, NOMINAL					
117-4-2	20604AD4K1	127-4-2	20604B4K1	517-4-2	20604M4K1	116-4-2	20605AD4K1	126-4-2	20605B4K1	516-4-2	20605M4K1
9117-4-1	20604AD4W1	9127-4-1	20604B4W1	9517-4-1	20604M4W1	9116-4-1	20605AD4W1	9126-4-1	20605B4W1	9516-4-1	20605M4W1
117-4-4	AD4K2	127-4-4	B4K2	517-4-4	M4K2	116-4-4	AD4K2	126-4-4	B4K2	516-4-4	M4K2
9117-4-2	AD4W2	9127-4-2	B4W2	9517-4-2	M4W2	9116-4-2	AD4W2	9126-4-2	B4W2	9516-4-2	M4W2
117-4-6	AD4K3	127-4-6	B4K3	517-4-6	M4K3	116-4-6	AD4K3	126-4-6	B4K3	516-4-6	M4K3
9117-4-3	AD4W3	9127-4-3	B4W3	9517-4-3	M4W3	9116-4-3	AD4W3	9126-4-3	B4W3	9516-4-3	M4W3
117-4-8	AD4K4	127-4-8	B4K4	517-4-8	M4K4	116-4-8	AD4K4	126-4-8	B4K4	516-4-8	M4K4
9117-4-4	AD4W4	9127-4-4	B4W4	9517-4-4	M4W4	9116-4-4	AD4W4	9126-4-4	B4W4	9516-4-4	M4W4
117-4-10	AD4K5	127-4-10	B4K5	517-4-10	M4K5	116-4-10	AD4K5	126-4-10	B4K5	516-4-10	M4K5
9117-4-5	AD4W5	9127-4-5	B4W5	9517-4-5	M4W5	9116-4-5	AD4W5	9126-4-5	B4W5	9516-4-5	M4W5
117-4-12	AD4K6	127-4-12	B4K6			116-4-12	AD4K6	126-4-12	B4K6		
9117-4-6	AD4W6	9127-4-6	B4W6			9116-4-6	AD4W6	9126-4-6	B4W6		
117-4-14	AD4K7	127-4-14	B4K7			116-4-14	AD4K7	126-4-14	B4K7		
9117-4-7	AD4W7	9127-4-7	B4W7			9116-4-7	AD4W7	9126-4-7	B4W7		
5/32 RIVET DIAMETER, NOMINAL						5/32 RIVET DIAMETER, NOMINAL					
117-5-4	20604AD5K2	127-5-4	20604B5K2	517-5-4	20604M5K2	116-5-4	20605AD5K2	126-5-4	20605B5K2	516-5-4	20605M5K2
9117-5-2	20604AD5W2	9127-5-2	20604B5W2	9517-5-2	20604M5W2	9116-5-2	20605AD5W2	9126-5-2	20605B5W2	9516-5-2	20605M5W2
117-5-8	AD5K4	127-5-8	B5K4	517-5-8	M5K4	116-5-8	AD5K4	126-5-8	B5K4	516-5-8	M5K4
9117-5-4	AD5W4	9127-5-4	B5W4	9517-5-4	M5W4	9116-5-4	AD5W4	9126-5-4	B5W4	9516-5-4	M5W4
117-5-12	AD5K6	127-5-12	B5K6	517-5-12	M5K6	116-5-12	AD5K6	126-5-12	B5K6	516-5-12	M5K6
9117-5-6	AD5W6	9127-5-6	B5W6	9517-5-6	M5W6	9116-5-6	AD5W6	9126-5-6	B5W6	9516-5-6	M5W6
3/16 RIVET DIAMETER, NOMINAL						3/16 RIVET DIAMETER, NOMINAL					
117-6-4	20604AD6K2	127-6-4	20604B6K2	517-6-4	20604M6K2	116-6-4	20605AD6K2	126-6-4	20605B6K2	516-6-4	20605M6K2
9117-6-2	20604AD6W2	9127-6-2	20604B6W2	9517-6-2	20604M6W2	9116-6-2	20605AD6W2	9126-6-2	20605B6W2	9516-6-2	20605M6W2
117-6-8	AD6K4	127-6-8	B6K4	517-6-8	M6K4	116-6-8	AD6K4	126-6-8	B6K4	516-6-8	M6K4
9117-6-4	AD6W4	9127-6-4	B6W4	9517-6-4	M6W4	9116-6-4	AD6W4	9126-6-4	B6W4	9516-6-4	M6W4
117-6-12	AD6K6	127-6-12	B6K6	517-6-12	M6K6	116-6-12	AD6K6	126-6-12	B6K6	516-6-12	M6K6
9117-6-6	AD6W6	9127-6-6	B6W6	9517-6-6	M6W6	9116-6-6	AD6W6	9126-6-6	B6W6	9516-6-6	M6W6
117-6-16	AD6K8	127-6-16	B6K8			116-6-16	AD6K8	126-6-16	B6K8		
9117-6-8	AD6W8	9127-6-8	B6W8			9116-6-8	AD6W8	9126-6-8	B6W8		
117-6-20	AD6K10	127-6-20	B6K10			116-6-20	AD6K10	126-6-20	B6K10		
9117-6-10	AD6W10	9127-6-10	B6W10			9116-6-10	AD6W10	9126-6-10	B6W10		
117-6-24	AD6K12	127-6-24	B6K12			116-6-24	AD6K12	126-6-24	B6K12		
9117-6-12	AD6W12	9127-6-12	B6W12			9116-6-12	AD6W12	9126-6-12	B6W12		
1/4 RIVET DIAMETER, NOMINAL						1/4 RIVET DIAMETER, NOMINAL					
117-8-4	20604AD8K2	127-8-4	20604B8K2	517-8-4	20604M8K2					516-8-4	20605M8K2
9117-8-2	20604AD8W2	9127-8-2	20604B8W2	9517-8-2	20604M8W2					9516-8-2	20605M8W2
117-8-8	AD8K4	127-8-8	B8K4	517-8-8	M8K4	116-8-8	20605AD8K4	126-8-8	20605B8K4	516-8-8	M8K4
9117-8-4	AD8W4	9127-8-4	B8W4	9517-8-4	M8W4	9116-8-4	20605AD8W4	9126-8-4	20605B8W4	9516-8-4	M8W4
117-8-12	AD8K6	127-8-12	B8K6	517-8-12	M8K6	116-8-12	AD8K6	126-8-12	B8K6	516-8-12	M8K6
9117-8-6	AD8W6	9127-8-6	B8W6	9517-8-6	M8W6	9116-8-6	AD8W6	9126-8-6	B8W6	9516-8-6	M8W6
117-8-16	AD8K8	127-8-16	B8K8	517-8-16	M8K8	116-8-16	AD8K8	126-8-16	B8K8	516-8-16	M8K8
9117-8-8	AD8W8	9127-8-8	B8W8	9517-8-8	M8W8	9116-8-8	AD8W8	9126-8-8	B8W8	9516-8-8	M8W8
117-8-20	AD8K10	127-8-20	B8K10	517-8-20	M8K10	116-8-20	AD8K10	126-8-20	B8K10	516-8-20	M8K10
9117-8-10	AD8W10	9127-8-10	B8W10	9517-8-10	M8W10	9116-8-10	AD8W10	9126-8-10	B8W10	9516-8-10	M8W10
117-8-24	AD8K12	127-8-24	B8K12	517-8-24	M8K12	116-8-24	AD8K12	126-8-24	B8K12	516-8-24	M8K12
9117-8-12	AD8W12	9127-8-12	B8W12	9517-8-12	M8W12	9116-8-12	AD8W12	9126-8-12	B8W12	9516-8-12	M8W12
117-8-28	AD8K14	127-8-28	B8K14	517-8-28	M8K14	116-8-28	AD8K14	126-8-28	B8K14	516-8-28	M8K14
9117-8-14	AD8W14	9127-8-14	B8W14	9517-8-14	M8W14	9116-8-14	AD8W14	9126-8-14	B8W14	9516-8-14	M8W14
				517-8-32	M8K16					516-8-32	M8K16
				9517-8-16	M8W16					9516-8-16	M8W16

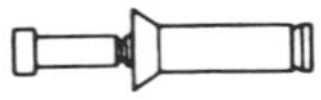
CR 516

CR 517

CR 9516

CR 9517

Monel with steel stem; no finish (can also be furnished with cadmium plating on special order).

DZUS FASTENERS

IN SELECTING LENGTH OF FASTENER–
ADD N+P+Q+H – SUBTRACT SPRING DEFLECTION.
FOR INSTALLATION WITHOUT GROMMET – OMIT "N".

INSTALLATION TOOLS

Available only in sets as indicated in table below.

PART NUMBER	FOR TYPE FASTENER	SET CONSISTS OF
A-4	A4, AJ4	4 tools, 2 blocks
A-5	A5, AJ5	4 tools, 2 blocks
A-6½	A6½, AJ6½	4 tools, 2 blocks
F-5	FA5	3 tools, 3 blocks
F-6½	FA6½	3 tools, 3 blocks

TYPE "A" FASTENER
OVAL HEAD SHORT UNDERCUT

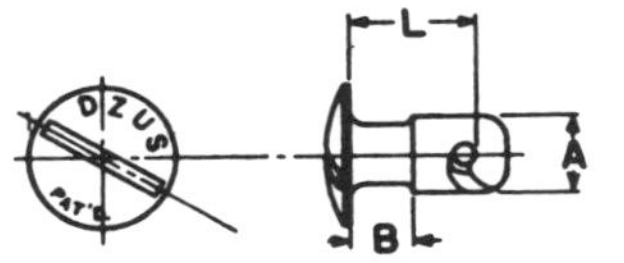

TYPE "AJ" FASTENER
OVAL HEAD LONG UNDERCUT

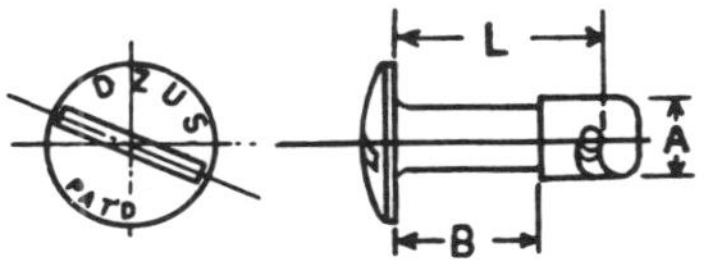

TYPE "FA" FASTENER
FLUSH HEAD ROUNDED EDGE

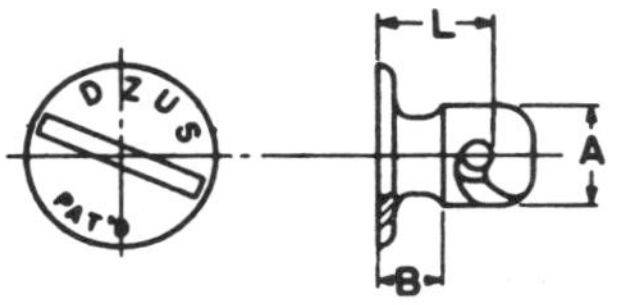

PART NUMBER	DIMENSIONS (Inches) A	B	L	USES GROMMET	USES SPRING
A4-25	.250	.080	.250	GA4-250	S4-200, S4-225
A4-30			.300		
A4-35			.350		
A5-30	.3125	.100	.300	GA5-312, GA5-375	S5A-200, S5A-225, S5A-250
A5-35			.350		
A5-40			.400		
A5-45			.450		
A5-50			.500		
A6½-40	.4062	.125	.400	GA6½-375, GA6½-500	S6A-250, S6A-275
A6½-50			.500		
AJ4-25	.250	.075	.250	GA4-250	S4-200, S4-225
AJ4-35		.150	.350		
AJ4-50		.250	.500		
AJ5-30	.3125	.100	.300	GA5-312, GA5-375	S5A-200, S5A-225, S5A-250
AJ5-35		.125	.350		
AJ5-40		.190	.400		
AJ5-45		.190	.450		
AJ5-50		.250	.500		
AJ6½-40	.4062	.188	.400	GA6½-375, GA6½-500	S6A-250, S6A-275
AJ6½-45		.188	.450		
AJ6½-50		.250	.500		
AJ6½-55		.300	.550		
AJ6½-80		.450	.800		
AJ6½-90		.550	.900		
AJ6½-100		.650	1.000		
FA5-35	.3125	.188	.350	GF5-175	S5A-300, S5A-325
FA5-40			.400		
FA5-45			.450		
FA5-50			.500		
FA6½-45	.4062	.250	.450	GF6½-218, GF6½-250	S6A-400, S6A-425
FA6½-50			.500		
FA6½-55			.550		
FA6½-60			.600		
FA6½-65			.650		

WING TYPE FASTENERS

Dimensions A, B, and L are same as standard slotted head fasteners. Example: AW4-25 dimensions are the same as those of A4-25. Wing in locked position usually aligns with the spring rivets. Use same size grommets and springs for wing type as are specified for standard slotted head type.

PART NUMBER	DIMENSIONS (Inches) P	Q	R
AW4-25	.875	.3125	.050
AW4-30			
AW4-35			
AW5-30	1.125	.4375	.062
AW5-35			
AW5-40			
AW5-45			
AW5-50			
AJW5-30	1.125	.4375	.062
AJW5-35			
AJW5-40			
AJW5-45			

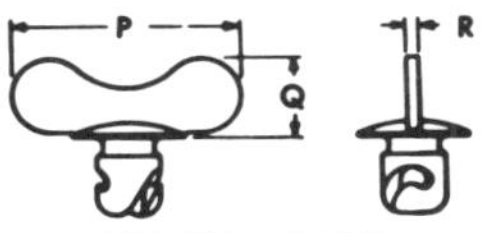

TYPE AW and AJW

TYPE S SPRINGS

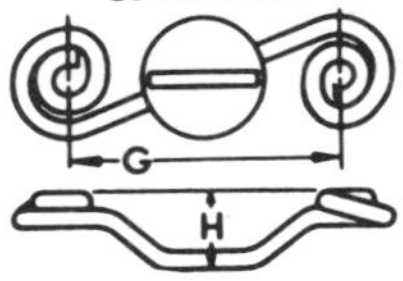

PART NUMBER	DIMENSIONS (Inches) H	G
S4-200	.200	.750
S4-250	.250	
S5A-200	.200	1.000
S5A-225	.225	
S5A-300	.300	
S5A-325	.325	
S6A-225	.225	1.375
S6A-250	.250	
S6A-275	.275	
S6A-400	.400	
S6A-425	.425	

TYPES GA AND GF GROMMETS

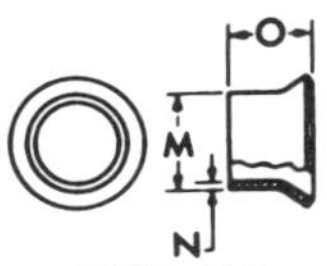

TYPE GA

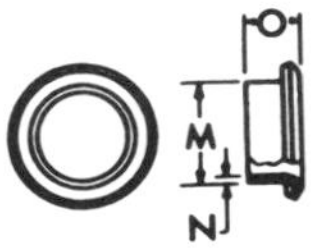

TYPE GF

PART NUMBER	DIMENSIONS (Inches) O	M	N	MAX. DEPTH OF PANEL
GA3-175	.175	.2187	.015	.025
GA3-200	.200	.2187	.015	.050
GA4-250	.250	.3125	.025	.050
GA5-312	.312	.375	.028	.062
GA5-375	.375	.375	.028	.125
GA6½-375	.375	.500	.035	.093
GA6½-500	.500	.500	.035	.200
GF5-175	.175	.375	.028	.062
GF6½-250	.250	.500	.035	.093
GSA6½T-750	.750	.500	.035	.600

CLAMPS

ADEL No. 755 CLIP

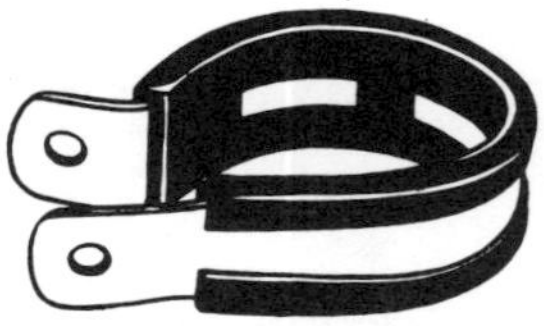

Line-support clip for tubing or conduit; has synthetic rubber cushion bonded to inner surface and edges. Cushion assures tight fit, eliminates vibration, thus preventing line abrasions. Made of 24S-T aluminum alloy in accordance with Federal specification QQ-A-362, Type T. Additional sizes of No. 755 clips, and also many other styles of Adel clips, are available from stock or on special order.

ORDERING TABLE

ADEL NUMBER*	DIA. OF TUBE (IN.)	ADEL NUMBER*	DIA. OF TUBE (IN.)
755-2-2-6	1/8	755-8-2-8	1/2
755-2-2-8	1/8	755-9-2-8	9/16
755-3-2-6	3/16	755-10-2-8	5/8
755-3-2-8	3/16	755-12-2-8	3/4
755-4-2-6	1/4	755-14-2-8	7/8
755-4-2-8	1/4	755-14-4-8	7/8
755-4-4-6	1/4	755-15-2-8	15/16
755-5-2-6	5/16	755-16-2-8	1
755-5-2-8	5/16	755-18-2-8	1 1/8
755-6-2-6	3/8	755-20-2-8	1 1/4
755-6-2-8	3/8	755-24-2-8	1 1/2
755-7-2-6	7/16	755-26-2-8	1 5/8
755-7-2-8	7/16	755-42-2-8	2 5/8
755-8-2-6	1/2		

*2nd dash number indicates cushion thickness in 32nds.
3rd dash number indicates width of metal in 16ths.

NATIONAL AIRCRAFT STANDARDS COMMITTEE

AIRCRAFT INDUSTRIES ASSOCIATION OF AMERICA, INC., 610 SHOREHAM BUILDING, WASHINGTON 5, D. C.

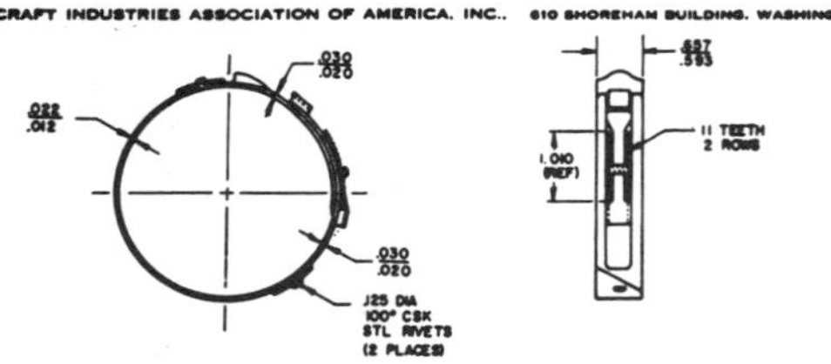

PART NO.	HOSE O.D. MIN.	HOSE O.D. MAX.	PART NO.	HOSE O.D. MIN.	HOSE O.D. MAX.	PART NO.	HOSE O.D. MIN.	HOSE O.D. MAX.
-40	2.375	2.500	-82	5.000	5.125	-124	7.625	7.750
-42	2.500	2.625	-84	5.125	5.250	-126	7.750	7.875
-44	2.625	2.750	-86	5.250	5.375	-128	7.875	8.000
-46	2.750	2.875	-88	5.375	5.500	-130	8.000	8.125
-48	2.875	3.000	-90	5.500	5.625	-132	8.125	8.250
-50	3.000	3.125	-92	5.625	5.750			
-52	3.125	3.250	-94	5.750	5.875			
-54	3.250	3.375	-96	5.875	6.000			
-56	3.375	3.500	-98	6.000	6.125			
-58	3.500	3.625	-100	6.125	6.250			
-60	3.625	3.750	-102	6.250	6.375			
-62	3.750	3.875	-104	6.375	6.500			
-64	3.875	4.000	-106	6.500	6.625			
-66	4.000	4.125	-108	6.625	6.750			
-68	4.125	4.250	-110	6.750	6.875			
-70	4.250	4.375	-112	6.875	7.000			
-72	4.375	4.500	-114	7.000	7.125			
-74	4.500	4.625	-116	7.125	7.250			
-76	4.625	4.750	-118	7.250	7.375			
-78	4.750	4.875	-120	7.375	7.500			
-80	4.875	5.000	-122	7.500	7.625			

② MATERIAL: ENDS-SAE 1064 STEEL—TEMPERED; CENTER-SAE 1080 STEEL, SPECIFICATION MIL-S-7952

② ~~[illegible]~~

② EXAMPLE: NAS 398-40- HOSE CLAMP, FOR A HOSE WITH A MAX. O. D. OF 2.500
~~[illegible]~~

② FINISH: PHOSPHATE PLUS ZINC CHROMATE PRIMER.
PHOSPHATE COATING IN ACCORDANCE WITH SPEC. MIL-C-490, GRADE I.
ZINC CHROMATE COATING IN ACCORDANCE WITH SPEC. MIL-P-6889, TYPE I.

APPROVAL DATE APRIL, 1947 REVISION ① 11-15-54 ② 31 Oct. 1956

TITLE	CLASSIFICATION STANDARD PART
CLAMP-RATCHET, THREE PIECE	NAS 398

AERO-SEAL HOSE CLAMPS

Tangential type worm drive clamps. Similar to AN737 clamps (page 79) but provide a greater range between minimum and maximum diameters to allow for variations in wall thickness and hose diameters. For use on fuel, oil, air and coolant lines. Special 10-pitch hardened steel worm produces true tangential pull and uniform clamping pressure; no hose distortion; no leakage under high pressures.

Aero-Seal clamps are available in all stainless steel (QS200-) or with stainless steel band and saddle and cadmium plated carbon steel screw assembly (QS100-).

EXAMPLES OF PART NUMBERS for varieties of Aero-Seal clamps having a range of 1 1/16" to 2":
- QS100-M24S (stainless band; carbon, screwdriver slot screw)
- QS100-M24W (stainless band; carbon, thumb grip screw)
- QS200-M24S (all stainless; screwdriver slot screw)
- QS200-M24W (all stainless; thumb grip screw)

WORM ENDS SCREW-DRIVER	WORM ENDS THUMB GRIP	RANGE OF DIA. MIN. (IN.)	RANGE OF DIA. MAX. (IN.)
-M6S	-M6W	7/16	25/32
-M8S	-M8W	1/2	29/32
-M10S	-M10W	9/16	1 1/16
-M12S	-M12W	11/16	1 1/4
-M16S	-M16W	13/16	1 1/2
-M20S	-M20W	13/16	1 3/4
-M24S	-M24W	1 1/16	2
-M28S	-M28W	1 5/16	2 1/4
-M32S	-M32W	1 9/16	2 1/2
-M36S	-M36W	1 13/16	2 3/4
-M40S	-M40W	2 1/16	3
-M44S	-M44W	2 5/16	3 1/4
-M48S	-M48W	2 9/16	3 1/2
-M52S	-M52W	2 13/16	3 3/4
-M56S	-M56W	3 1/16	4
-M60S	-M60W	3 5/16	4 1/4
-M64S	-M64W	3 9/16	4 1/2

SCREWDRIVER SLOT or THUMB GRIP

CHOICE OF ENDS ON WORM

For commercial and industrial use, the screwdriver-operated type of hose clamp is generally preferred to the thumb grip type. *However, "Aero-Seals" are available either way, and at the same price.* The screwdriver type worm has a cup surrounding the slotted head so the screwdriver can be engaged easily in dark or difficult locations and will not slip out to cause damage.

AN737 CLAMPS

AN737 Hose Clamps are for use on fuel, oil and coolant lines. Installation can be made while hose is in place; clamps may be tightened repeatedly while in service. No lock wire needed. All parts of these clamps are made of stainless steel except the thumbscrews which are cadmium plated carbon steel. All clamps meet current MIL and AN specifications.

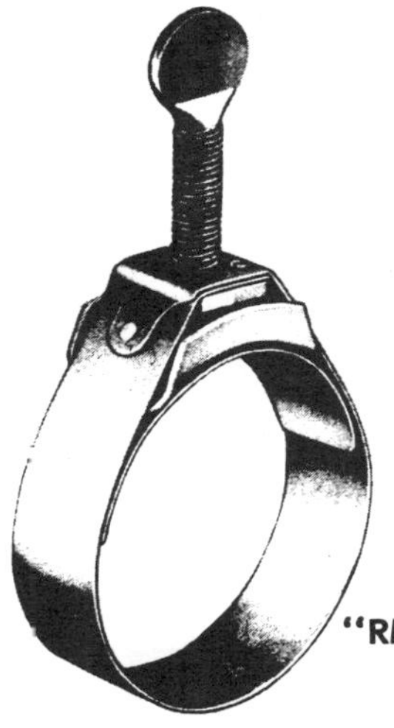

"RM" Type

AN737RM Radial-Machine Screw Type
Clamp band is 5/8" wide and .020" thick for all sizes. The band is welded to the nut by certified spot-welding methods. This unit, when assembled to the bridge and its flexible bridge extender, provides a clamp which combines superior strength with uniform clamping action. Order by AN part number, per table below. Example: AN737RM22 is a single size, radial-machine screw type clamp having a minimum I.D. of 7/16".

AN737TW Tangential-Worm Screw Type
The one-piece curved housing is spot-welded to the recessed band on certified equipment to form an integral unit. Contour serrations give greater band strength and are a perfect mate for the buttress thread of the screw for smooth clamping action. Order by AN part number per table below. Example: AN737TW22 is a single size, tangential-worm screw type clamp having a minimum I.D. of 7/16".

"TW" Type

Clamps are easily installed or removed with hose in place

ORDERING TABLE AN737 Hose Clamps

DASH NO.	MIN. I.D. OF CLAMP (INCHES)	MAX. I.D. OF CLAMP (INCHES)	FOR USE WITH HOSE I.D.	
			NON SELF-SEALING	SELF-SEALING
– 22	7/16	11/16	1/4	......
– 24	1/2	3/4	5/16	......
– 26	9/16	13/16	3/8	......
– 30	11/16	15/16	1/2	......
– 34	13/16	1 1/16	5/8	......
– 38	15/16	1 3/16	3/4	......
– 44	61/64	1 3/8	......	5/8
– 46	1	1 7/16	1	......
– 48	1 5/64	1 1/2	......	3/4
– 56	1 21/64	1 3/4	......	1
– 58	1 3/8	1 13/16	1 1/4	......
– 66	1 5/8	2 1/16	1 1/2	1 1/4
– 74	1 7/8	2 5/16	1 3/4	1 1/2
– 82	2 1/8	2 9/16	2	......
– 91	2 13/32	2 27/32	......	2
– 98	2 5/8	3 1/16	2 1/2	......
–107	2 29/32	3 11/32	......	2 1/2
–114	3 1/8	3 9/16	3	......

NOTE: *The above table shows only those Single Size clamps covered by the AN737 Standard Drawing.*

STRUT VALVES

AN812-1
High Pressure Valve designed for shock struts and accumulators to permit inflation and gauging. Less AN809-1 core and AN813-1B valve cap.

AN6287-1
For use in high pressure shock absorber struts, hydraulic pressure accumulators, surge cylinders and high pressure pneumatic systems. Permits inflation, deflation and gauging of air cushions in these units. Supersedes the AN812-1 valve. Although not required on most aircraft, for design purposes it has been the current model since 1948. Consists of stainless steel stem, core, and MS 20813-1 cap. FOR 3000 LB. SYSTEMS.

MS28889-1
Essentially the same as the AN6287-1 strut valve. For design purposes, the MS28889-1 superseded the AN6287-1 in 1955 and is functionally and dimensionally interchangeable with the AN6287-1 with the exception of over-all and mounting base lengths. For 5,000 lb. systems.

(AN-812-1)
For pressures up to 1500 p.s.i.

(AN-6287-1)
For pressures up to 3000 p.s.i.

TUBE FITTINGS

AN standard fittings for use with aircraft tubing. Made in three basic types: flared, universal, and screw thread; and in combinations of these types. Flared tube fittings consisting of a nut, sleeve, and body form a leakproof connection . . . no soldering necessary. Screw thread fittings and universal fittings employ a soft brass or aluminum gasket to make a leakproof connection.

Available in brass, steel, or aluminum alloy as indicated in the tables accompanying each illustration. Popular sizes only are listed on these pages. Most fittings can be ordered in larger sizes. Price and delivery information will be furnished on request.

Order tube fittings by AN number plus dash number for size and material desired. Example:

AN804-5 (Tee, 5/16" tube size, steel)

AN806-4D (Plug, 1/4" tube size, aluminum alloy)

See page 90 for AN Drawing

AN775 DASH NUMBER		TUBE
STEEL	ALUM. ALLOY	O.D.
–4	–4D	1/4"
–5	–5D	5/16"
–6	–6D	3/8"
–8	–8D	1/2"
–10	–10D	5/8"

AN775 BOLT
Universal

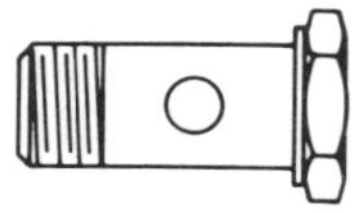

AN776 DASH NUMBER		TUBE
STEEL	ALUM. ALLOY	O.D.
–4	–4D	1/4"
–5	–5D	5/16"
–6	–6D	3/8"
–8	–8D	1/2"
–10	–10D	5/8"

AN776
ELBOW, 90°
Universal

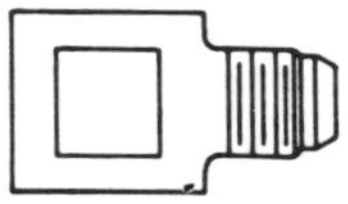

See page 91 for AN Drawing

AN777 DASH NUMBER		TUBE
STEEL	ALUM. ALLOY	O.D.
–4	–4D	1/4"
–5	–5D	5/16"
–6	–6D	3/8"
–8	–8D	1/2"
–10	–10D	5/8"

AN777
ELBOW, 75°
Universal

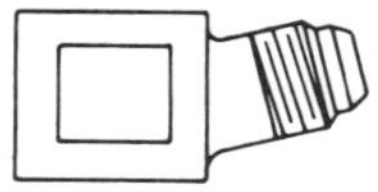

AN778 DASH NUMBER		TUBE
STEEL	ALUM. ALLOY	O.D.
–4	–4D	1/4"
–5	–5D	5/16"
–6	–6D	3/8"
–8	–8D	1/2"
–10	–10D	5/8"

AN778
ELBOW, 45°
Universal

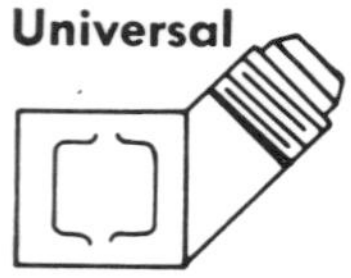

Shown here are those AN and MS fittings commonly used on aircraft.

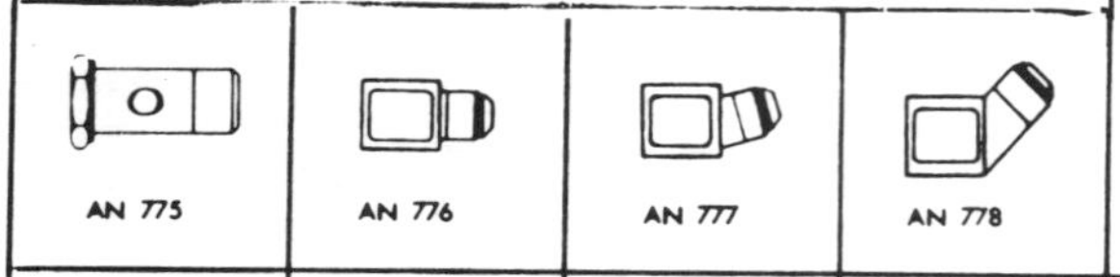

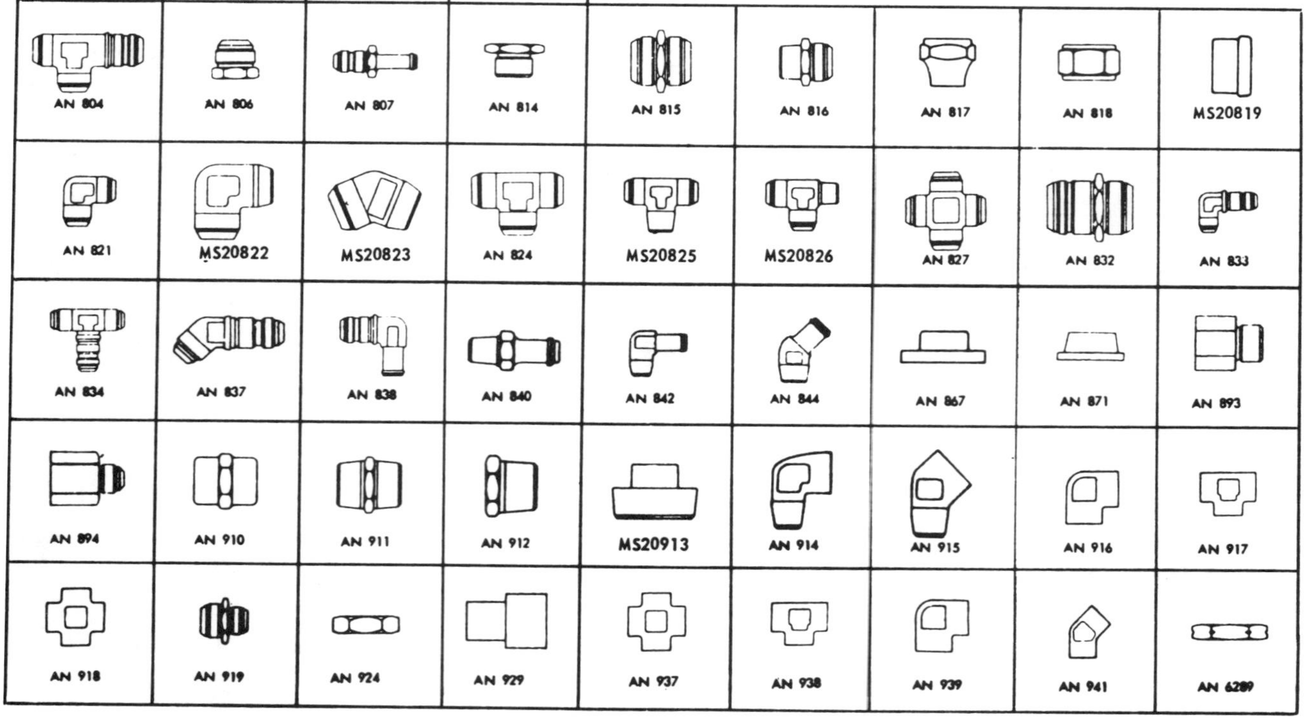

TUBE FITTINGS

AN and MS standard fittings for use with aircraft tubing. Made in three basic types: flared, universal, and screw thread; and in combinations of these types. Flared tube fittings consisting of a nut, sleeve, and body form a leakproof connection . . . no soldering necessary. Screw thread fittings and universal fittings employ a soft brass or aluminum gasket to make a leakproof connection. Available in brass, steel, or aluminum alloy as indicated in the tables accompanying each illustration. Popular sizes only are listed on these pages. Most fittings can be ordered in larger sizes.

Order tube fittings by AN number plus dash number for size and material desired. Example:
AN804-5 (Tee, 5/16" tube size, steel)
AN806-4D (Plug, 1/4" tube size, aluminum alloy)

FLARELESS TUBE FITTINGS

Van Dusen carries a complete line of the new Government approved Flareless Tube Fittings, which have been widely accepted in the aviation industry. Write or call our nearest branch office for complete information.

AN804 TEE — Flared Tube with Bulkhead on Run

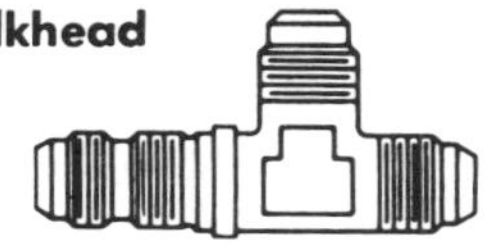

AN804 DASH NUMBER		TUBE
STEEL	ALUM. ALLOY	O.D.
–2	–2D	1/8"
–3	–3D	3/16"
–4	–4D	1/4"
–5	–5D	5/16"
–6	–6D	3/8"
–8	–8D	1/2"
–10	–10D	5/8"
–12	–12D	3/4"
–16	–16D	1"

AN806 PLUG Flared Tube

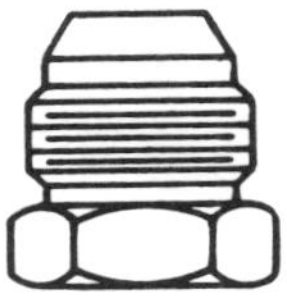

AN806 DASH NUMBER		TUBE
STEEL	ALUM. ALLOY	O.D.
–2	–2D	1/8"
–3	–3D	3/16"
–4	–4D	1/4"
–5	–5D	5/16"
–6	–6D	3/8"
–8	–8D	1/2"
–10	–10D	5/8"
–12	–12D	3/4"
–16	–16D	1"

AN814 PLUG AND BLEEDER Screw Thread

AN814 DASH NUMBER		TUBE
STEEL	ALUM. ALLOY	O.D.
–2	–2D	1/8"
–3	–3D	3/16"
–4	–4D	1/4"
–5	–5D	5/16"
–6	–6D	3/8"
–8	–8D	1/2"
–10	–10D	5/8"
–12	–12D	3/4"
–16	–16D	1"

AN815 UNION Flared Tube

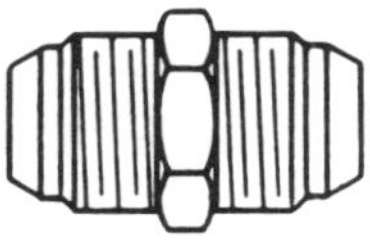

AN815 DASH NUMBER		TUBE
STEEL	ALUM. ALLOY	O.D.
–2	–2D	1/8"
–3	–3D	3/16"
–4	–4D	1/4"
–5	–5D	5/16"
–6	–6D	3/8"
–8	–8D	1/2"
–10	–10D	5/8"
–12	–12D	3/4"
–16	–16D	1"

AN816 NIPPLE Flared Tube and Pipe Thread

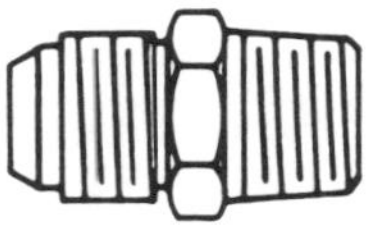

AN816 DASH NUMBER		TUBE	PIPE
STEEL	ALUM. ALLOY	O.D.	THREAD
–2	–2D	1/8"	1/8"
–3	–3D	3/16"	1/8"
–4	–4D	1/4"	1/8"
–5	–5D	5/16"	1/8"
–6	–6D	3/8"	1/4"
–8	–8D	1/2"	3/8"
–10	–10D	5/8"	1/2"
–12	–12D	3/4"	3/4"
–16	–16D	1"	1"

AN817 NUT Sleeve, Coupling

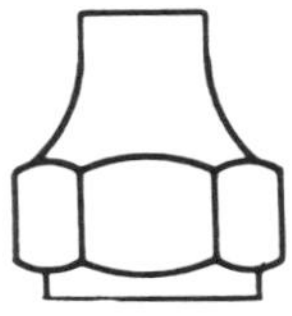

AN817 DASH NUMBER		TUBE
STEEL	ALUM. ALLOY	O.D.
–2	–2D	1/8"
–3	–3D	3/16"
–4	–4D	1/4"
–5	–5D	5/16"
–6	–6D	3/8"
–8	–8D	1/2"
–10	–10D	5/8"
–12	–12D	3/4"
–16	–16D	1"

AN818 NUT Coupling

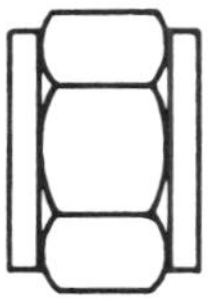

AN818 DASH NUMBER		TUBE
STEEL	ALUM. ALLOY	O.D.
–2	–2D	1/8"
–3	–3D	3/16"
–4	–4D	1/4"
–5	–5D	5/16"
–6	–6D	3/8"
–8	–8D	1/2"
–10	–10D	5/8"
–12	–12D	3/4"
–16	–16D	1"

MS20819 SLEEVE Coupling

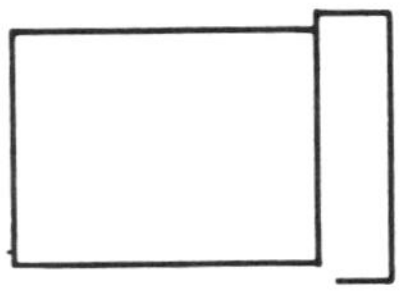

MS20819 DASH NUMBER		O.D.
STEEL	ALUM. ALLOY	TUBE
–2	–2D	1/8"
–3	–3D	3/16"
–4	–4D	1/4"
–5	–5D	5/16"
–6	–6D	3/8"
–8	–8D	1/2"
–10	–12D	5/8"
–12	–12D	3/4"
–16	–16D	1"

AN821 ELBOW, 90° Flared Tube

AN821 DASH NUMBER		O.D.
STEEL	ALUM. ALLOY	TUBE
–2	–2D	1/8"
–3	–3D	3/16"
–4	–4D	1/4"
–5	–5D	5/16"
–6	–6D	3/8"
–8	–8D	1/2"
–10	–10D	5/8"
–12	–12D	3/4"
–16	–16D	1"

TUBE FITTINGS

AN837 ELBOW, 45°
Flared Tube, Bulkhead and Universal

AN837 DASH NUMBER		TUBE O.D.
STEEL	ALUM. ALLOY	
−2	−2D	1/8"
−3	−3D	3/16"
−4	−4D	1/4"
−5	−5D	5/16"
−6	−6D	3/8"
−8	−8D	1/2"
−10	−10D	5/8"
−12	−12D	3/4"
−16	−16D	1"

AN893
BUSHING
Screw Thread Reducer

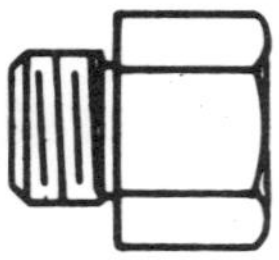

AN893 DASH NUMBER		TUBE O.D.	
STEEL	ALUM. ALLOY	SMALL END	LARGE END
−2	−2D	1/4"	3/8"
−3	−3D	1/4"	1/2"
−7	−7D	5/16"	3/8"
−8	−8D	5/16"	1/2"
−12	−12D	3/8"	1/2"
−121	−121D	3/8"	5/8"
−151	−151D	1/2"	5/8"
−16	−16D	1/2"	3/4"
−17	−17D	1/2"	1"

AN894
BUSHING
Screw Thread Expander

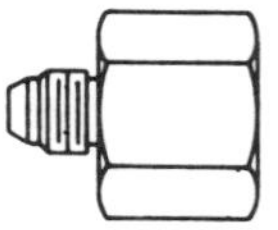

AN894 DASH NUMBER		TUBE O.D.	
STEEL	ALUM. ALLOY	LARGE END	SMALL END
−3-2	D3-2	3/16"	1/8"
−4-3	D4-3	1/4"	3/16"
−5-4	D5-4	5/16"	1/4"
−6-5	D6-5	3/8"	5/16"
−8-5	D8-5	1/2"	5/16"
−8-6	D8-6	1/2"	3/8"
−10-6	D10-6	5/8"	3/8"
−10-8	D10-8	5/8"	1/2"
−12-8	D12-8	3/4"	1/2"

AN919
REDUCER
External Thread, Flared Tube

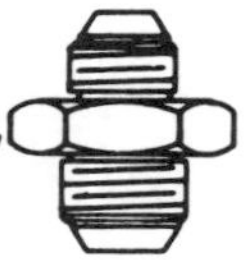

AN919 DASH NUMBER		TUBE O.D.	
STEEL	ALUM. ALLOY	LARGE END	SMALL END
−2	−2D	1/4"	3/16"
−3	−3D	5/16"	1/4"
−6	−6D	3/8"	1/4"
−7	−7D	3/8"	5/16"
−11	−11D	1/2"	5/16"
−12	−12D	1/2"	3/8"
−14	−14D	5/8"	3/8"
−15	−15D	5/8"	1/2"
−19	−19D	3/4"	1/2"

AN924
NUT
Flared Tube, Bulkhead and Universal Fitting

AN924 DASH NUMBER		TUBE O.D.
STEEL	ALUM. ALLOY	
−2	−2D	1/8"
−3	−3D	3/16"
−4	−4D	1/4"
−5	−5D	5/16"
−6	−6D	3/8"
−8	−8D	1/2"
−10	−10D	5/8"
−12	−12D	3/4"
−16	−16D	1"

AN929
CAP (Assembly)
Pressure Seal Flared Tube Fitting

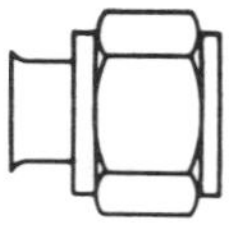

AN929 DASH NUMBER	TUBE O.D.
−2	1/8"
−3	3/16"
−4	1/4"
−5	5/16"
−6	3/8"
−8	1/2"
−10	5/8"
−12	3/4"
−16	1"

AN937
CROSS
Internal Screw Thread

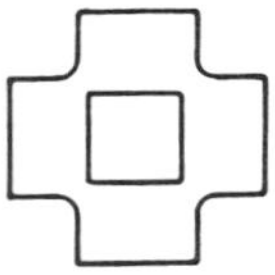

AN937 DASH NUMBER		TUBE O.D.
STEEL	ALUM. ALLOY	
−4	−4D	1/4"
−5	−5D	5/16"
−6	−6D	3/8"
−8	−8D	1/2"
−10	−10D	5/8"
−12	−12D	3/4"
−16	−16D	1"
−20	−20D	1 1/4"

AN938
TEE
Internal Screw Thread

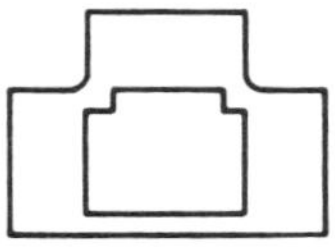

AN938 DASH NUMBER		TUBE O.D.
STEEL	ALUM. ALLOY	
−4	−4D	1/4"
−5	−5D	5/16"
−6	−6D	3/8"
−8	−8D	1/2"
−10	−10D	5/8"
−12	−12D	3/4"
−16	−16D	1"
−20	−20D	1 1/4"

AN939
ELBOW, 90°
Internal Screw Thread

AN939 DASH NUMBER		TUBE O.D.
STEEL	ALUM. ALLOY	
−4	−4D	1/4"
−5	−5D	5/16"
−6	−6D	3/8"
−8	−8D	1/2"
−10	−10D	5/8"
−12	−12D	3/4"
−16	−16D	1"
−20	−20D	1 1/4"

PIPE FITTINGS

Pipe fittings approved for use on aircraft. Made of aluminum alloy, brass, and steel as indicated in the tables accompanying each illustration. All applicable Government specifications and standard drawings are adhered to in the manufacture of these fittings. Fittings are available in sizes larger than those listed in the following tables. Price and delivery information will be furnished on request.

When ordering fittings use AN or MS number that corresponds with desired fitting and add dash number for size and material -- example: AN915-6 (Elbow, 45°, ¾" pipe size, brass)

AN867 DASH NUMBER		PIPE SIZE
STEEL	ALUM. ALLOY	
−2S	−2	¼"
−3S	−3	⅜"
−1S	−1	⅛"
−4S	−4	½"
−6S	−6	¾"
−8S	−8	1"
−10S	−10	1¼"
−12S	−12	1½"

AN867 FLANGE
Pipe Thread, Welding

AN910 COUPLING

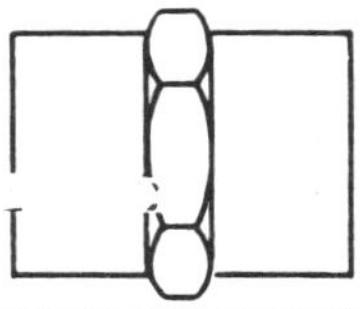

AN910 DASH NUMBER		PIPE SIZE
BRASS	ALUM. ALLOY	
−1	−1D	⅛"
−2	−2D	¼"
−3	−3D	⅜"
−4	−4D	½"
−6	−6D	¾"
−8	−8D	1"

AN911 HEX. SHOULDER NIPPLE

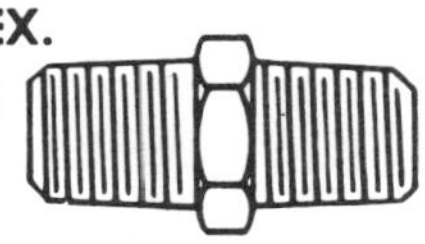

AN911 DASH NUMBER		PIPE SIZE
BRASS	ALUM. ALLOY	
−1	−1D	⅛"
−2	−2D	¼"
−3	−3D	⅜"
−4	−4D	½"
−6	−6D	¾"
−8	−8D	1"

AN912 BUSHING Reducer

AN912 DASH NUMBER		PIPE SIZE
BRASS	ALUM. ALLOY	
−1	−1D	¼" x ⅛"
−2	−2D	⅜" x ¼"
−3	−3D	⅜" x ⅛"
−4	−4D	½" x ⅜"
−5	−5D	½" x ¼"
−7	−7D	¾" x ½"
−8	−8D	¾" x ⅜"
−10	−10D	1" x ¾"

MS20913 PLUG
Square Head

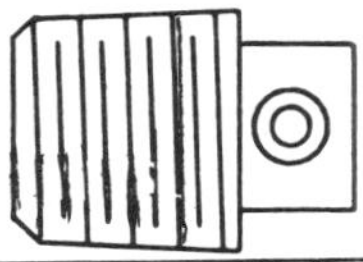

MS20913 DASH NUMBER			PIPE SIZE
BRASS	ALUM. ALLOY	STEEL	
−1	−1D	−1S	⅛"
−2	−2D	−2S	¼"
−3	−3D	−3S	⅜"
−4	−4D	−4S	½"
−6	−6D	−6S	¾"
−8	−8D	−8S	1"
−10	−10D	−10S	1¼"

AN914 ELBOW, 90°
Internal and External Thread

AN914 DASH NUMBER		PIPE SIZE
BRASS	ALUM. ALLOY	
−1	−1D	⅛"
−2	−2D	¼"
−3	−3D	⅜"
−4	−4D	½"
−6	−6D	¾"
−8	−8D	1"

AN915 ELBOW, 45°
Internal and External Thread

AN915 DASH NUMBER		PIPE SIZE
BRASS	ALUM. ALLOY	
−1	−1D	⅛"
−2	−2D	¼"
−3	−3D	⅜"
−4	−4D	½"
−6	−6D	¾"
−8	−8D	1"

AN916 ELBOW, 90°
Internal Thread

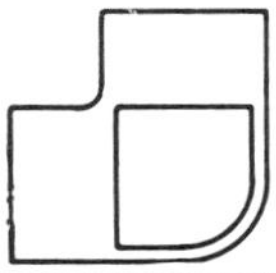

AN916 DASH NUMBER		PIPE SIZE
BRASS	ALUM. ALLOY	
−1	−1D	⅛"
−2	−2D	¼"
−3	−3D	⅜"
−4	−4D	½"
−6	−6D	¾"
−8	−8D	1"

AN917 TEE
Internal Thread

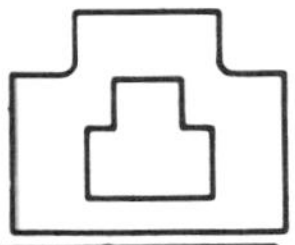

AN917 DASH NUMBER		PIPE SIZE
BRASS	ALUM. ALLOY	
−1	−1D	⅛"
−2	−2D	¼"
−3	−3D	⅜"
−4	−4D	½"
−6	−6D	¾"

AN918 CROSS
Internal Thread

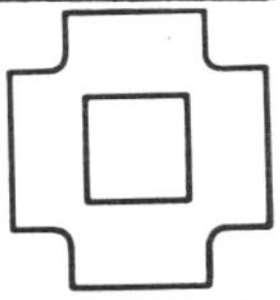

AN918 DASH NUMBER		PIPE SIZE
BRASS	ALUM. ALLOY	
−1	−1D	⅛"
−2	−2D	¼"
−3	−3D	⅜"
−4	−4D	½"
−6	−6D	¾"
−8	−8D	1"

When ordering hose fittings use AN number plus dash number that corresponds to desired size and material — example: AN 840-6D (Hose nipple, aluminum alloy, 1/4" pipe thread for 3/8" I.D. hose).

HOSE FITTINGS

AN807 ADAPTOR
Hose to Universal

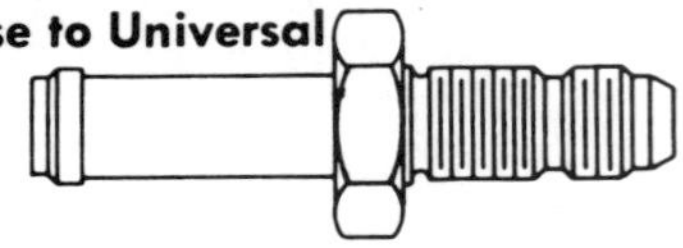

AN807 DASH NUMBER		HOSE I.D.	TUBE O.D.
BRASS	ALUM. ALLOY		
–4	–4D	1/4"	1/4"
–6	–6D	3/8"	3/8"
–8	–8D	1/2"	1/2"
–10	–10D	5/8"	5/8"
–12	–12D	3/4"	3/4"
–16	–16D	1"	1"
–20	–20D	1 1/4"	1 1/4"
–24	–24D	1 1/2"	1 1/2"

AN838 ELBOW, 90°
Hose to Universal

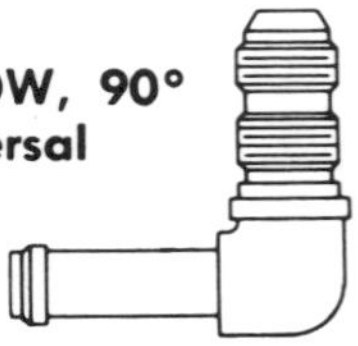

AN838 DASH NUMBER		HOSE I.D.	TUBE O.D.
BRASS	ALUM. ALLOY		
–4	–4D	1/4"	1/4"
–6	–6D	3/8"	3/8"
–8	–8D	1/2"	1/2"
–10	–10D	5/8"	5/8"
–12	–12D	3/4"	3/4"
–16	–16D	1"	1"
–20	–20D	1 1/4"	1 1/4"
–24	–24D	1 1/2"	1 1/2"

AN840
HOSE NIPPLE
Pipe Thread

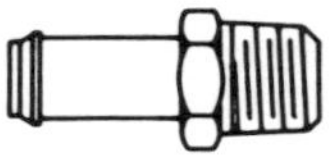

AN840 DASH NO.		HOSE I.D.	PIPE THREAD
BRASS	ALUM.		
–4	–4D	1/4"	1/8"
–6	–6D	3/8"	1/4"
–8	–8D	1/2"	3/8"
–10	–10D	5/8"	1/2"
–11	–11D	3/4"	5/8"
–12	–12D	3/4"	3/4"
–16	–16D	1"	3/4"
–17	–17D	1"	1"

SAFETY WIRE

MS20995 SAFETY WIRE is manufactured to MS20995 Standard Drawing. Available in 1-pound and 5-pound spools. Order by part number, as per table below (specify spool weight).

EXAMPLE: 20 FIVE-POUND SPOOLS MS20995-F20 (Signifies 100 lbs. of .020 diameter, carbon steel safety wire packaged 5 lbs. per spool).

NOTE: MS20995 supersedes the old AN995 standard

ORDERING TABLE - Military Specification Wire

MATERIAL	Nominal Wire Diameter (inches)				
	.020	.025	.032	.040	.041
	MS20995 DASH NUMBER				
MONEL Nickel-copper alloy, specification QQ-N-281, Class A, annealed	-NC20	—	-NC32	-NC40	—
CARBON STEEL Zinc coated, specification QQ-W-461, FS1005, 1010 or 1015 annealed	-F20	-F25	-F32	—	-F41
CORROSION-RESISTANT STEEL Specification QQ-W-423, Form 1, Composition FS302, or FS304, Cond. A	-C20	-C25	-C32	—	-C41

O-RINGS

O-RINGS

Military specifications for O-Rings designate specific requirements for the compound, identification, etc. These requirements are summarized in the MILITARY SPECIFICATION INDEX

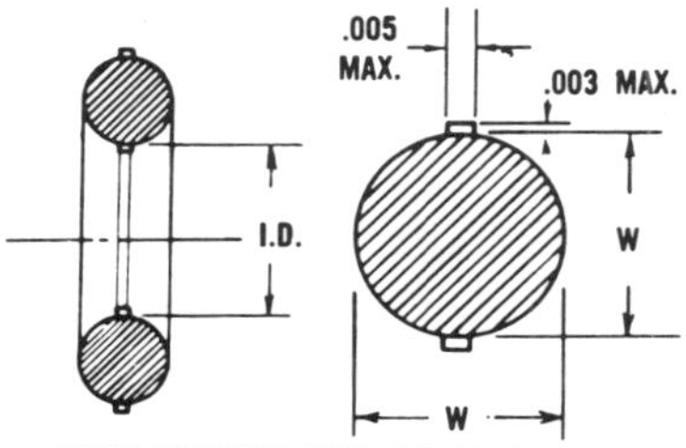

CROSS SECTIONAL VIEW FOR ALL O-RINGS

BACK-UP RINGS

MIL-P-5514D requires the use of back-up rings for all static and dynamic installations over 1500 psi pressure. Back-up rings are recommended for all installations, regardless of pressures, because they prolong the life of the O-Ring and allow for variations in clearances. (Extrusion of O-Rings into the clearance gap is the main cause of O-Ring failure).

When pressure is from one direction only back-up rings may be used on the low pressure side only. However, the use of two back-up rings is strongly recommended in all installations to provide standardization of groove d[illegible]ons and prevent incorrect installation of a s[illegible]g.

Leather back-up rings are inferior to Teflon and are never allowed in Type II or III systems (over 160°F), Leather must be used as a continuous ring while Teflon rings may be scarf-cut or spiral for ease of installation. Unlike Teflon, leather must be initially lubricated. Leather can only be installed with the smooth side against the O-Ring.

Teflon back-up rings are resistant to all hydraulic fluids and fuels and have an exceptional service life. The low friction of Teflon prevents spiral failure of O-Rings and contributes to prolonged life and improved performance, even in rotary seals.

INSTALLATION

The general specification for installation and gland design for O-Rings in military hydraulic systems is MIL-P-5514D. Use of this specification is mandatory for the Departments of the Army, the Navy and the Air Force. The following information on installation and back-up rings is in accordance with this specification.

O-Ring installation should be designed so that threads, ports or other sharp corners will not damage the O-Ring during installation or performance. Components of the mechanism must satisfactorily resist corrosion during normal service life. The following are maximum surface roughness height ratings (as defined in MIL-STD-10):

Surfaces over which O-Rings slide	16
Groove diameter	32
Groove sides without back-up rings	32
Groove sides with back-up rings	63

MILITARY SPECIFICATION INDEX

DRAWING NUMBER	SPECIFICATION NUMBER	DESCRIPTION
AN6227B	MIL-P-5516 Class B	PACKING, O-RING, HYDRAULIC, AIRCRAFT – (MIL-H-5606 petroleum base hydraulic fluid, –65 to +160°F) (–54 to +71°C).
AN6230B	MIL-P-5516 Class B	GASKET, O-RING, HYDRAULIC, AIRCRAFT – (MIL-H-5606 petroleum base hydraulic fluid, –65 to +160°F) (–54 to +71°C).
MS9058	AMS3651B	BACK-UP,RING, BOSS CONNECTION – Teflon back-up ring for use with MS28778 O-Rings. Dash numbers correspond. (See MS28778 on page 92 for dash number).
MS28774	MIL-R-8791B	RETAINER, BACK-UP PACKING, SINGLE TURN, TETRAFLUOROETHYLENE – Teflon back-up rings for use with MS29513 O-Rings. Dash numbers correspond. (See Cross Reference Chart beginning on page 95).
MS28777	MIL-R-5521	RING, HYDRAULIC FITTING, GASKET BACK-UP – Leather back-up ring to use with MS28778 O-Ring. (MS9058 Teflon equivalents are normally used).
MS28778	MIL-G-5510A	PACKING, PREFORMED, STRAIGHT THREAD TUBE FITTING BOSS – O-Ring for +160°F hydraulic fluid systems per MIL-H-5606.
MS28782	MIL-R-8791B	RETAINER, PACKING, BACK-UP – Teflon back-up rings for use with AN6227B O-Rings. Dash numbers correspond. (See Cross Reference Chart beginning on page 95).
MS28783	MIL-R-8791B	RING, GASKET, BACK-UP – Teflon back-up ring to use with AN6230B O-Rings. Dash numbers correspond. (See Cross Reference Chart beginning on page 95).
MS29513	MIL-P-5315A	PACKING O-RING, HYDROCARBON FUEL RESISTANT – For use in MIL-F-5572, MIL-F-5616 and MIL-F-5624 fuels from –65° to +200°F.

AN100

MS20115

AN100 THIMBLES

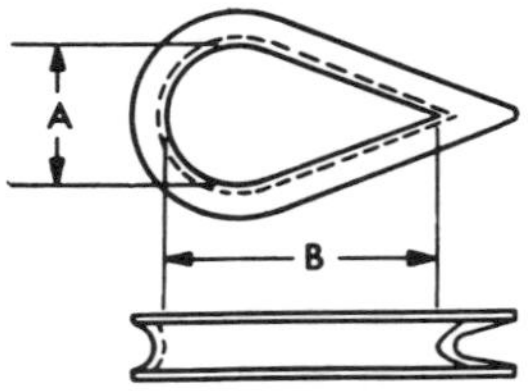

Available in cadmium-plated steel or in corrosion-resisting steel. Thimbles are made in conformance with Government procurement specification MIL-T-5677 and standard drawing AN100. When ordering thimbles use AN number plus dash number for thimble desired. Example? AN100-6 (Carbon-steel thimble for $\frac{3}{16}$ inch cable).

ORDERING AND WEIGHT TABLE

Approximate Weights of AN100 Thimbles (Pounds per Thimble)

DASH NUMBER		FITS CABLE SIZE	DIMENSIONS (INCHES)		WEIGHT
CARBON STEEL	COR. RES. STEEL		A	B	
–3	–C3	1/16-5/64	.35	.70	.0030
–4	–C4	3/32-7/64-1/8	.35	.70	.0043
–5	–C5	5/32	.40	.80	.0065
–6	–C6	3/16	.50	1.00	.0090
–7	–C7	7/32	.60	1.20	.0143
–8	–C8	1/4	.70	1.40	.0163
–9	–C9	9/32	.80	1.60	.0262
–10	–C10	5/16	.90	1.80	.0295
–12	–C12	3/8	1.00	2.00	.0770
–14	–C14	7/16	1.125	2.250	.1220
–16	–C16	1/2	1.250	2.500	.1670

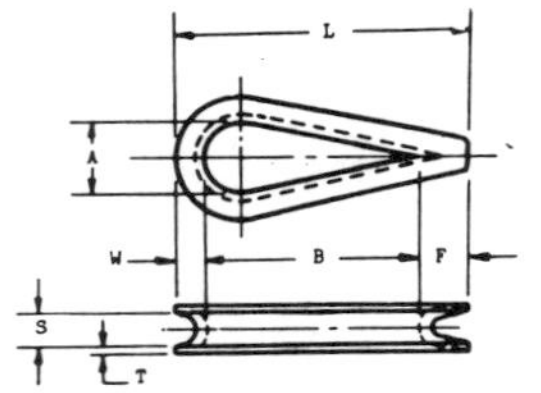

AN PART NO.			CABLE DIA	A	B	F MIN	L APPROX	S +1/64 -0	T	W +1/64 -0
LOW CARBON STEEL	CORROSION RESISTING STEEL	PHOSPHOR BRONZE								
AN100-3	AN100C-3	AN100B-3	1/16 5/64	.350	45/64	3/16		3/32	.032±.003	5/64
AN100-4	AN100C-4	AN100B-4	3/32 7/64 1/8	.350	45/64	7/32		9/64	.032±.003	5/64
AN100-5	AN100C-5	AN100B-5	5/32	.400	51/64	7/32		11/64	.032±.003	7/64
AN100-6	AN100C-6	AN100B-6	3/16	.500	1	5/16		13/64	.032±.003	9/64
AN100-7	AN100C-7	AN100B-7	7/32	.600	1-13/64	3/8		15/64	.032±.003	5/32
AN100-8	AN100C-8	AN100B-8	1/4	.700	1-13/32	13/32		17/64	.032±.003	11/64
AN100-9	AN100C-9	AN100B-9	9/32	.800	1-39/64	13/32		19/64	.040±.004	13/64
AN100-10	AN100C-10	AN100B-10	5/16	.900	1-51/64	7/16		21/64	.040±.004	7/32
AN100-12	AN100C-12	AN100B-12	3/8	1.000	2	5/8		25/64	.060±.004	17/64
AN100-14	AN100C-14	AN100B-14	7/16	1.125	2-1/4	13/16		29/64	.080±.004	21/64
AN100-16	AN100C-16	AN100B-16	1/2	1.250	2-1/2	1		33/64	.080±.004	13/32
(a) AN100-18			9/16	1.500 (MIN)	2-3/4 (MIN)		4	37/64	(1) .156±.016	
(a) AN100-20			5/8	1.750 MIN	3-1/4 MIN		5-3/8	41/64	(1) .168±.019	

(a) FOR -18 AND -20, THE FINISH REQUIREMENT ONLY, OF THE PROCUREMENT SPECIFICATION SHALL APPLY. THE MATERIAL SHALL BE OF COMMERCIAL WROUGHT STEEL SUITABLE FOR THE PURPOSE.

MATERIAL: SEE PROCUREMENT SPECIFICATION.

FINISH: SEE PROCUREMENT SPECIFICATION.

DIMENSIONS IN INCHES. UNLESS OTHERWISE SPECIFIED, TOLERANCES: FRACTIONS ±1/64, DECIMALS ±.010.

APPROVED 12 Nov 42 REVISED (1) 20 Mar 51

PROCUREMENT SPECIFICATION	AIR FORCE–NAVY AERONAUTICAL STANDARD	AN100
(1) MIL-T-5677	THIMBLE - WIRE CABLE	

MS20115 CABLE SHACKLES

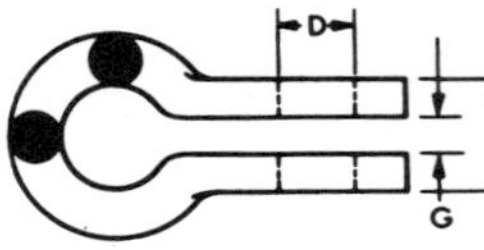

Available in carbon steel, cadmium plated per QQ-P-416 and in corrosion resistant steel (MIL-S-7720). Plain dash numbers are for corrosion resistant steel; add "F" before dash number for non-corrosion resistant (carbon) steel. Examples:

MS20115-6 (Corr.Resist. Steel, 3/16 dia. cable)
MS20115F6 (Carbon steel, 3/16 dia. cable)

NOTE: MS20115 Shackles supersede the old AN115 Shackles.

DASH NUMBER		D	F	G	USED WITH	
Cor. Res. Steel	Carbon Steel	dia.			Thimble	Clevis Pin
-2	F2	.190	.218	.093	AN100-3	MS20392-2-9
-3	F3	.190	.254	.108	AN100-4	MS20392-2-11
-4	F4	.190	.383	.195	AN100-4	MS20392-2-11
-5	F5	.250	.406	.202	AN100-5	MS20392-3-15
-6	F6	.313	.543	.260	AN100-6	MS20392-4-17
-7	F7	.318	.625	.296	AN100-7	MS20392-5-19
-8	F8	.375	.688	.313	AN100-8	MS20391-5-19
-9	F9	.438	.719	.327	AN100-9	MS20392-6-25
-10	E10	.438	.765	.348	AN100-10	MS20392-7-27
-12	F12	.500	.830	.380	AN100-12	MS20392-7-27

FED. SUP CLASS 4030

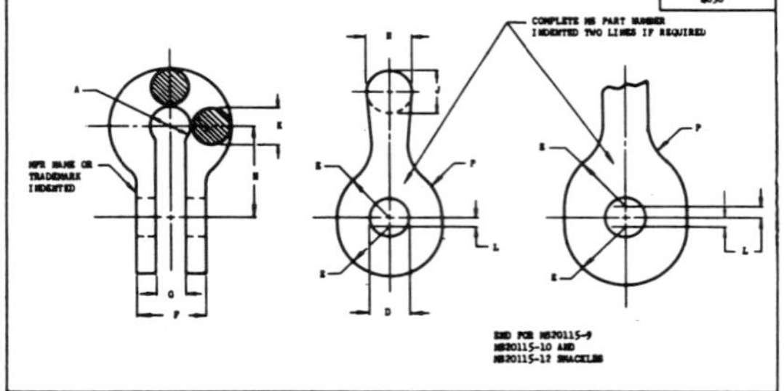

DASH NUMBERS			NOMINAL WIRE ROPE DIA (REF)	MINIMUM BREAKING STRENGTH LB	A +.016 -.000 DIA	D +.005 -.000 DIA	E RAD	F +.010 -.005	G +.007 -.003	H ±.016	J ±.016	K ±.016	L	M ±.016	P ±.016 RAD
CORROSION RESISTANT STEEL	NON-CORROSION RESISTANT STEEL	NICKLE COPPER ALUMINUM ALLOY													
-2	F2	K2	1/16	480	.188	.190	.172	.218	.093	.156	.156	.156	.031	.562	.375
-3	F3	K3	3/32	920	.219		.219	.254	.108						
-4	F4	K4	1/8	2 000	.250		.274	.383	.195	.250	.250	.219			
-5	F5	K5	5/32	2 800	.281	.250	.344	.406	.202					.750	.438
-6	F6	K6	3/16	4 200	.375	.313	.391	.543	.260	.312	.312	.281	.062	.812	.500
-7	F7	K7	7/32	5 600	.406		.453	.625	.296	.344	.344	.312		.875	
-8	F8	K8	1/4	7 000	.438	.375	.484	.688	.313	.406	.359	.359		1.000	.406
-9	F9	K9	9/32	8 000	.500	.438	.578	.719	.327	.438	.391	.391			
-10	F10	K10	5/16	9 800	.594		.633	.765	.348	.469	.422	.422	.094	1.125	.500
-12	F12	K12	3/8	14 000	.625	.500	.750	.830	.380	.562	.500	.500	.125	1.250	

MATERIAL: CORROSION RESISTANT STEEL, MIL-S-7720, CHROME STEEL, MIL-S-6758, MIL-S-8503, MIL-S-8707; NICKEL-COPPER-ALUMINUM ALLOY, QQ-N-286
FINISH: CADMIUM PLATED, QQ-P-416 OR ZINC PLATED, QQ-Z-325

DIMENSIONS IN INCHES: UNLESS OTHERWISE SPECIFIED, TOLERANCES: DECIMALS ±.010.
HOLES TO BE CO-AXIAL WITHIN ±.005

MILITARY STANDARD MS20115 (ASG)

SWAGEABLE TERMINALS

MS20663C BALL AND DOUBLE SHANK

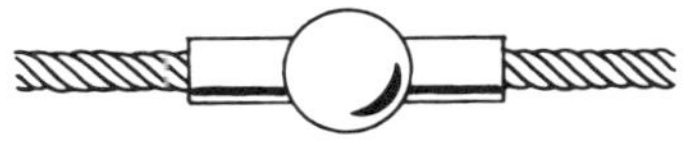

DASH NO.	CABLE DIA. (INCH)	HOLE SIZE (INCH)
-2	1/16	.073
-3	3/32	.104
-4	1/8	.139
-5	5/32	.169
-6	3/16	.201
-7*	7/32	.233
-8*	1/4	.264

MS20664C BALL AND SHANK

The swaging terminals on this page provide the best means of attaching fittings to cable. These MS terminals are made of corrosion-resisting stainless steel (FED-STD-66) to Procurement Specification MIL-T-781. Polished finish.

MINIMUM BREAKING STRENGTH

Properly swaged to the cable, these terminals will maintain the minimum rated breaking strength of that cable. (See page 99 for cable breaking strengths.) For example, 1/8" dia. 7 x 19 stainless steel cable has a breaking strength of 1,760 lbs. Therefore, all "-4" terminals swaged to that cable would carry that same 1,760 lbs. breaking strength.

MS21259 STUD END

DASH NUMBER		SIZE AND THREAD	CABLE DIAMETER (INCH)
RIGHT HAND	LEFT HAND		
-2RH	-2LH	6-40	1/16
-3RH	-3LH	10-32	3/32
-4RH	-4LH	1/4-28	1/8
-5RH	-5LH	1/4-28	5/32
-6RH	-6LH	5/16-24	3/16
-7RH	-7LH	3/8-24	7/32
-8RH	-8LH	3/8-24	1/4
-9RH	-9LH	7/16-20	9/32
-10RH	-10LH	1/2-20	5/16

HOW TO ORDER

Designate terminal desired by proper MS number and add dash number for size that corresponds with cable diameter being used. Examples:

MS21259-2RH (Stud End, RH thread, for 1/16" cable)
MS20667-6 (Fork End, for 3/16" cable)
MS20668-9 (Eye End, for 9/32" cable)
MS21260-L3RH (Turnbuckle End, Long, RH thread, for 3/32" cable)

MS21259 and MS21260 cable terminals are available in either right or left hand threads are listed in the Ordering Tables accompanying the illustrations. MS21260 terminals are offered in long or short varieties.

MS20667 FORK END

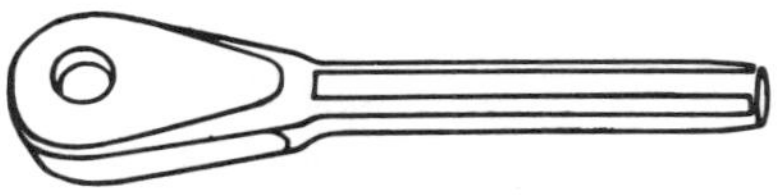

MS21260 TURNBUCKLE END

DASH NUMBER		SIZE AND THREAD	CABLE DIAMETER (INCH)
RIGHT HAND	LEFT HAND		
MS21260 TURNBUCKLE END - LONG			
-L2RH	-L2LH	6-40	1/16
-L3RH	-L3LH	10-32	3/32
-L4RH	-L4LH	1/4-28	1/8
-L5RH	-L5LH	1/4-28	5/32
-L6RH	-L6LH	5/16-24	3/16
-L7RH	-L7LH	3/8-24	7/32
-L8RH	-L8LH	3/8-24	1/4
-L9RH	-L9LH	7/16-20	9/32
-L10RH	-L10LH	1/2-20	5/16
MS21260 TURNBUCKLE END - SHORT			
-S2RH	-S2LH	6-40	1/16
-S3RH	-S3LH	10-32	3/32
-S4RH	-S4LH	1/4-28	1/8
-S5RH	-S5LH	1/4-28	5/32
-S6RH	-S6LH	5/16-24	3/16

TIE RODS AND TERMINALS
We can also supply custom-made tie rods, both Round Drawn (to Specification AN701 thru AN708) and Streamline (to Specification AN671 thru AN682), as well as AN665 tie rod terminals. Contact your nearest Van Dusen office for further information.

STREAMLINE

ROUND DRAWN

AN665 TERMINALS

MS20668 EYE END

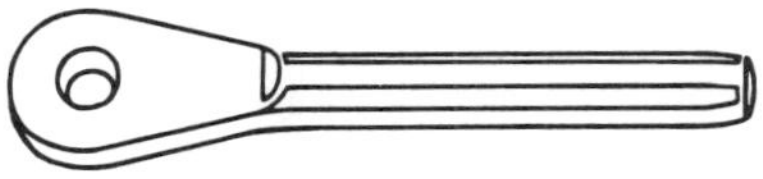

DASH NUMBER	CABLE DIAMETER (INCH)	PIN HOLE DIAMETER (INCH)
–2	1/16	.188
–3	3/32	.188
–4	1/8	.188
–5	5/32	.250
–6	3/16	.313
–7	7/32	.313
–8	1/4	.375
–9	9/32	.438
–10	5/16	.438

MS TURNBUCKLES (CLIP-LOCKING)

MS Standard Drawings for clip-locking turnbuckles supersede various AN Drawings for conventional (lockwire type) turnbuckle parts and NAS Drawings for clip-locking turnbuckle parts. Refer to the cross reference tables on this page for AN and NAS equivalents.

See pages 105 and 106 for MS Standard Drawings

Clip-Locking Turnbuckles utilize two locking clips instead of lockwire for safetying. The turnbuckle barrel and terminals are slotted lengthwise to accomodate the the locking clips. After the proper cable tension is reached the barrel slots are aligned with the terminal slots and the clips are inserted. The curved end of the locking clips expand and latch in the vertical slot in the center of the barrel.

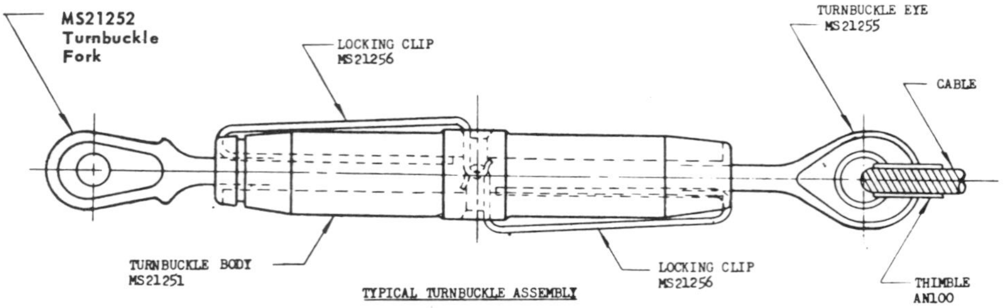

TYPICAL TURNBUCKLE ASSEMBLY

MS21251 TURNBUCKLE BARREL

Supersedes AN155 and NAS649 barrels. MS21251 items can replace AN155 items of like material and thread, but the AN155 items cannot replace the MS21251 items. MS21251 items are interchangeable with the NAS649 items of like material and thread. MS21251 barrels are available in brass (QQ-B-637, composition 2 or MIL-T-6945), steel (cadmium plated to QQ-P-416, type 2, class 3) or aluminum alloy (anodized to MIL-A-8725). The cross reference table shows equivalent items made of brass.

MS21251 DASH NO.	ROPE DIA.	THREAD SIZE	AN155 DASH NO.	NAS649 DASH NO.	USES MS21256 CLIP DASH NO.
B2S	1/16	6-40	B8S	B8S	-1
B2L	1/16	6-40	B8L	B8L	-2
B3S	3/32	10-32	B16S	B16S	-1
B3L	3/32	10-32	B16L	B16L	-2
B5S	5/32	1/4-28	B32S	B32S	-1
B5L	5/32	1/4-28	B32L	B32L	-2
B6S	3/16	5/16-24	B46S	B46S	-1
B6L	3/16	5/16-24	B46L	B46L	-2
B8L	1/4	3/8-24	B80L	B80L	-2
B9L	9/32	7/16-20	B125L	B125L	-3
B10L	5/16	1/2-20	B175L	B175L	-3

MS21256 TURNBUCKLE CLIP

Made of corrosion resistant steel wire, QQ-W-423, composition FS302, condition B. These are NOT interchangeable with the NAS651 clips. Available in 3 sizes: MS21256-1, -2 and -3. For applications, see the MS21251 Turnbuckle Barrel Cross Reference Chart elsewhere on this page.

MS21260 SWAGED STUD END

These clip-locking terminals are available in corrosion resistant steel and in cadmium plated carbon steel. MS21260 items can replace AN669 items of the same dash numbers, but the AN669 items cannot always replace the MS21260 items.

TERMINALS

MS items can replace AN items of like thread except for the -22 and -61 sizes, but the AN items cannot replace the MS items. MS items are interchangeable with the NAS items of like thread except for the -22 and -61 sizes. These MS terminals are available only in steel, cadmium plated to QQ-P-416, type 2, class 3. Available with right-hand (R) or left-hand (L) threads.

MS21252 TURNBUCKLE FORK supersedes AN161 and NAS645 forks.

MS21254 PIN EYE supersedes AN165 and NAS648 eyes.

MS21255 CABLE EYE supersedes AN170 and NAS647 eyes.

MS21260 SWAGED STUD END supersedes AN669 studs.

MS21252 MS21254 MS21255 DASH NOS.		WIRE ROPE DIA.	THREAD SIZE	AN161 AN165 AN170 DASH NOS.		NAS645 NAS648 NAS647 DASH NOS.	
RH THD	LH THD			RH THD	LH THD	RH THD	LH THD
-2RS	-2LS	1/16	6-40	-8RS	-8LS	-8RS	-8LS
-2RL*	-2LL*	1/16	6-40	---	---	---	---
-3RS	-3LS	3/32	10-32	-16RS	-16LS	-16RS	-16LS
-3RL	-3LL	3/32	10-32	-16RL	-16LL	-16RL	-16LL
-5RS	-5LS	5/32	1/4-28	-32RS	-32LS	-32RS	-32LS
-5RL	-5LL	5/32	1/4-28	-32RL	-32LL	-32RL	-32LL
-6RS	-6LS	3/16	5/16-24	-46RS	-46LS	-46RS	-46LS
-6RL	-6LL	3/16	5/16-24	-46RL	-46LL	-46RL	-46LL
-8RL	-8LL	1/4	3/8-24	-80RL	-80LL	-80RL	-80LL
-9RL	-9LL	9/32	7/16-20	-125RL	-125LL	-125RL	-125LL
-10RL	-10LL	5/16	1/2-20	-175RL	-175LL	-175RL	-175LL

*MS21254 and MS21255 eyes only; MS21252 fork not made in this size.

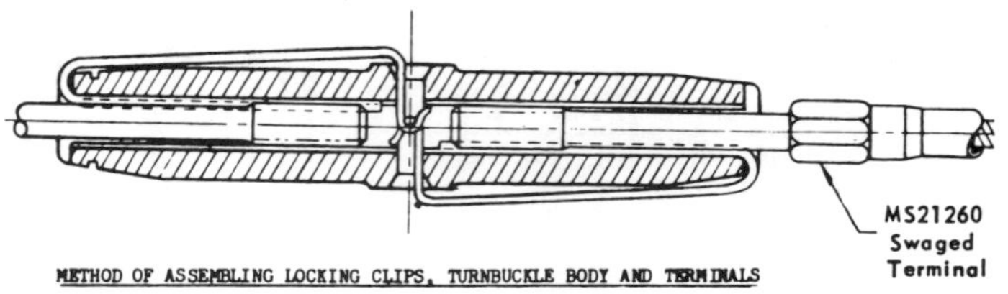

METHOD OF ASSEMBLING LOCKING CLIPS, TURNBUCKLE BODY AND TERMINALS

PULLEYS

(See next page for MS Drawings)

PHENOLIC PULLEYS are machined from fabric-reinforced, high impact phenolic sheets exceeding specification requirements. Bearing ring bonds bearing to pulley. Bearing lubricated with MIL-G-3278 grease (a grease retainer shield also protects the bearing. Sheave material will not corrode tin or zinc coated carbon steel cable. Fungus resistant sheave per MIL-P-7034.

ALUMINUM PULLEYS are machined to exacting tolerances; lighter in weight than specification requirements and superior in strength. Sheave material conforms to QQ-268-T4 and QQ-A-355-T4; anodized to specification AN-QQ-A-696.

NOTE: 1. All "MS" Pulleys conform to Procurement Spec. MIL-P-7034.

2. AN and MS pulleys of like dash numbers are universally, functionally, and dimensionally interchangeable.

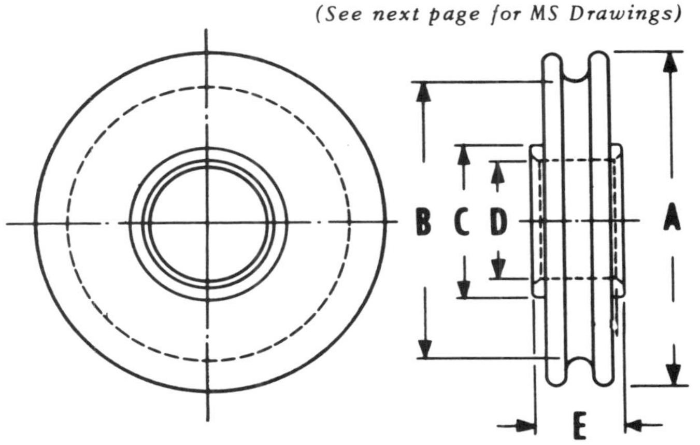

MS 20219

SECONDARY CONTROL PULLEY. MS 20219A1 and MS20219A3 available in aluminum only. MS20219-2, -4 and -5 available in phenolic or aluminum. Examples of part numbers:

MS20219-2 Pulley with phenolic sheave.

MS20219A2 Pulley with aluminum sheave.

MS PART NUMBER	OLD AN PART NUMBER	BEARING NO.	CABLE SIZE	A diam. ± .005	B diam. ±.005	C diam. Ref.	D diam. -.0005	E -.005	ALLOWABLE LIMIT LOAD ON PULLEY
MS20219A1	AN219A1	P4K	1/16 and 3/32	1.312	1.000	.423	.2500	.438	480
MS20219-2	AN219-2	P4K		1.750	1.438	.423	.2500	.438	480
MS20219A3	AN219A3	P10K		1.750	1.438	.769	.6250	.438	480
MS20219-4	AN219-4	P4K		2.625	2.312	.423	.2500	.438	920
MS20219-5	AN219-5	P10K		2.625	2.312	.769	.6250	.438	920

MS 20220

FLIGHT CONTROL PULLEY. All parts listed available in either phenolic or aluminum. Examples of part numbers:

MS20220-2 Pulley with phenolic sheave

MS20220A2 Pulley with aluminum sheave

NOTE: "D" diameter on all MS20220 pulleys is .3145 inches.

MS PART NUMBER	OLD AN PART NO.	BEARING NO.	CABLE SIZE	A Dia. —.010	B Dia. —.010	C -.030 Dia.	E -.005	ALLOWABLE LIMIT LOAD ON PULLEY
MS20220-1	AN220-1	P5K	1/8,	1.755	1.255	.465	.625	500
MS20220-2	AN220-2	PD5K	5/32,	3.005	2.505	.515	.625	1680
MS20220-3	AN220-3	PD5K	AND	4.255	3.755	.515	.625	2500
MS20220-4	AN220-4	PD5K	3/16	5.505	5.005	.515	.625	2500

MS 20221

HEAVY DUTY CONTROL PULLEY. MS 20221A1 available in aluminum only. MS 20221-2 and -3 furnished in phenolic or aluminum. Examples of part numbers:

MS20221-2 Pulley with phenolic sheave

MS20221A2 Pulley with aluminum sheave

MS PART NUMBER	OLD AN PART NO.	BEARING NO.	CABLE SIZE	A DIA. —.010	B DIA. —.010	C DIA.	D DIA.	E —.005	ALLOWABLE LIMIT LOAD ON PULLEY
MS20221A1	AN221A1	PD5K	3/16, 7/32, AND 1/4	2.630	2.005	.515—.030	.3145—.0025	.625	2800
MS20221-2	AN221-2	P8		4.130	3.505	.800—.050	.505—.005	.750	4900
MS20221-3	AN221-3	P8		5.630	5.005	.800—.050	.505—.005		7000

MS24566

ANTI-FRICTION BEARING PULLEY. Note: AN210-1A, -2A, -3A and -4A pulleys to Spec. AN-P-60 are available, subject to availability of bearings. AN 210-A pulleys are dimensionally the same as the "B" pulleys, but have metal shielded bearings rather than removable composition seal type bearings. All MS24566 pulleys available in phenolic or aluminum. Examples of part numbers:

MS24566-1B **Pulley with phenolic** sheave

MS24566A1B **Pulley with aluminum** sheave

MS PART NUMBER	OLD AN PART NUMBER	CABLE SIZE	B DIA. +.000 -.010	A DIA. +.000 -.010	D +.0000 -.0005 DIA.	E +.000 -.005	MAXIMUM ALLOWABLE DESIGN LOAD LB.	BEARING NO.
MS24566-1B	AN210-1B	1/16, 5/64, 3/32	.972	1.250	.1900	.297	185	KP3AK
MS24566-2B	AN210-2B	1/16, 5/64, 3/32	2.222	2.500	.1900	.297	500	KP3K
MS24566-3B	AN210-3B	1/8, 5/32, 3/16	1.510	2.000	.2500	.484	450 (a)	KP4K
MS24566-4B	AN210-4B	1/8, 5/32, 3/16	3.010	3.500	.2500	.484	1200	KP4K
MS24566-5B	AN210-5B	3/16, 7/32, 1/4	4.374	5.000	.3750	.620	2000	KP6
MS24566-6B	AN210-6B	3/16, 7/32, 1/4	5.374	6.000	.3750	.620	2500	KP6

WISCO FILLER CAPS & ADAPTERS

Designed and tooled primarily for critical aircraft and missle applications, these filler components are also used in automotive and ground equipment. All components shown here are approved for MIL-H-3136, fuel; MIL-L-6082, oil (piston engine); MIL-E-5559, ethylene glycol; MIL-A-6091, alcohol ethyl. Temperature range: -65F to 215F.

Modified designs are also available for diester-based synthetic lubricants, such as MIL-L-7808.
1 1/2" components shown here are for 15 psi service; 2" and 3" components are for 10 psi service. Higher pressure units in 2" size are available on request.

1½ INCH COMPONENTS

Repair parts available are: USAF 39A4241 Gasket and USAF 39A4240 or MS 29523 Hook.

42B3581 Assembly (not shown) is composed of a 42A3589 cap and a 39B4233 adapter.

42A3589
Cap, Basic Unit
20 gauge (.036") deep drawn steel.
Cadmium plated finish.
Basic unit in this group.

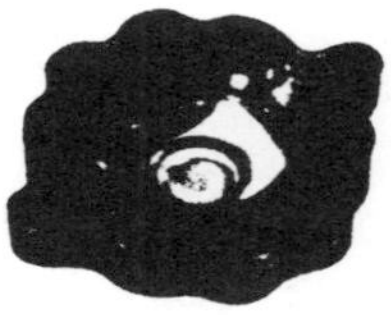

39B4234
Adapter
.060 deep drawn steel.
Cadmium plated finish.
Available without head and in shorter length.
For use with cap shown above.

39B4233
Adapter
.060 deep drawn steel.
Cadmium plated finish.
Available without head or thread or safety chain.
For use with cap shown above.

M704 CAP: Locking bayonet type, slightly different from 42A3589 cap in design. Has vent hole to permit entrance of air as tank is drained. Provided with a synthetic rubber gasket. Fastens to filler neck by safety chain.
M-142 ALUMINUM TANK FLANGE: Designed for welding to aluminum tanks. Inner surface threaded to receive 39B4233 Filler Neck.

M-142 ALUMINUM TANK FLANGE — **M-704 CAP**

2 INCH COMPONENTS

NAF1025-1 Assembly is a NAF1025-2 cap plus a NAF1025-3 neck
NAF1025-5 is a NAF1025-2 plus a NAF1025-4
NAF1025-8 is a NAF1025-2 plus a NAF1025-6
NAF1025-9 is a NAF1025-2 plus a NAF1025-7

NAF1025-2
Cap, Basic Unit
Cap: 18 gauge Body: 20 gauge.
Cadmium plated finish—corrosion resistant.
Spring loaded—dip stick can be added.
Basic unit in this group.

NAF1025-3
Filler Neck
20 gauge deep drawn steel.
Cadmium plated finish—corrosion resistant.
12" length.
For use with cap shown above.

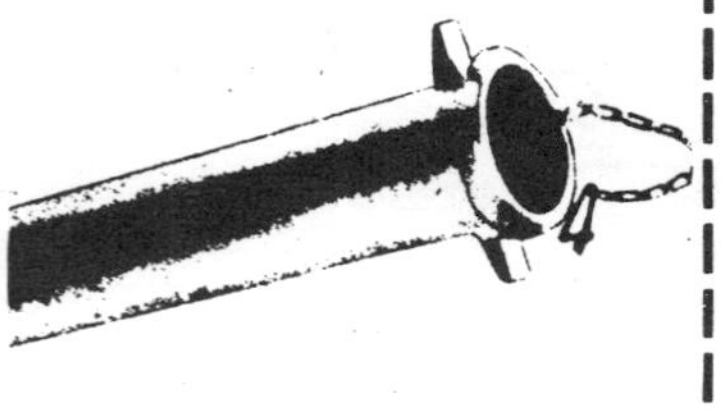

NAF1025-3 as shown, becomes:
NAF1025-4 without chain
NAF1025-6 without lugs
NAF1025-7 without chain or lugs

3 INCH COMPONENTS

Standard Cap and Adapter assemblies are: 39B5435, 40B1718, and 45B11359. **Repair Parts:** All parts for this cap assembly are available as repair parts.

39B5435
Cap, Basic Unit
.188 thickness deep drawn steel.
Cadmium plated finish.
Basic unit in this group.

This cap will fit the following standard adapters: AN4122-1A, AN4122-2A, AN4122-3A, AN4122-4A, AN4123-1, AN4123-2, 39B5235, 39B5294, 39B5438, and 45B11358.

39B5295
Adapter
18 gauge (.048) deep drawn steel.
Cadmium plated finish.
Over-all height ⅞" or 1¼".
Furnished with or without chain.
For use with cap shown above.

AN4122-2A
Adapter (curved)
18 gauge deep drawn steel.
Over-all diameter 7".
Cadmium plated finish.
For use with cap shown above.

AN4123-1
Adapter (8 hole flange)
.125 thickness deep drawn steel.
Bolt circle 4$\frac{5}{16}$" over-all height 1¼".
Cadmium plated finish.
For use with cap shown above.

AN4123-2
Adapter (12 hole adapter)
.188 thickness deep drawn steel.
Bolt circle 4¾" over-all height 1¼".
Cadmium plated finish.
For use with cap shown above.

AN4122-1A
Adapter
18 gauge (.048) deep drawn steel.
Over-all diameter 7" over-all height ¾".
Cadmium plated finish.
For use with cap shown above.

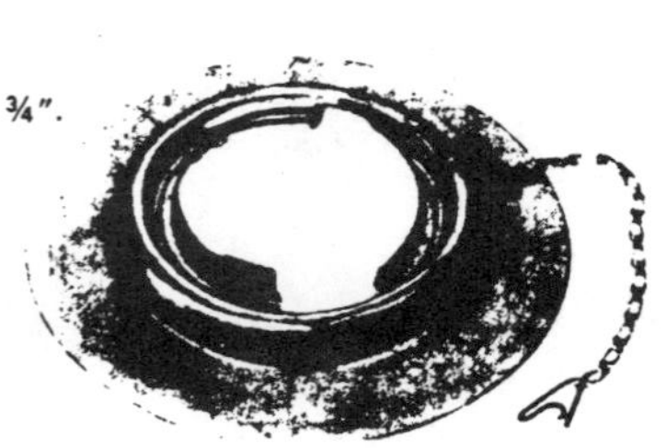

NATIONAL AEROSPACE STANDARDS NUMERICAL INDEX

NOTICE

National Aerospace Standards are proposed as guides to the aerospace industry in serving the public interest and the national defense program. The standards therein are designed to eliminate misunderstandings between manufacturers and purchasers, to facilitate interchangeability and improvement of products, and to assist purchaser in expeditiously selecting and obtaining the proper product for a particular need.

Symbols Used to Denote Status of National Aerospace Standards
*Inactive for new design
**New revisions and amendments issued since date of last index
†New specifications or standards issued since date of last index
XCanceled

Number	Current Issue	Title
NAS1X	12/41	Rivet–Countersunk Head 100°, Sup. by AN426
NAS2*	11/50	Lubrication Fittings, Sup by AN944
NAS3*	4/45	Plate–Bolt, Locking
NAS4*	1/46	End–Rod, Control Resistance Welding Type, X1020 Steel, See NAS5
NAS5	9/63	Rod End, Threaded, Alloy Steel, Flash Welding Type (ANA Bul. 147)
NAS6*	12/44	Manifold–Single Port, De-Icer
NAS7*	12/44	Manifold–Double Port, De-Icer
NAS8*	12/44	Manifold–Triple Port, De-Icer
NAS9*	12/44	Manifold–Four Port, De-Icer
NAS10*	8/45	Standard Form for National Aircraft Standard Drawings
NAS11*	4/57	Valve Assembly–Flapper
NAS12*	10/47	Screws–Non-Losable, Alum Alloy, #8-32 and #10-32
NAS13*	10/47	Seal–Control Tube, Bellows Type
NAS14*	10/47	Retainer–Bellows Type Seal, Control Tube
NAS15*	10/47	Joint–Universal, Male
NAS16*	10/47	Joint–Universal, Female
NAS17*	1/50	Panel–Connector, Electrical, For #6 Posts, Sup. by AN3435
NAS18*	1/50	Panel–Connector, Electrical, For #8 and #10 Posts, Sup. by AN3435
NAS19*	1/50	Panel–Connector, Electrical, For 1/4, 5/16 and 3/8 Posts, Sup. by AN3435
NAS20*	1/50	Screw–Post, Connector Panel, Electrical, Sup. by AN3432
NAS21	1/63	Strip, Terminal Board Mounting Nut Insulating
NAS22*	1/50	Strip–Insulating, Base Connector Panel, Electrical, Sup. by AN3437
NAS23	1/63	Terminal–Tank Strap, Plain (ANA Bul. 147)
NAS24	1/63	Terminal–Tank Strap, Forked (ANA Bul. 147)
NAS25	5/68	Trunnion–Tank Strap, Threaded
NAS26	5/68	Trunnion–Tank Strap, Unthreaded
NAS27	5/68	Turnbuckle–Tank Strap
NAS28	5/68	Bolt–Tee Head
NAS29(4)	1/63	Strap Assembly–Tank (ANA Bul. 147)
NAS30*	9/56	Surface Roughness Designation, See MIL-STD-10
NAS31(2)*	2/58	Conversion Table–Surface Rougnness Designation, See MIL-STD-10
NAS34*	11/52	Pulley–3/32 Dia. Engine Control Cable, Anti-Friction
NAS35*	10/52	Lock–Cable, Pulley, 3/32 Dia. Engine Control Cable
NAS36	1/63	Rod End, Spherical Bearing 3/16 Bore x 1/4–28 Interchangeable Type (ANA Bul. 147–AF Only)
NAS37*	11/52	End–Rod, Spherical Bearings 3/16 x 5/16-24 Interchangeable Type
NAS38	10/62	Specification–Rod End, Spherical Bearing, Interchangeable Type (ANA Bul. 147–AF Only)
NAS39X	7/45	Terminal–Electrical, Mech., Sup. by AN659
NAS40*	7/54	Hinge–Half, Continuous, Extruded Aluminum Alloy
NAS41	7/54	Pin–Continuous Hinge
NAS42	11/64	Spacers–Rivet (ANA Bul. 147)
NAS43	11/64	Spacers–Screw and Bolt (ANA Bul. 147)
NAS44	10/62	Busbar–Electrical (ANA Bul. 147)
NAS45*	1/50	Panel Assembly–Electrical Connector, Sup. by AN3436
NAS46(4)*	12/44	Code–Wire Identification, Sup. by MIL-W-5088
NAS48X	12/45	Leveling Points, Sup. by MIL-M-6756
NAS49(2)*	10/62	Fitting, Engine Cable Control, Quick Disconnect
NAS50(2)*	7/56	Rings–Internal Retainer, See NAS669
NAS51(2)*	7/56	Rings–External Retainer, See NAS670
NAS52*	10/47	Conduit/or Assembly/–Electrical, Shielded, Flexible
NAS53*	6/49	Bolt–Close Tolerance, Al. Stl. & Alum. Al. #10-32, Sup. by AN173–AN186
NAS54*	6/49	Bolt–Close Tolerance, Al. Stl. & Alum. Al. 1/4-28, Sup. by AN173–AN186
NAS55*	6/49	Bolt–Close Tolerance, Al. Stl. & Alum. Al. 5/16-24, Sup. by AN173–AN186
NAS56*	6/49	Bolt–Close Tolerance, Al. Stl. & Alum. Al. 3/8-24, Sup. by AN173–AN186
NAS57*	6/49	Bolt–Close Tolerance, Al. Stl. & Alum. Al. 7/16-20, Sup. by AN173–AN186
NAS58*	6/49	Bolt–Close Tolerance, Al. Stl. & Alum. Al. 1/2-20, Sup. by AN173–AN186
NAS59*	6/49	Bolt–Close Tolerance, Al. Stl. & Alum. Al. 9/16-18, Sup. by AN173–AN186
NAS60*	6/49	Bolt–Close Tolerance, Al. Stl. & Alum. Al. 5/8-18, Sup. by AN173-AN186
NAS62*	6/49	Bolt–Close Tolerance, Al. Stl. & Alum. Al. 3/4-16, Sup. by AN173–AN186
NAS64*	6/49	Bolt–Close Tolerance, Al. Stl. & Alum. Al. 7/8-14, Sup. by AN173–AN186
NAS66*	6/49	Bolt–Close Tolerance, Al. Stl. & Alum. Al. 1-14, Sup. by AN173–AN186

Number	Current Issue	Title
NAS67	11/66	Guide–Fastener, Low Form, Cowl, Dzus Type (ANA Bul. 147)
NAS68•	10/47	Guide–Fastener, Cowl, Dzus Type
NAS69•	10/47	Guide–Fastener, Cowl, Dzus Type, Dimpled Rivet Holes
NAS70•	7/54	Washer–Plain, Use AN960
NAS71(13)X	4/44	Symbols–Aircraft Wiring Diagram, Sup. by AND10378
NAS72(2)	10/62	Bushing–Clamp-Up, Steel, Chrome Plated (ANA Bul. 147)
NAS73(2)	11/66	Bushing–Clamp-Up, Steel, Cadmium Plated (ANA Bul. 147)
NAS74(2)	10/62	Bushing–Clamp-Up, Bronze (ANA Bul. 147)
NAS75	11/66	Bushing–Plain, Press-Fit, Steel (ANA Bul. 147)
NAS76	11/66	Bushing–Plain, Press-Fit, Bronze (ANA Bul. 147)
NAS77	10/68	Bushing–Flanged, Press Fit, Steel & Bronze
NAS78(2)	1/63	Fitting, Self-Sealing, Quick Disconnect Tubing
NAS79(2)X	7/45	Standard Indentors and Nests for Applying Mechanical Terminals
NAS80•	7/46	Bolt–100° Flush Head Close Tolerance 1/4-28, See NAS334
NAS81•	7/46	Bolt–100° Flush Head Close Tolerance 5/16-24, See NAS335
NAS82•	7/46	Bolt–100° Flush Head Close Tolerance 3/8-24, See NAS336
NAS83•	7/46	Bolt–100° Flush Head Close Tolerance 7/16-20, See NAS337
NAS84•	7/46	Bolt–100° Flush Head Close Tolerance 1/2-20, See NAS338
NAS85•	7/46	Bolt–100° Flush Head Close Tolerance 9/16-18, See NAS339
NAS86•	7/46	Bolt–100° Flush Head Close Tolerance 5/8-18, See NAS340
NAS90•	1/50	End–Rod Control, Threaded, Sup. by AN490
NAS91•	10/46	Rod Assembly–Control 1/4 Solid, Steel, Adjustable Clevis Ends
NAS92•	10/46	Rod Assembly–Control 1/4 Solid, Steel, Adjustable Clevis and Bearing Ends
NAS93•	10/46	Rod Assembly–Control, 1/4 Solid, Steel, Adjustable Bearing Ends
NAS94•	5/48	Rod–Control, 1/4 Solid, Steel, Sup. by NAS354
NAS95•	10/46	Rod Assembly–Control, Welded Tube, 3/8 O.D., Steel, Fixed and Adjustable Clevis Ends
NAS96•	10/46	Rod Assembly–Control, Welded Tube, 3/8 O.D., Steel, Fixed Clevis and Adjustable Bearing Ends
NAS97•	10/46	Welded Assembly–Tube 3/8 O.D., Steel, Clevis and Threaded Rod Ends, Use NAS359
NAS98•	10/46	Rod Assembly–Control, Welded Tube, 3/8 O.D., Steel, 1/4 Adjustable Clevis Ends
NAS99•	10/46	Rod Assembly–Control, Welded Tube, 3/8 O.D., Steel, 1/4 Adjustable Clevis and Bearing Ends
NAS100•	10/46	Rod Assembly–Control, Welded Tube, 3/8 O.D., Steel, 1/4 Adjustable Bearing Ends
NAS101•	10/46	Welded Assembly–Tube, 3/8 O.D., Steel, Threaded Rod Ends, Use NAS358
NAS102•	10/46	Tube–Control, 3/8 O.D., and 1/2 O.D., Steel, Use NAS360
NAS103•	10/46	Rod Assembly–Control, Welded Tube, 1/2 O.D., Steel, 1/4 Adjustable Clevis Ends
NAS104•	10/46	Rod Assembly–Control, Welded Tube, 1/2 O.D., Steel, 5/16 Adjustable Clevis Ends
NAS105•	10/46	Rod Assembly–Control, Welded Tube, 1/2 O.D., Steel, 1/4 Adjustable Clevis and Bearing Ends
NAS106•	10/46	Rod Assembly–Control, Welded Tube, 1/2 O.D., Steel, 5/16 Adjustable Clevis and Bearing Ends
NAS107•	10/46	Rod Assembly–Control, Welded Tube, 1/2 O.D., Steel, 1/4 Adjustable Bearing Ends
NAS108•	10/46	Rod Assembly–Control, Welded Tube, 1/2 O.D., Steel, 5/16 Adjustable Bearing Ends
NAS109•	10/46	Welded Assembly–Tube, 1/2 O.D., Steel, 1/4 Threaded Rod Ends, Use NAS358
NAS110•	10/46	Welded Assembly–Tube, 1/2 O.D., Steel, 5/16 Threaded Rod Ends, Use NAS358
NAS111•	10/46	Rod Assembly–Control, Riveted Tube, 3/8 O.D., Alum. Alloy, 1/4 Adjustable Clevis Ends
NAS112•	10/46	Rod Assembly–Control, Riveted Tube, 3/8 O.D., Alum. Alloy, 1/4 Adjustable Clevis and Bearing Ends
NAS113•	10/46	Rod Assembly–Control, Riveted Tube, 3/8 O.D., Alum. Alloy, 1/4 Adjustable Bearing Ends
NAS114•	10/46	Riveted Assembly–Tube, 3/8 O.D., Alum. Alloy, Threaded Rod Ends, Use NAS355
NAS115•	10/46	Tube–Control, 3/8 O.D. and 1/2 O.D., Alum. Alloy, Use NAS357
NAS116•	10/46	Rod Assembly–Control, Riveted Tube, 1/2 O.D., Alum. Alloy, 1/4 Adjustable Clevis Ends
NAS117•	10/46	Rod Assembly–Control, Riveted Tube, 1/2 O.D., Alum. Alloy, 1/4 Adjustable Clevis and Bearing Ends
NAS118•	10/46	Rod Assembly–Control, Riveted Tube, 1/2 O.D., Alum. Alloy, 1/4 Adjustable Bearing Ends
NAS119•	10/46	Riveted Assembly–Tube, 1/2 O.D., Alum. Alloy, Threaded Rod Ends, Use NAS355
NAS120	1/63	Knob Control 3/4 Spherical (ANA Bul. 147–AF Only)
NAS121	1/63	Knob–Control, 1 Inch Spherical (ANA Bul. 147–AF Only)
NAS122•	12/45	Knob-Control, 1-1/4 Spherical, Plastic, Use NAS198 or NAS199
NAS123	1/63	Knob–Control, 1 Inch Dia. x 5/8 Oval, Plastic (ANA Bul. 147–AF Only)
NAS124	1/63	Knob–Control, 1-1/4 Dia. x 13/16 Horizontal Semi-Round Plastic (ANA Bul. 147–AF Only)
NAS125	1/63	Knob–Control, 1-1/8 Dia. x 1-3/16 Angular Semi-Round, Plastic (ANA Bul. 147–AF Only)
NAS126	1/63	Knob, 1-1/8 Cubical Plastic (ANA Bul. 147–AF Only)
NAS127	1/63	Button–1/2 Dia., Luminous Letter (ANA Bul. 147)
NAS128	1/63	Filler-Knob Lever Slot (ANA Bul. 147)
NAS129	10/61	Data–Std. Knob, Engineering Reference (ANA Bul. 147)
NAS130X	4/43	Angles–Equal Legs, Extruded, 24S Aluminum Alloy, Sup. by AND10133
NAS131(2)X	4/43	Angles–Unequal Legs, Extruded, 24S Aluminum Alloy, Sup. by AND10134
NAS132X	4/43	Angles–Bulb, Extruded, 24S Aluminum Alloy, Sup. by AND10135
NAS133(2)X	4/43	Tees–Plain, Extruded, 24S Aluminum Alloy, Sup. by AND10136
NAS134(4)X	4/43	Channels–Plain, Extruded, 24S Aluminum Alloy, Sup. by AND10137
NAS135(2)X	4/43	Zees–Equal Legs, Extruded, 24S Aluminum Alloy, Sup. by AND10138
NAS136(2)X	4/43	Zees–Unequal Legs, Extruded, 24S Aluminum Alloy, Sup. by AND10139
NAS137X	4/43	I-Shapes–Plain, Extruded, 24S Aluminum Alloy, Sup. by AND10140
NAS138X	4/43	Tees–Bulb, Extruded, 24 S Aluminum Alloy, Sup. by AND10141
NAS139•	8/45	Stud–Coarse Thread, Sup. by NAS183
NAS140•	8/45	Stud–Fine Thread, Sup. by NAS184
NAS141	10/61	Tapped Holes for Studs in Aluminum Alloy and Soft Metals
NAS142	10/61	Tapped Holes for Studs in Steel and Hard Metals (Incl. Brass)
NAS143•	9/54	Washers–4130 Or Equiv., 125,000 to 145,000 PSI, Countersunk And Plain Types, See MS20002
NAS144–158(2)	11/66	Bolt–Internal Wrenching–Steel 1/4-28 thru 1-1/8-12
NAS159(4)	3/67	Bolt–Internal Wrenching–Aircraft

Number	Current Issue	Title
NAS160(2)*	5/46	Wheel–Control, Pilot's, Military Aileron, Use NAS348
NAS161*	10/61	Cap Assembly–Military Control Wheel Hub
NAS162	8/64	Cap–Military Control Wheel Hub
NAS163*	11/61	Insignia–Military Control Wheel Hub Cap
NAS164	8/64	Insert–Military Control Wheel Hub Cap
NAS165x	11/55	Switch–Pilots Control Wheel Trigger
NAS166*	5/46	Switch (Bomb Release & Radio)–Military Control Wheel, Use NAS350
NAS167x	11/55	Cap–Pilots Control Wheel Trigger Switch Cavity
NAS168	10/61	Knob, One Inch Dia. Knurled Plastic (ANA Bul. 147–AF Only)
NAS169(7)*	10/47	Specification–Rolled Formed Sections
NAS170	1/67	Clevis–Engine Control Rod, Adjustable
NAS171	6/49	Clevis–Rod End for Resistance Welding (ANA Bul. 147)
NAS172	1/67	Bolt–Internal Wrenching–Steel, 1–1/4–12
NAS174	1/67	Bolt–Internal Wrenching–Steel, 1–3/8–12
NAS176	1/67	Bolt–Internal Wrenching–Steel, 1–1/2–12
NAS177	9/67	Pin–100° C'Sunk Head Hi-Shear Rivet
NAS178	9/67	Pin–Flat Head Hi-Shear Rivet
NAS179	4/58	Collar–Hi-Shear Rivet (ANA Bul. 147)
NAS180(6)*	6/45	Angles–Equal Leg, Rolled Form Aluminum Alloy
NAS181*	10/47	Bearings–Plain Spherical (Interchangeable)
NAS182	9/45	Bearing–Plain Spherical (ANA Bul. 147)
NAS183	1/67	Stud–Coarse Thread
NAS184	1/67	Stud–Fine Thread
NAS185*	10/47	Spacers–Split, Screw & Bolt
NAS187*	10/47	Fairing–Tube Outlet
NAS188	3/58	Pad–Tank Strap, See AN8031 (ANA Bul. 147–AF Only)
NAS190x	6/66	Dimensioning–Decimal Usage For, see NAS380
NAS191	12/61	Terminal Board–Assembly–Molded-In Screw Electrical (ANA Bul. 147–AF Only)
NAS192	12/61	Post, Binding–Terminal Board, Electrical (ANA Bul. 147–AF Only)
NAS193(5)x	7/45	Specification–Solderless (Pressure) Terminals For Electric Cables
NAS194x	5/52	Magnesium Alloy–Standard Thicknesses for Sheet and Plate, Sup. By ANA Bul. 408
NAS195x	5/52	Steel Sheet And Strip–Preferred Thicknesses
NAS197(3)x	3/54	Insulation–Space and Air Duct Thermal
NAS198	12/61	Knob–Control, 1.25 Spher., Plastic (ANA Bul. 147–AF Only)
NAS199	12/61	Knob and Letter Assy.–Control, 1.25 Spherical Plastic (ANA Bul. 147–AF Only)
NAS200*	11/52	Screws–Flush Head, 100° Countersunk, Phillips Recess, Low Carbon Steel and Brass
NAS201*	11/52	Screws–Flush Head, 100° Countersunk, Frearson Recess, Low Carbon Steel and Brass
NAS202*	11/52	Screws–Round Head, Phillips Recess, Low Carbon Steel and Brass
NAS203*	11/52	Screws–Round Head, Frearson Recess, Low Carbon Steel and Brass
NAS204*	11/47	Screws–100° Flat Head, Phillips Recess, Alum. Alloy, Bronze and Alloy Steel #8-32, Sup. by AN509
NAS205*	11/47	Screws–100° Flat Head, Phillips Recess, Alum. Alloy, Bronze and Alloy Steel #10-32, Sup. by AN509
NAS206*	11/47	Screws–100° Flat Head, Phillips Recess, Alum. Alloy, Bronze and Alloy Steel, 1/4-28, Sup. by AN509
NAS207*	11/47	Screws–100° Flat Head, Phillips Recess, Alum. Alloy, Bronze and Alloy Steel, 5/16-24, Sup. by AN509
NAS208*	11/47	Screws–100° Flat Head, Phillips Recess, Alum. Alloy, Bronze and Alloy Steel, 3/8-24, Sup. by AN509
NAS209*	11/47	Screws–100° Flat Head, Phillips Recess, Alum. Alloy, Bronze and Alloy Steel, 7/16-20, Sup. by AN509
NAS210*	11/47	Screws–100° Flat Head, Phillips Recess, Alum. Alloy, Bronze and Alloy Steel, 1/2-20, Sup. by AN509
NAS211*	11/47	Screws–100° Flat Head, Phillips Recess, Alum. Alloy, Bronze and Alloy Steel, 9/16-18, Sup. by AN509
NAS212*	11/47	Screws–100° Flat Head, Frearson Recess, Alum. Alloy, Bronze and Alloy Steel, #8-32, Sup. by AN509
NAS213*	11/47	Screws–100° Flat Head, Frearson Recess, Alum. Alloy, Bronze and Alloy Steel, #10-32, Sup. by AN509
NAS214*	11/47	Screws–100° Flat Head, Frearson Recess, Alum. Alloy, Bronze and Alloy Steel, 1/4-28, Sup. by AN509
NAS215*	11/47	Screws–100° Flat Head, Frearson Recess, Alum. Alloy, Bronze and Alloy Steel, 5/16-24, Sup. by AN509
NAS216*	11/47	Screws–100° Flat Head, Frearson Recess, Alum. Alloy, Bronze and Alloy Steel, 3/8-24, Sup. by AN509
NAS217*	11/47	Screws–100° Flat Head, Frearson Recess, Alum. Alloy, Bronze and Alloy Steel, 7/16-20, Sup. by AN509
NAS218*	11/47	Screws–100° Flat Head, Frearson Recess, Alum. Alloy, Bronze and Alloy Steel, 1/2-20, Sup. by AN509
NAS219*	11/47	Screws–100° Flat Head, Frearson Recess, Alum. Alloy, Bronze and Alloy Steel, 9/16-18, Sup. by AN509
NAS220–227(3)*	11/64	Screws, Brazier Head, Phillips Recess, See MS27039
NAS228*	3/58	Screws–Brazier Head, Frearson Recess, Alum. Alloy, Bronze and Alloy Steel, 8-32, See NAS220, NAS623 or NAS1402
NAS229*	3/58	Screws–Brazier Head, Frearson Recess, Alum. Alloy, Bronze and Alloy Steel, 10-32, See NAS221, NAS623 or NAS1403
NAS230*	3/58	Screws–Brazier Head, Frearson Recess, Alum. Alloy, Bronze and Alloy Steel, 1/4-28, See NAS222, NAS623 or NAS1404
NAS231*	3/58	Screws–Brazier Head, Frearson Recess, Alum. Alloy, Bronze and Alloy Steel, 5/16-24, See NAS223, NAS623 or NAS1405
NAS232*	3/58	Screws–Brazier Head, Frearson Recess, Alum. Alloy, Bronze and Alloy Steel, 3/8-24, See NAS224, NAS623 or NAS1406
NAS233*	9/53	Screws–Brazier Head, Frearson Recess, Alum. Alloy, Bronze and Alloy Steel, 7/16-20
NAS234*	9/53	Screws–Brazier Head, Frearson Recess, Alum. Alloy, Bronze and Alloy Steel, 1/2-20
NAS235*	9/53	Screws–Brazier Head, Frearson Recess, Alum. Alloy, Bronze and Alloy Steel, 9/16-18
NAS240*	10/47	Nut–Flared Tube Coupling Plastic
NAS241x	1/46	Ring–Flared Tube Fitting, Steel
NAS242*	10/47	Nut–Flared Tube Fitting Plastic Lock
NAS243*	10/47	Cap–Flared Tube Fitting Plastic
NAS244*	10/47	Cap–Rigid Tube or Hose Fitting Synthetic
NAS245*	10/47	Nipple–Flared Tube & Pipe Thread Plastic
NAS246*	10/47	Union–Flared Tube, Plastic
NAS247*	10/47	Reducer–Flared Tube, Plastic
NAS248*	10/47	Union–Bulkhead Universal and Flared Tube Plastic
NAS249*	10/47	Nipple–Hose and Flared Tube Plastic
NAS250*	10/47	Union–Hose, Flared Tube, Bulkhead & Universal, Plastic
NAS251*	10/47	Nipple–Pipe Thread and Hose, Plastic
NAS252*	10/47	Union–Hose Bulkhead, Plastic
NAS253*	10/47	Elbow–90° Flared Tube and Pipe Thread, Plastic
NAS254*	10/47	Elbow–90° Bulkhead Universal and Flared Tube, Plastic
NAS255*	10/47	Elbow–Hose, Flared Tube, Bulkhead & Universal, 90°, Plastic
NAS256*	10/47	Elbow–Hose and Pipe Thread, 90°, Plastic
NAS257*	10/47	Elbow–Hose, Bulkhead, 90°, Plastic
NAS258*	10/47	Elbow–90°, Female Screw Thread Plastic Tee–Female Screw Thread Cross–Female Screw Thread
NAS259*	10/47	Elbow–45° Flared Tube and Pipe Thread Plastic
NAS260*	10/47	Elbow–45° Universal and Flared Tube Plastic
NAS261*	10/47	Elbow–45° Universal Cone and Hose Plastic
NAS262*	10/47	Elbow–45° Pipe Thread and Hose Plastic
NAS263*	10/47	Elbow–45° Bulkhead Hose, Plastic

Number	Current Issue	Title
NAS264•	10/47	Elbow–45° Female Straight Threads Plastic
NAS265•	10/47	Tee–Flared Tube Plastic
NAS266•	10/47	Tee–Flared Tube, Run Universal Side Plastic
NAS267•	10/47	Tee–Flared Tube, Run Pipe Thread Side Plastic
NAS268•	10/47	Tee–Pipe Thread & Flared Tube Run and Side Plastic
NAS269•	10/47	Tee–Pipe Thread and Hose Run and Side Plastic
NAS270•	10/47	Tee–Hose Run Pipe Thread Side Plastic
NAS271•	10/47	Tee–Hose Plastic
NAS272•	10/47	Cross–Flared Tube Plastic
NAS273•	10/47	Cross–Hose Pipe Thread Bottom Plastic
NAS274•	10/47	Cross–Hose Plastic
NAS275•	10/47	Plug and Bleeder –Screw Thread Plastic
NAS276•	10/47	Plug–Pipe Thread Plastic
NAS277x	1/46	Protector–Flared Tube, Plastic
NAS278(16)•	10/47	Specification for Tube Fittings Molded from Thermosetting Plastic Materials
NAS279(14)•	10/47	Specification for Tubing Fabricated from Thermoplastic Materials
NAS280(9)•	10/47	Specification for the Installation of Plastic Fittings and Tubing
NAS281•	10/47	Tubing–Thermoplastic
NAS287	12/61	Terminal–Chain To Cable, (For Swagging) (ANA Bul. 147)
NAS288	12/61	Terminal–Chain to Turnbuckle (ANA Bul. 147)
NAS289x	11/52	Bi-Planar Block–Chain
NAS290•	3/52	Bearing–Heavy Duty Inner and Outer Races, Non-Separable, Single Row Needle, Sup. by NAS505
NAS291•	11/52	Bearing–Heavy Duty, Outer Race Retainment Separable Inner Race, Single Row Needle, Sup. by NAS515
NAS292•	3/52	Bearing–Roller Type, Heavy Section Outer Race and Inner Race Non-Separable, Single Row Needle, Sup. by NAS502
NAS293•	3/52	Bearing–Roller Type, Heavy Section Outer Race and Inner Race Non-Separable, Double Row Needle, Sup. by NAS503
NAS294•	3/52	Bearings–Heavy Duty Self-Aligning, Inner and Outer Races Non-Separable, Single and Double Row Needle, Sup. by NAS504
NAS295	12/61	Bearing–Thin Shell, Open and Closed Ends, Regular and Extra Capacity, Single Row, Small Needle (ANA Bul. 147)
NAS296•	3/52	Bearing–Thin Shell with Non-Separable Inner Race, Single Row Small Needle, Sup. by NAS506
NAS302–310	12/61	Cable Assembly Swaged Type, Type I Terminals (ANA Bul. 147)
NAS312–320	12/61	Cable Assembly Swaged Type, Type I and Type II Terminals (ANA Bul. 147)
NAS322–330	12/61	Cable Assembly Swaged Type, Type II Terminals (ANA Bul. 147)
NAS333–340(3)	1/67	Bolt–100° Close Tolerance, High Strength (ANA Bul. 147–AF Only)
NAS341(2)	3/46	Angles–Equal Leg, Extruded Aluminum, Supplementary Series (ANA Bul. 147)
NAS342(4)	3/46	Angles–Unequal Leg, Extruded Aluminum, Supplementary Series (ANA Bul. 147)
NAS343	3/46	Angles–Bulb, Extruded Aluminum, Supplementary Series (ANA Bul. 147)
NAS344(3)	3/46	Tees–Extruded Aluminum, Supplementary Series (ANA Bul. 147)
NAS345(5)	3/46	Channels–Extruded Aluminum, Supplementary Series (ANA Bul. 147)
NAS346(2)	3/46	Zees–Equal Leg, Extruded Aluminum, Supplementary Series (ANA Bul. 147)
NAS347(2)	3/46	Zees–Unequal Leg, Extruded Aluminum, Supplementary Series (ANA Bul. 147)
NAS348(2)•	4/57	Wheel–Pilots Control
NAS349x	5/54	Wheel Assembly–Pilots Control
NAS350•	3/62	Switch–Pilots Control Wheel Push Button
NAS351•	9/61	Nut–Pilot's Control Wheel Attaching
NAS352•	9/61	Shaft End–Pilot's Control Wheel
NAS353(8)•	9/61	Specification–Nut, Aircraft Self-Locking, For 160,000 to 180,000 PSI Tension Bolts
NAS354	11/66	Rod–Control Solid–Steel (ANA Bul. 147)
NAS355	12/64	Tube Assembly–Control, Aluminum Alloy with Riveted Threaded Rod Ends (ANA Bul. 147)
NAS356	12/64	Tube Assembly–Control, Aluminum Alloy with Riveted Clevis and Threaded Rod Ends (ANA Bul. 147)
NAS357	12/64	Tube–Control, Aluminum Alloy (ANA Bul. 147)
NAS358	11/66	Tube Assembly–Control Steel with Welded Threaded Rod-Ends (ANA Bul. 147)
NAS359	11/66	Tube Assembly–Control Steel with Welded Clevis and Threaded Rod Ends (ANA Bul. 147)
NAS360	9/61	Tube–Control, Steel (ANA Bul. 147)
NAS361	1/64	Tube Assembly–Control, Steel, With Flash Welded and Threaded Rod-Ends (ANA Bul. 147)
NAS362	1/64	Tube Assembly–Control, Steel, With Flash Welded Clevis and Threaded Rod-Ends (ANA Bul. 147)
NAS363–369(2)	1/64	Control Rod and Tube Assembly, 1/4-28UNF-3A Thread (ANA Bul. 147)
NAS370–373	1/64	Control Rod and Tube Assembly, 5/16-24UNF-3A Thread (ANA Bul. 147)
NAS375(38)x	5/54	Aircraft Equipment Testing; Specification For
NAS376	1/64	Pulley–Secondary Control, For 1/16 and 3/32 Dia Wire Rope (ANA Bul. 147–AF Only)
NAS377•	7/52	Pulley–Flight Control for 1/8, 5/32, and 3/16 Dia. Cable, See AN220
NAS378•	7/52	Pulley–Heavy Duty Control for 3/16, 7/32, and 1/4 Dia. Cable, See AN221
NAS379(6)x	1/64	Specification–Pulleys: Grease Lubricated Anti-Friction Bearing Control, Use MIL-P-7034
NAS380(12)	4/67	NAS Documents Preparation and Maintenance
NAS381•	11/50	Storage Batteries–12-Volt, Lead Acid
NAS382•	11/52	Bushing–Plain Bearing Pulley
NAS383•	11/52	Pulleys–Plain Bearing
NAS384	1/64	Screw–100° Oval Recessed Head, Sheet Metal (ANA Bul. 147–AF Only)
NAS385	1/64	Screw–82° Oval Recessed Head, Sheet Metal (ANA Bul. 147–AF Only)
NAS386	1/64	Screw–82° Oval Recessed Head, (Wood) (ANA Bul. 147–AF Only)
NAS387	1/64	Screw–100° Oval Recessed Head, Mach. (Coarse & Fine Thread) (ANA Bul. 147–AF Only)
NAS388	8/47	Screw–82° Oval Recessed Head–Mach. (Coarse & Fine Thread) (ANA Bul. 147–AF Only)
NAS389x	11/52	Screw–82° Oval Recessed Head–Self Tapping Machine
NAS390	11/66	Washer–Flush Type Finishing (ANA Bul. 147)
NAS391	11/66	Washer–Countersunk Type Finishing (ANA Bul. 147–AF Only)
NAS392x	11/55	Bearing–Brake and Wheel Location, 8" and 10" Wheel
NAS393x	11/55	Bearing–Brake and Wheel Location, 6" Wheel
NAS394•	4/57	Axle–Landing Gear Wheel 1-1/2 O.D.
NAS395	12/54	Nut–"U" Type Sheet Spring (ANA Bul. 147)
NAS396	12/54	Nut–"U" Type Countersunk Sheet Spring (ANA Bul. 147)
NAS397	4/59	Clamp–Ratchet, One Piece (ANA Bul. 147)
NAS398	10/56	Clamp–Ratchet, Three Piece (ANA Bul. 147–AF Only)
NAS400	7/47	Connection–Airplane Ground Air Conditioning 5 Inch
NAS401	7/47	Connection–Airplane Ground Air Conditioning 8 Inch

Number	Current Issue	Title
NAS402	10/47	Adapter–Fuel and Oil Tank Filler Cap, Design and Installation of (For Aircraft)
NAS403	10/47	Plug and Receptacle–Fuel Nozzle Grounding, Design and Installation of (For Aircraft)
NAS404*	2/58	Jack Pads–Design and Installation of (For Aircraft), See MS33559
NAS405*	9/61	Receptacle–External Power Connection Direct Current, Design and Installation of (For Aircraft)
NAS406*	9/61	Plug–External Power Connection Design of (For Aircraft)
NAS407X	11/55	Valve and Adapter–Low Pressure Oxygen Connection, Installation of (For Aircraft)
NAS408X	11/55	Valve and Adapter–Low Pressure Oxygen Connection, Design of (For Aircraft)
NAS409*	9/61	Coupling–External CO_2 Connection Design & Installation of (For Aircraft)
NAS416X	11/55	Flange–Low Pressure Fluid Connection (3 Bolt)
NAS417	9/61	Flange–Low Pressure Fluid Connection (4 Bolt)
NAS418	9/53	Tube–Low Pressure Fluid Connection
NAS419*	4/57	Connection–Low Pressure Fluid (4 Bolt)
NAS420*	11/52	Connection–Low Pressure Fluid (3 Bolt)
NAS421X	11/55	Gasket–Fluid Connection, 3 & 4 Bolt
NAS422*	4/57	Sleeve–Cable Turnbuckle Lock
NAS423X	11/52	Master and Ignition Switch–Single Engine
NAS424	2/61	Coupling–Tube, Flared Fitting, Female Thread (ANA Bul. 147)
NAS425*	12/49	Fork Turnbuckle, Sup. by AN162
NAS426*	12/49	Terminal–Fork End Cable (For Swaging), Sup. by AN658
NAS427	11/66	Pin–Pulley Guard (ANA Bul. 147)
NAS428	9/61	Bolt–Adjusting Crowned Hexagon Head (ANA Bul. 147)
NAS429*	5/51	Bearing–Sintered, Plain
NAS430*	5/51	Bearing–Super Sintered, Plain
NAS431*	5/51	Bearing–Ball, Double Row Rod End Type
NAS432*	5/51	Bearing–Ball, Double Row Rod End Type
NAS433*	5/51	Bearing–Ball, Bellcrank
NAS434*	5/51	Clamp–Loop Type, Uncushioned Sheet Metal Nut
NAS435*	5/51	Clamp–Loop Type–Cushioned Sheet Metal Nut
NAS436*	5/51	Clamp–Harness
NAS437*	5/51	Clamp–Wire Bundle
NAS438*	5/51	Clip–Box Cover Fastener
NAS439*	5/51	Clip–Trim Lining
NAS440*	5/51	Cutouts–Flush Latch
NAS441*	5/51	Fuse–Electric Safety
NAS442*	5/51	Latch–Flush Type
NAS443*	5/51	Nut–Self-Locking Internal Wrenching
NAS444*	5/51	Nut–Double Lug Anchor Type
NAS445*	5/51	Nut–Offset Anchor Type
NAS446	9/61	Nut–Flat Type Sheet Spring (ANA Bul. 147)
NAS447*	5/51	Nut Plate–Self-Locking, One Lug
NAS448*	3/50	Nut Plate–Self-Locking, Angle
NAS449*	5/51	Nut–Corner Anchor Type
NAS450*	5/51	Nut Plate–Self-Locking, Plain
NAS451*	5/51	Plug Button
NAS452*	5/51	Rivnut–100° Csk. Head
NAS453*	5/51	Rivnut–Flat Head
NAS454*	5/51	Rivet–Blind Csk. Head (Self Plugging Type)
NAS455*	5/51	Rivet–Blind Brazier Head (Self Plugging Type)
NAS456*	5/51	Ring–Locking, Rosan
NAS457*	5/51	Sleeve–Threaded, Rosan (Insert)
NAS458(6)*	9/53	Vendor Parts–Contractor Requirements For Drawings, Data Lists, And Specification Of Specification For
NAS459(8)*	9/61	Protection of Vendor Parts and Accessories Against Corrosion Specification for, See MIL-F-7179
NAS460	11/66	Washer–Tab Type (ANA Bul. 147–AF Only)
NAS461	3/62	Gage–100° Countersink Screw Head Flushness (ANA Bul. 147–AF Only)
NAS462	3/62	Gaging–100° Countersink Screw Head Flushness (ANA Bul. 147–AF Only)
NAS463(2)	11/62	Shim, Anchor Nut, Plain (ANA Bul. 147)
NAS464(3)	11/66	Bolt–Shear, Close Tolerance (ANA Bul. 147)
NAS465	8/49	Standard Dimensions and Tolerances for Rod End Shank Outside Diameters and Tube End Inside and Outside Diameters
NAS466*	2/58	Fitting–Lubrication, See MS15000
NAS467	3/62	Bolt–Hex Head, Brass, Electrical (ANA Bul. 147)
NAS468*	5/51	Bearing–Ball, Single Row, Unshielded
NAS469*	5/51	Bearing–Ball, Single Row, Rod End Type
NAS470*	5/51	Bearing–Sintered, Flanged
NAS471*	5/51	Bead–Insulator
NAS472*	5/51	Bracket–Angle Speed Nut
NAS473*	5/51	Bracket–Angle Speed Nut
NAS474*	5/51	Bracket–Angle Speed Nut
NAS475*	5/51	Bracket–Angle Speed Nut
NAS476*	5/51	Clamp–Cushioned Lug Type
NAS477*	5/51	Clamp–Cushioned Lug Type
NAS478*	5/51	Clamp–Conduit
NAS479*	5/51	Clip–Bulb Angle or Straight Channel
NAS480*	5/51	Clip–Tubular Speed
NAS481*	5/51	Clip–Electric Fuse
NAS482*	5/51	Clip–Push On
NAS483*	5/51	Cock–Valve Drain
NAS484*	5/51	Fuse–Type 4AB
NAS485*	5/51	Hose–Oil and Coolant
NAS486*	5/51	Lamp–Sealed Beam
NAS487	11/66	Nut–Instrument Mount (ANA Bul. 147–AF Only)
NAS488	3/62	Nut–"J" Type Sheet Spring (ANA Bul. 147)
NAS489*	5/51	Nut–Latching Type
NAS490*	5/51	Shock Mount (Insulator Vibration)
NAS491*	5/51	Shock Mount–Cylindrical (300 Series)
NAS492*	5/51	Stud–Trimount (Tapered Type)
NAS493*	5/51	Stud
NAS494*	5/51	Terminal–Ball Type Cable
NAS495(2)*	3/58	Bolt–Internal Wrenching (High Tensile and High Fatigue Strength), See MS20004-MS20024
NAS496(14)	11/66	Bolts–Internal Wrenching
NAS497*	12/52	Fitting–Lubricator 1/8 Inch Drive, Flush Type, Sup. by NAS516
NAS498(7)**	5/69	Bolts–Shear
NAS499(10)*	11/62	Fasteners, Panel, Fast Operating, See MIL-F-5591
NAS500	11/66	Shim Anchor Nut–Countersunk (ANA Bul. 147)
NAS501(3)	2/68	Bolt–Stabilized–Non-Magnetic Corr Res Steel
NAS502	5/60	Bearing–Needle, Non-Separable, Heavy-Duty, Track-Roller, Single-Row (ANA Bul. 147–AF Only)
NAS503	5/60	Bearing–Needle, Non-Separable, Heavy-Duty, Track-Roller, Double-Row (ANA Bul. 147–AF Only)
NAS504	5/60	Bearing–Needle, Heavy-Duty, Self-Aligning, Non-Separable (ANA Bul. 147–AF Only)
NAS505	5/60	Bearing–Needle, Heavy-Duty, Non-Separable, Single-Row (ANA Bul. 147–AF Only)
NAS506	5/60	Bearing–Needle, Thin-Shell, Non-Separable, Single-Row (ANA Bul. 147–AF Only)
NAS507(6)	5/52	Airframe Castings–Specification for Classification and Inspection Of
NAS508*	12/62	Rivet–Universal Head Monel, See MS20615
NAS509(2)	4/65	Nut, Drilled Jam (ANA Bul. 147)
NAS510X	1/56	Link Assembly–Cable Take-Up
NAS511	11/66	Link, Cable Take-Up (ANA Bul. 147)
NAS512(7)*	5/63	Specification–Coupling; Flexible Fuel for One Inch Through Four Inch Tube Sizes, See MIL-C-25014
NAS513(2)	12/61	Washer–Rod End Locking (ANA Bul. 147)
NAS514(2)	8/63	Screw, Machine, 100°, Flat Head, Full Threaded, Alloy Steel (ANA Bul. 147)
NAS515	12/61	Bearing–Needle, Heavy Duty, Separable Inner Race, Single Row (ANA Bul. 147–AF Only)
NAS516	11/66	Fitting, Lubrication–1/8 Inch Drive, Flush Type (ANA Bul. 147)

Number	Current Issue	Title
NAS517(2)	11/66	Screw–100° Close Tolerance Flat Head 160,000 PSI (ANA Bul. 147–AF Only)
NAS518(2)	12/61	Gage–100° Flat Head Flushness (ANA Bul. 147–AF Only)
NAS519	3/53	Stylus–100° Flat Head Gage (ANA Bul. 147–AF Only)
NAS520	3/53	Coupling–Flexible, Low Pressure, 450° Air Line (ANA Bul. 147)
NAS521	3/53	Coupling–Flexible, Low Pressure, 250° Air Line (ANA Bul. 147)
NAS522(7)	3/53	Coupling–Flexible, Low Pressure, Air Line (ANA Bul. 147–AF Only)
NAS523(11)	5/67	Code–Rivet
NAS524x	12/61	Screw Threads–Standard NAS, See Spec. MIL-S-7742
NAS525(2)	3/62	Rivet–100° Countersunk Head Hi-Shear Close Tolerance Head and Shank (ANA Bul. 147)
NAS526(2)	3/62	Flushness Gage and Stylus for 100° Countersunk Flush Fasteners (ANA Bul. 147–AF Only)
NAS527(6)	3/62	Inspection Procedure for Flush Fasteners (ANA Bul. 147–AF Only)
NAS528	3/62	Collar (High Strength) High–Shear Rivet (ANA Bul. 147)
NAS529(2)	3/62	Rivet–Flat Head, Hi-Shear, Close Tolerance Shank (ANA Bul. 147)
NAS530*	3/58	Pulley–Secondary Control for 1/16 & 3/32 Dia. Cable 1 Dia., See MS20219
NAS531(9)	9/53	Charts–Rockwell Hardness
NAS532*	2/60	Guard, Switch Toggle, Three Position, See MS25214
NAS533(3)	10/53	Tubing Assembly–Flexible Transparent Vinyl (ANA Bul. 147)
NAS534	3/62	Ring, Flexible Tubing End (ANA Bul. 147)
NAS535	3/62	Washer, Step-Up Flexible Tubing Adapter (ANA Bul. 147)
NAS536	10/53	Washer–Retention Flexible Tubing Adapter (ANA Bul. 147)
NAS537	11/66	Bushing–Plain Press Fit–Undersize Inside Diameter
NAS538	10/68	Bushing–Flanged, Press Fit–Undersize Inside Diameter
NAS539	4/64	Knob Assembly, Transilluminated Pointer (ANA Bul. 147–AF Only)
NAS540(3)	5/64	Involute Serrations, 45° Pressure Angle, Loose Fit
NAS541(2)	5/64	Involute Serrations–Internal, 45° Pressure Angle, Loose Fit
NAS542(2)	5/64	Involute Serrations–External, 45° Pressure Angle, Loose Fit
NAS543	4/64	Switch–One Hole Mounting, Single Pole Toggle (ANA Bul. 147–AF Only)
NAS544	5/64	Switch–One Hole Mounting, Double Pole Toggle (ANA Bul. 147–AF Only)
NAS545*	2/58	Nut Switch Mounting, See MS25082–8
NAS546*	5/57	Ring–Locking, Use MS25081
NAS547(9)	7/59	Fastener, Rotary, Quick-Operating High Strength (ANA Bul. 147–AF Only)
NAS548*	5/64	Screw–100° Head, Type B Tapping, Frearson, Phillips, and Slotted Recess, See MS21207
NAS549	2/64	Washer–Flat, Resin Fiber (ANA Bul. 147)
NAS550	4/68	Fittings, Installation of Universal Elbow
NAS551	4/64	Bolt–Universal Fitting (ANA Bul. 147–AF Only)
NAS552	8/61	Elbow–Universal 90° (ANA Bul. 147–AF Only)
NAS553	8/61	Elbow–Universal 15° (ANA Bul. 147–AF Only)
NAS554	8/61	Elbow–Universal 45° (ANA Bul. 147–AF Only)
NAS555	8/61	Tee–Universal (ANA Bul. 147–AF Only)
NAS556(11)*	11/62	Specification for Spot and Seam Welding of Commercially Pure Titanium, See MIL-W-6858
NAS557	2/68	Grommet, Plastic–Split
NAS558	11/66	Keys–Square and Rectangular (ANA Bul. 147)
NAS559(2)	8/61	Lock–Rod End (Key Type) (ANA Bul. 147–AF Only)
NAS560(2)	8/61	Screw–Machine, Non Magnetic High Temperature, Structural, 100° Flush Head (ANA Bul. 147–AF Only)
NAS561(2)	11/66	Pins–Spring, Slotted & Coiled, Heavy Duty (ANA Bul. 147–AF Only)
NAS562(2)	11-67	Cam Follower, Needle Bearing, High Strength Stud
NAS563–572(2)	8/66	Bolt–Full Threaded, Fully Identified Head (ANA Bul. 147–AF Only)
NAS573(2)	8/65	Clamp Assembly–Electronic Equipment (ANA Bul. 147)
NAS574(2)	1/62	Pin, Rear Mounting–Electronic Equipment (ANA Bul. 147–AF Only)
NAS575(2)	4/57	Thread Length for Bolts & Screws
NAS576	10/61	Thread Length for Self-Locking Nuts
NAS577(2)**	3/69	Nut, Self-Locking, Barrel–Floating
NAS578	2/64	Retainer–Floating Barrel Nut (ANA Bul. 147–AF Only)
NAS579(3)	8/54	Involute Spline 30° Full Fillet
NAS580(2)	8/54	Involute Spline–Hole (30° Full Fillet)
NAS581(2)	8/54	Involute Spline–Shaft (30° Full Fillet)
NAS582(2)	10/61	Swaging Process–Tube to Rod End
NAS583–590(2)	11/66	Bolt, 100° Head "Hi-Torque", Close Tolerance, 160,000 PSI (ANA Bul. 147–AF Only)
NAS591(2)	8/61	Coupling–Rigid Tube Connector, Aluminum (ANA Bul. 147–AF Only)
NAS592	8/61	Ring–Rigid Tube Connector, Aluminum (ANA Bul. 147–AF Only)
NAS593	8/61	Nut–Rigid Tube Connector, Aluminum (ANA Bul. 147–AF Only)
NAS594(2)	8/61	Coupling–Rigid Tube Connector, Corrosion Resistant Steel (ANA Bul. 147–AF Only)
NAS595	8/61	Ring–Rigid Tube Connector, Corrosion Resistant Steel (ANA Bul. 147–AF Only)
NAS596	8/61	Nut–Rigid Tube Connector, Corrosion Resistant Steel (ANA Bul. 147–AF Only)
NAS597(12)	12/64	Fittings–Rigid Tube Connector, Flared Tube (ANA Bul. 147–AF Only)
NAS598(3)*	12/57	Tolerances–Aluminum and Magnesium Alloy Extruded Rod, Bar, and Shapes, See Fed. Std. 245
NAS599	10/61	Plate, Identification–Hydraulic Cylinder (ANA Bul. 147–AF Only)
NAS600–606	10/61	Screw, Machine–Aircraft, Pan Head, Phillips Recess, Full Threaded, Alloy Steel (ANA Bul. 147–AF Only)
NAS607(2)	10/61	Pin, Straight, Headless–(Dowel) (ANA Bul. 147–AF Only)
NAS608(2)*	4/62	Screws–Socket Head Cap, Plain and Self Locking, See NAS1351 and NAS1352
NAS609(2)*	4/62	Screws–Socket Head Cap (Non-Preferred Thread), See NAS1351 and NAS1352
NAS610–616*	2/59	Screw–Machine, Aircraft, Pan Head, Reed & Prince Recess Full Threaded, Alloy Steel, See NAS600–NAS606
NAS617	11/60	Packing, Preformed, Straight Thread Tube Fitting Boss, Synthetic Lubricant Resistant (ANA Bul. 147)
NAS618(7)	6/58	Fastener–Recommended Shank, Hole and Head-to-Shank Fillet Radius Limits For (ANA Bul. 147–AF Only)
NAS619(13)	4/56	Handbooks–Helicopter Inspection Requirements
NAS620	11/62	Washer, Flat–Reduced Outside Diameter (ANA Bul. 147)
NAS621(9)	6/66	Fasteners, Titanium Alloy, Procurement Specification (ANA Bul. 147–AF Only)
NAS622	10/61	Hook, Support–Electronic Equipment Clamp, (ANA Bul. 147–AF Only)
NAS623(2)	4/65	Screw, Machine–Aircraft, Pan Head, Phillips Recess, Short Thread, 160,000 PSI Alloy Steel (ANA Bul. 147–AF Only)
NAS624–644(2)**	3/69	Bolt–Twelve Point, External Wrenching–180,000 PSI

Number	Current Issue	Title
NAS645*	10/64	Fork, Turnbuckle, Clip Locking, See MS21252
NAS646*	10/64	Fork, Turnbuckle (For Bearing), Clip Locking, See MS21253
NAS647*	10/64	Eye, Turnbuckle (For Cable), Clip Locking, See MS21255
NAS648*	10/64	Eye, Turnbuckle (For Pin), Clip Locking, See MS21254
NAS649*	10/64	Barrel, Turnbuckle, Clip Locking, See MS21251
NAS650*	11/64	Terminal, Turnbuckle Cable, Clip Locking, See MS21260
NAS651*	10/64	Clip, Turnbuckle Locking, See MS21256
NAS652	1/62	Coupling–Rigid Tube Bulkhead Connector (ANA Bul. 147–AF Only)
NAS653–658(2)	10/68	Bolt, Close Tolerance–Hexagon Head, Titanium, Short Thread, .190 to .500
NAS659(2)	5/61	Bearing–Rod End, External Thread, Self-Aligning, Anti-Friction (ANA Bul. 147)
NAS660(2)	5/61	Bearing–Rod End, Plain Shank, Self-Aligning, Anti-Friction (ANA Bul. 147)
NAS661(7)	1/62	Bearings, Ball and Roller, Rod End, Anti-Friction, Airframe (ANA Bul. 147–AF Only)
NAS662	10/64	Screw–Machine, Flathead 100° Plain and Self Locking (ANA Bul. 147–AF Only)
NAS663–668(2)	8/65	Bolt, 100° Flat Head Close Tolerance, Titanium Alloy (ANA Bul. 147)
NAS669(5)*	8/63	Ring–Internal Retainer, See MS16625
NAS670(5)*	8/63	Ring–External Retainer, See MS16624
NAS671	1/62	Nut–Plain Hexagon Small Pattern (ANA Bul. 147–AF Only)
NAS672(2)	6/66	Plating–High Strength Steels–Cadmium
NAS673–678(2)	10/68	Bolt, Close Tolerance–Hexagon Head, Titanium, .190 to .500
NAS679(2)	12/68	Nut, Self-Locking, Hexagon–Low Height
NAS680(2)	12/68	Nut, Self-Locking, Plate–Two Lug, Low Height, C'Bore
NAS681(2)	12/68	Nut, Self-Locking, Plate–Two Lug, Low Height, Csk
NAS682(2)	12/68	Nut, Self-Locking, Plate–One Lug, Low Height, C'Bore
NAS683(2)	12/68	Nut, Self-Locking, Plate–One Lug, Low Height, Csk
NAS684(2)	12/68	Nut, Self-Locking, Plate–Corner, Low Height, C'Bore
NAS685(2)	12/68	Nut, Self-Locking, Plate–Corner, Low Height, Csk
NAS686(2)**	3/69	Nut, Self-Locking, Plate–Two Lug, Low Height, C'Bore, Floating
NAS687(2)	12/68	Nut, Self-Locking, Plate–One Lug, Low Height, C'Bore Floating
NAS688–692(2)	12/68	Nut Assembly, Self-Locking, Gang Channel–Low Height, C'Bored
NAS693–695(2)	12/68	Nut Assembly, Self-Locking, Gang Channel–Low Height, Csk
NAS696(2)	12/68	Nut, Self-Locking, Plate–One Lug, Low Height, C'Bore, Reduced Rivet Spacing
NAS697(2)	12/68	Nut, Self-Locking, Plate–Two Lug, Low Height, C'Bore, Reduced Rivet Spacing
NAS698(2)	12/68	Nut, Self-Locking, Plate–Corner, Low Height, C'Bore, Reduced Rivet Spacing
NAS699	11/66	Nut–Cylinder Lock (ANA Bul. 147–AF Only)
NAS700(11)*	4/63	Capacitors, Fixed, Stacked Dielectric, Miniature, See MIL-C-5
NAS701(11)*	4/63	Capacitors, Fixed, Rolled Foil, Miniature, See MIL-C-14157 & MIL-C-25
NAS702(14)	2/61	Wire, Electrical, Insulated, Copper, Hook-Up, and General Purpose (For 105° C Service) (ANA Bul. 147–AF Only)
NAS703(15)	2/61	Wire, Electrical, Insulated, Copper, High Temperature, Hook-Up and General Purpose (For 200° C Service) (ANA Bul. 147–AF Only)
NAS704	2/64	Terminal, Stud, Turret–Hollow–Midget (ANA Bul. 147–AF Only)
NAS705	2/64	Terminal, Stud, Turret, Single End–Midget (ANA Bul. 147–AF Only)
NAS706	2/64	Terminal, Stud, Turret–Double End (ANA Bul. 147–AF Only)
NAS707	2/64	Terminal, Stud, Turret–Single End (ANA Bul. 147–AF Only)
NAS708	2/64	Terminal, Stud, Turret–Hollow (ANA Bul. 147–AF Only)
NAS709	2/64	Terminal, Stud, Turret–Double End–Midget (ANA Bul. 147–AF Only)
NAS710(58)	4/55	Resistors–Variable, Precision (ANA Bul. 147–AF Only)
NAS711(15)*	10/68	Rectifiers, Selenium, For Magnetic Amplifiers
NAS713(14)	7/58	Connectors–Electrical, for Printed Wiring (ANA Bul. 147–AF Only)
NAS714(2)	7/58	Connector–Electrical, Female, For Use With Printed Wiring (ANA Bul. 147–AF Only)
NAS715	7/58	Connector–Electrical, Male, For Use With Printed Wiring (ANA Bul. 147–AF Only)
NAS716*	4/63	Standard Power Supply Voltages, See MIL-STD-706
NAS717(16)*	4/63	Semiconductor Diodes, See MIL-S-19500
NAS718(13)	8/57	Cable–High Temperature, Coaxial (ANA Bul. 147–AF Only)
NAS719(2)	6/57	Dimensional System–Modular
NAS720	10/58	Screw–Pan Head, Assembled Washers–Electrical Terminal Attachment (Preferred) (ANA Bul. 147–AF Only)
NAS721(2)	1/62	Screw, Miniature, Fillister Head (ANA Bul. 147–AF Only)
NAS722(2)	1/62	Screw, Miniature, Pan Head (ANA Bul. 147–AF Only)
NAS723(2)	1/62	Screw, Miniature, 100 Deg Flat Head (ANA Bul. 147–AF Only)
NAS724(2)	1/62	Screw, Miniature, Binding Head (ANA Bul. 147–AF Only)
NAS725(8)	1/62	Miniature Screws (ANA Bul. 147–AF Only)
NAS726(24)	9/60	Gear–Spur, 20 Degree Involute, Low Inertia, 48 and Finer Pitch (ANA Bul. 147–AF Only)
NAS727(5)	9/60	Gear–Spur, 20 Degree Involute, Low Inertia, 48 and Finer Pitch (ANA Bul. 147–AF Only)
NAS728(139)	10/62	Test Methods for Electromagnetic Relays
NAS729(13)	8/63	Cable, Electrical, Flat Conductor, Flexible–300 Volts, Copper (ANA Bul. 147–AF Only)
NAS800(5)	11/47	Specification–Airline Life Rafts
NAS801(3)	9/52	Specification–Airline Life Vests
NAS802(4)	5/50	Specification–Aircraft Safety Belts
NAS804(5)	8/49	Specification–Parachutes
NAS805(2)	2/50	Specification–Flares, Landing
NAS806(4)*	1/56	Specification–Aircraft Seats and Berths, See NAS809
NAS807(10)	6/51	Specification–Twin Seaplane Floats
NAS808(6)	12/51	Specification–Aircraft Skis
NAS809(6)	1/56	Specification–Aircraft Seats and Berths
NAS810*	6/69	Protection of Fluid Lines and Equipment, See NAS850
NAS811*	6/69	Cap, Protective, Flareless Tube and Hose Assembly
NAS812*	6/69	Cap, Protective Flared Tube and Hose Assembly
NAS813(2)**	6/69	Cap–Protective, Electrical Connector
NAS814(2)**	6/69	Cap–Protective, Flared Fitting
NAS815(2)**	6/69	Plug–Protective, Flared Hose or Tube Assembly
NAS816(2)*	6/69	Cap or Plug–Protective, Plain Tube or Hole
NAS817(2)**	6/69	Cap–Protective, Flared Fitting
NAS818(2)**	6/69	Plug–Protective, Flared Tube, Hose Assembly or MS33649 Boss
NAS819(2)**	6/69	Cap–Protective, Flanged Fitting
NAS820(2)**	6/69	Plug–Protective, Electrical Connector
NAS821(4)	3/58	Inspection Requirements For Titanium Alloy Sheet
NAS822	6/58	Quantity Indicators & Controls–Fuel, Personal Aircraft

Number	Current Issue	Title
NAS823	11/58	Cast Surface Comparison Standard (ANA Bul. 147–AF Only)
NAS824(10)	1/64	Inspection, Ultrasonic, Wrought Metal
NAS850(2)	11/68	General Packaging Standard
NAS851(3)**	6/69	General Packaging Standard Indexes
NAS852	11/68	Quality Assurance, General Guidance
NAS853	11/68	Field Force, Protection For
NAS900	5/54	Mount–Optical Tooling, Spherical 2.2500 Inside Bore
NAS901	5/54	Spherical Mounts–Telescopes–Collimators–Targets For Optical Tooling
NAS902	5/54	Targets–Optical Tooling, 2-1/4 & 1-1/2 Outside Diameters
NAS903	5/54	Adapters–Optical Tooling For 2-1/4 & 1-1/2 Dia. Targets
NAS904	4/52	Base–Optical Tooling Spherical Mount Support
NAS905(3)x	10/62	Terminology for Masters Definitions and Preferred Terms
NAS906	5/54	Targets–Optical Tooling, Solid 2-1/4 & 1-1/2 Outside Diameters
NAS907(20)**	6/69	High Speed Steel Twist Drills 1/16" thru 1/2"
NAS908(4)	4/55	Glossary of Optical Tooling Terms
NAS909(22)	1/69	Milling Machines–General Purpose, Knee And Bed Types, Horizontal And Vertical
NAS910(17)	4/68	Horizontal Boring, Drilling and Milling Machine
NAS911(17)	9/60	Milling Machine–Airframe Skin, Non-Ferrous
NAS912(48)	11/68	Spar Mill, Numerically Controlled
NAS913(30)	10/68	Milling Machine–Numerically Controlled, Profiling and Contouring
NAS914(30)	12/68	Milling Machine, Tracer Controlled Profiling & Contouring–Vertical
NAS915x	10/62	Hand Knob, Knurled
NAS916x	10/62	Hand Knob, Large
NAS917x	10/62	Swinging "C" Washer
NAS918x	10/62	Pivot Screw, For Swing "C" Washer
NAS919x	10/62	Non-Precision Bushings
NAS920x	10/62	Lag Screw (For Profiling)
NAS921x	10/62	Washer–Head Thumb Screw
NAS922x	10/62	Strap Clamp
NAS923x	10/62	"C" Washer
NAS924*	3/57	Quarter-Turn Screw
NAS925x	10/62	Insert–Drop Hammer Die
NAS926x	10/62	Swing Bolt
NAS927x	10/62	Jig Pin, "L" Type
NAS928x	10/62	Jig Pin, "T" Type
NAS929(9)	5/58	Performance Requirement–Work Holder Airframe Spar Milling Machine
NAS930(23)	10/59	Stretch Presses–Airframe, Sheet and Extrusion
NAS931(7)	7/57	Specification for Dual Turning Machines
NAS932	12/62	Drills–Threaded Hexagonal Shank, Sizes #40 thru Letter 'F'
NAS933(13)x	9/63	Fusion Welding Machine, Circumferential–Longitudinal
NAS934(9)x	9/63	Fusion Welding Machine, Circumferential
NAS935(9)x	9/63	Fusion Welding Machine, Longitudinal
NAS936(23)	6/59	Milling Machine–Precision Bed-Type Horizontal and Vertical Spindles
NAS937	6/59	Drills–Procurement Specification Stepped Double Margin Sizes 5/32" thru 1/2"
NAS938(18)	6/66	Machine Axis and Motion Nomenclature
NAS939(8)	10/59	Press-Hydraulic, Type I, Single Action Ram with Die Cushion
NAS940(4)	10/59	Press–Hydraulic, Type II, Internal Expansion without Ram
NAS941	12/59	Identification Marking of Titanium Sheet
NAS942(48)	7/64	Strain-Gages, Bonded Resistance
NAS943(17)	11/63	Interchangeable Perforated Tape Standards Variable Block Format For Positioning and Straight Cut Numerically Controlled Equipment
NAS944(6)	4/60	Symbols, Hydraulic Test Equipment Drafting
NAS945	1/61	Cap–Protective, NAS591 Coupling
NAS946	1/61	Cap–Protective, NAS593 Nut
NAS947	1/61	Cap–Protective, Beaded Tube Assembly
NAS948(21)	10/64	Printed Circuit Board Drilling Machine, Numerically Controlled
NAS949(8)	5/61	Tube Bender–Precision-Thin Wall Aircraft Type
NAS950	2/60	Gasket, Dummy Stowage Receptacle, Electrical
NAS951(9)	8/62	Machine–Electrical Wire Processing, Numerically Controlled
NAS952(8)	12/63	Filament Winding Machine, Horizontal, Numerically Controlled
NAS953(12)	4/65	Numerical Control System
NAS954(22)	4/68	Numerically Controlled Horizontal Boring, Drilling and Milling Machine
NAS955(17)	11/63	Interchangeable Perforated Tape Standards Variable Block Format, Contouring and Contouring/Positioning Numerically Controlled Equipment
NAS956(17)	4/63	Tube Bender–Numerical Control, Precision Aircraft Type for Thin Wall, Ferrous and Non-Ferrous Alloys
NAS957(11)	4/63	Numerically Controlled Drafting Line Plotter
NAS958(10)	4/63	Shear Forming Equipment–Horizontal and Vertical
NAS959(13)	11/64	Manual Gas Metal Arc Welding Equipment
NAS960(19)	5/63	Drilling Machines, Numerically Controlled
NAS961(15)	12/64	Manual Gas Tungsten Arc (TIG) Fusion Welding Equipment
NAS962(16)	4/65	Machine Gas Tungsten Arc Welding Equipment
NAS963(31)	8/63	Numerically Controlled Horizontal and Vertical Jig Boring Machines
NAS964(13)	10/64	Machine Gas Metal Arc Welding Equipment
NAS965(6)	11/63	Drills–Special Application–Threaded Hexagonal Shank–Sizes No. 40 thru Letter "F"
NAS966(22)	3/68	Lathe, Precision–Numerically Controlled
NAS967(17)	10/64	Resistance Spot Welding Equipment–3 Phase
NAS968(6)	11/64	Interchangeable Digital Magnetic Tape Standards, Variable Block, Word Address Format, Contouring and Contouring/Positioning Equipment
NAS969(2)	11/64	Interchangeable Digital Magnetic Tape Cartridge–For Numerically Controlled Machines
NAS970(8)	11/66	Tool Holders–Basic for Numerically Controlled Machine Tools
NAS971(18)	12/64	Measuring Inspection Machine–Precision–Numerically Controlled
NAS972(9)	5/68	Glossary of Manufacturing Equipment Terms
NAS973(30)	11/64	Tool and Cutter Grinding Machine
NAS974(13)	5/65	Electrical Harness Assembling Machine, Numerically Controlled
NAS975(15)	6/66	Component Lead Cutting and Forming Equipment, Electronic Printed Circuit Board
NAS976(19)	8/65	Electron Beam Welding Machine
NAS977(22)	9/65	Milling Machine, Contour, Numerical Control, Multiaxis, Traveling Gantry/Table, Combination High and Low Speed
NAS978(19)	11/65	Numerically Controlled Combination Drilling, Boring, Milling and Tapping Machines
NAS979(59)	1/69	Uniform Cutting Tests–NAS Series Metal Cutting Equipment Specifications
NAS980(12)	2/66	Printed Circuit Board Soldering Machine–Automatically Controlled
NAS981(16)	3/67	Gas Tungsten Arc (GTA)-Spot Welding Equipment
NAS982(28)	11/67	Hot Forming, Sizing, and Straightening Press Specification
NAS983(16)	9/68	Standard Format for NAS 900 Series Equipment Specification
NAS1000	3/62	Tube End–Standard Dimensions For Inside Double Flare
NAS1001	10/56	Tube End–Standard Dimensions For Outside Double Flare

Number	Current Issue	Title
NAS1002	3/62	Stud–Electrical Terminal Block (ANA Bul. 147–AF Only)
NAS1003–1020(2)	3/67	Bolt–Machine, Hexagon Head, Non Magnetic, & Heat Resistant (ANA Bul. 147–AF Only)
NAS1021(2)**	3/69	Nut, Self-Locking, Hexagon–Regular Height
NAS1022(2)		Nut, Self-Locking, Hexagon–Low Height, Low Strength
NAS1023(2)	12/68	Nut, Self-Locking, Plate–Two Lug, Regular Height, Non-Csk
NAS1024(2)	12/68	Nut, Self-Locking, Plate–Two Lug, Regular Height, Csk
NAS1025(2)	12/68	Nut, Self-Locking, Plate–One Lug, Regular Height, Non-Csk
NAS1026(2)	12/68	Nut, Self-Locking, Plate–One Lug, Regular Height, Csk
NAS1027(2)	12/68	Nut, Self-Locking, Plate Corner, Regular Height, Non-Csk
NAS1028(2)	12/68	Nut, Self-Locking, Plate–Corner, Regular Height, Csk
NAS1029(2)	12/68	Nut, Self-Locking, Plate–Two Lug, Regular Height, Bevel Seat
NAS1030(2)	12/68	Nut, Self-Locking, Plate–One Lug, Regular Height, Non-Csk, Bevel Seat
NAS1031(2)	12/68	Nut, Self-Locking, Plate–Two Lug, Regular Height, Floating
NAS1032(2)	12/68	Nut, Self-Locking, Plate–One Lug, Regular Height, Non-Csk, Floating
NAS1033(2)	12/68	Nut, Self-Locking, Plate–Right Angle, Regular Height, C-Bore, Floating
NAS1034–1038(2)	12/68	Nut Assembly, Self-Locking, Gang Channel–Regular Height, Non-Csk
NAS1039–1041(2)	12/68	Nut Assembly, Self-Locking, Gang Channel–Regular Height, Csk
NAS1042	4/60	Shackle–Anchor Forged Carbon Steel Round Pin (For Ground Support Equipment) (ANA Bul. 147–AF Only)
NAS1043	4/60	Shackle–Chain Forged Carbon Steel Screw Pin (For Ground Support Equipment) (ANA Bul. 147–AF Only)
NAS1044	4/60	Shackle–Chain, Forged Carbon Steel, Round Pin (For Ground Support Equipment) (ANA Bul. 147–AF Only)
NAS1045	11/66	Thimble–Heavy, Wrought Steel, Wire Cable (For Ground Support Equipment) (ANA Bul. 147–AF Only)
NAS1046	7/56	Clip–Wire Rope, Forged Carbon Steel (For Ground Support Equipment) (ANA Bul. 147–AF Only)
NAS1047	7/59	Turnbuckle Assembly, Forged Carbon Steel, Jaw and Jaw (For Ground Support Equipment) (ANA Bul. 147–AF Only)
NAS1048	7/56	Chain–Welded, Alloy Steel (For Ground Support Equipment) (ANA Bul. 147–AF Only)
NAS1049	2/63	Sling Link, Welded, Alloy Steel, Pear Shaped (For Ground Support Equipment) (ANA Bul. 147–AF Only)
NAS1050	2/63	Grab Hook, Forged Alloy Steel, Direct Eye (For Ground Support Equipment) (ANA Bul. 147–AF Only)
NAS1051	2/63	Safety Hook, Forged Carbon Steel, Snap, Direct Eye (For Ground Support Equipment) (ANA Bul. 147–AF Only)
NAS1052	2/63	S Hook, Alloy Steel (For Ground Support Equipment) (ANA Bul. 147–AF Only)
NAS1053	2/63	Eyebolt Assembly, Forged Carbon Steel, Shoulder Nut (For Ground Support Equipment) (ANA Bul. 147–AF Only)
NAS1054(2)	9/67	Rivet, Protruding Head Hi-Shear, Sup. NAS178
NAS1055(2)	9/67	Rivet, Hi-Shear 100° Flush Close Tolerance Head, Sup. NAS177
NAS1056	1/69	Spacer, Sleeve-Rivet, High Temperature, Non Magnetic
NAS1057	1/69	Spacer, Sleeve-Screw And Bolt, High Temperature, Non Magnetic
NAS1058(8)	2/63	Fasteners, Blind
NAS1059(2)	2/63	Fastener–Blind, High Strength

Number	Current Issue	Title
NAS1060	1/57	Bus–High Temperature, Terminal Block (ANA Bul. 147–AF Only)
NAS1061*	10/67	Washer–High Temperature–Lock, Spring, See MS35338
NAS1062	3/62	Strip–High Temperature Terminal Block Nut Insulating (ANA Bul. 147–AF Only)
NAS1063	12/61	Strip–High Temperature Terminal Block, Base Insulating (ANA Bul. 147–AF Only)
NAS1064(2)*	10/67	Terminal Block–Electrical, High Temperature, See MS27212
NAS1065	12/61	Stud–High Temperature Electrical Terminal Block (ANA Bul. 147–AF Only)
NAS1066*	10/67	Terminal Block Assembly–Electrical High Temperature, See MS27212
NAS1067(2)	12/68	Nut, Self-Locking, Plate–One Lug, Regular Height, Side-by-Side
NAS1068(2)	12/68	Nut, Self-Locking, Plate–Two Lug, Low Height, C'Bore, Floating, Reduced Rivet Spacing
NAS1069(24)	6/59	Tension Fatigue Test Procedure for Aeronautical Fasteners
NAS1070	4/64	Washer–Plain, Oversize, for Aluminum Terminals (ANA Bul. 147)
NAS1071	12/61	Switch–Toggle, Double Pole, Three Position (ANA Bul. 147–AF Only)
NAS1072	10/61	Guard, Channel, Toggle Switch (ANA Bul. 147–AF Only)
NAS1073	10/61	Guard, Two-Position Toggle-Switch Lock (ANA Bul. 147–AF Only)
NAS1074*	10/64	Turnbuckle Assemblies, Clip Locking, Assembly and Safetying of, See MS33736
NAS1075	6/58	Bearing–Rod End, 3/8-24 External Thread, Self-Aligning, Anti-Friction (ANA Bul. 147)
NAS1076(4)	10/61	Lubricant, High Gear Loading, All Weather (ANA Bul. 147–AF Only)
NAS1077	1/61	Fitting End, Hose, Standard Dimensions for Flareless Tube
NAS1078(2)	2/64	Recess and Recess Gage Dimensions–Torq-Set
NAS1079	2/64	Driver–Control Dimensions, Torque–Set
NAS1080(6)**	6/69	Collar, Swage Locking (For Pull Type or Stump Type Lockbolts)
NAS1081(2)	2/63	Setscrew, Self-Locking 250° F (ANA Bul. 147–AF Only)
NAS1082	9/59	Insignia–National (ANA Bul. 147–AF Only)
NAS1083–1088(2)	8/65	Bolt, 100° Flat Head, Close Tolerance, Short Thread Titanium Alloy (ANA Bul. 147–AF Only)
NAS1089	10/63	Streamer, Warning (ANA Bul. 147–AF Only)
NAS1090	6/64	Hook, Chain, S (ANA Bul. 147–AF Only)
NAS1091	10/63	Streamer Assembly, Warning (ANA Bul. 147–AF Only)
NAS1092*	2/59	Terminal Board Assembly–Electrical, See MS25123
NAS1093(2)*	2/59	Terminal Board–Electrical, See MS25123
NAS1094*	2/59	Strip–Terminal Board Base Insulating, See MS25123
NAS1095(11)	11/60	Flareless Fitting Connections Fabrication and Assembly
NAS1096	1/67	Screw, Hex Head, Recessed, Full Thread (ANA Bul. 147–AF Only)
NAS1097	7/68	Rivet, Solid–100° Flush Shear Head, Aluminum Alloy
NAS1098*	12/62	Nut, Universal Tube Fitting, See NAS1410
NAS1099	2/64	Washer–Bevel 9-1/2° (ANA Bul. 147–AF Only)
NAS1100(2)	5/68	Screw, Machine–Pan Head, Full Thread, Torq-Set (ANA Bul. 147-AF Only)
NAS1101(2)	5/68	Screw, Machine–Flat Fillister Head, Full Thread, Torq-Set
NAS1102(2)	5/68	Screw, Machine–Flat 100 Deg Head, Full Thread, Torq-Set (ANA Bul. 147-AF Only)
NAS1103–1120(3)	9/68	Bolt, Shear–Hexagon Head Modified, Short Thread
NAS1111(2)X	1/61	Screw, 82° Flat Head, Torq-Set, Full Thread, See NAS1393

Number	Current Issue	Title
NAS1121–1128(3)	5/68	Screw, Machine–Flat Fillister Head, Close Tol, Short Thd, Torq-Set
NAS1131–1138(3)	5/68	Screw, Machine–Pan Head, Close Tol, Short Thd, Torq-Set
NAS1141–1148(3)	5/68	Screw, Machine–Pan Head Modified, Close Tol, Short Thd, Torq-Set
NAS1151–1158(4)	5/68	Screw, Machine–Flat 100 Deg Head, Close Tol, Short Thd, Torq-Set
NAS1161–1168(3)	7/68	Screw, Self-Locking–Flat 100 Deg Head, Shear, Torq-Set
NAS1169(2)	10/62	Washer–Dimpled–100° (ANA Bul. 147–AF Only)
NAS1170	10/62	Fastener, Chain (ANA Bul. 147–AF Only)
NAS1171–1178(3)	5/68	Screw, Self-Locking–Pan Head, Shear, Torq-Set
NAS1179(15)	3/61	Oxygen Mask Assembly, Passenger
NAS1181–1188(3)	7/68	Screw, Self-Locking–Flat Fillister Head Close Tol, Torq-Set
NAS1189(2)	5/68	Screw, Self-Locking–Flat 100 Deg Head, Full Thread
NAS1190(2)	7/68	Screw, Self-Locking–Pan Head Full Thread
NAS1191(2)	5/68	Screw, Self-Locking–Flat Fillister Head, Full Thread
NAS1192(5)	4/59	Performance Specification for Hard Anodic Coatings on Aluminum Alloys
NAS1193(2)	10/64	Locking Device, Positive Index (ANA Bul. 147)
NAS1194(3)x	3/60	Aluminum Alloy, 7075; Bare and Clad Plate, Extruded Bar, Rod, & Shapes: Rolled Bar and Shapes: Controlled Stretched
NAS1195(2)	4/63	Shim–Miniature Anchor Nut (ANA Bul. 147–AF Only)
NAS1196	2/60	Receptacle–Dummy Storage, Electrical Connector (ANA Bul. 147–AF Only)
NAS1197	7/60	Washer–5052 Aluminum Alloy (ANA Bul. 147)
NAS1198	9/67	Rivet–Solid Universal Head, A286 Corroston Resistant Steel
NAS1199	9/67	Rivet–Solid–100° Flush Head, A286 Corrosion Resistant Steel
NAS1200	9/67	Rivet–Solid–100° Flush Shear Head–A286 Corrosion Resistant Steel
NAS1201	3/62	Chain–Bead, Components and Assembly (ANA Bul. 147–AF Only)
NAS1202–1210(3)	3/62	Bolt, 100° Close Tolerance Head and Shank, 160,000 PSI Short Thread (ANA Bul. 147–AF Only)
NAS1211(2)	6/61	Fitting, Tie Down Strap End (ANA Bul. 147–AF Only)
NAS1212(2)	6/61	Strap Assembly–Tie Down (ANA Bul. 147–AF Only)
NAS1213(2)	6/61	Strap Assembly–Tie Down Buckle (ANA Bul. 147–AF Only)
NAS1214(2)*	2/58	Fitting End–Standard Dimensions For Flareless Tube Connections and Gasket Seal, See MS33514
NAS1215(2)*	2/58	Fitting End–Standard Dimensions for Bulkhead Flareless Tube Connections, See MS33515
NAS1216(2)	4/65	Screw, Pan Head, "Hi Torque" Recess, Full Thread (ANA Bul. 147–AF Only)
NAS1217(3)	4/65	Screw, Pan Head, "Hi Torque" Recess, Short Thread (ANA Bul. 147–AF Only)
NAS1218(3)	4/65	Screw, Pan Head, "Hi Torque" Recess, Long Thread (ANA Bul. 147–AF Only)
NAS1219(2)	4/65	Screw, 100° Csk Head, "Hi Torque" Recess, Full Thread (ANA Bul. 147–AF Only)
NAS1220(3)	4/65	Screw, 100° Csk Head, "Hi Torque" Recess, Short Thread (ANA Bul. 147–AF Only)
NAS1221(3)	4/65	Screw, 100° Csk Head, "Hi Torque" Recess, Long Thread (ANA Bul. 147–AF Only)
NAS1222*	4/62	Inserts, Screw Threads Helical Coil, Screw Locking, Cres, Coarse and Fine Threads, See MS21209
NAS1223–1235(4)	12/68	Bolt, Shear–Hexagon Head, Self Locking
NAS1236	3/62	Bolt–Universal, Tube End–Flareless (ANA Bul. 147–AF Only)
NAS1237	1/59	Elbow–Universal 90° Flareless (ANA Bul. 147–AF Only)
NAS1238	1/59	Elbow–Universal 75°, Flareless (ANA Bul. 147–AF Only)
NAS1239	1/59	Elbow–Universal 45°, Flareless (ANA Bul. 147–AF Only)
NAS1240	1/59	Tee–Universal, Flareless (ANA Bul. 147–AF Only)
NAS1241	4/63	Rivet–Solid 100° Flush Head, Aluminum Alloy, MS20426 Repair (ANA Bul. 147–AF Only)
NAS1242	4/63	Rivet–Solid, Universal Head, Aluminum Alloy, MS20470 Repair (ANA Bul. 147–AF Only)
NAS1243–1250	11/66	Bolt, .0156 Oversize Shank, 100° Head "Hi Torque", Close Tolerance, 160,000 PSI, Short Thread (ANA Bul. 147–AF Only)
NAS1251(2)	12/68	Bolt, Ring–Swivel
NAS1252	11/64	Washer–7075 Aluminum Alloy Flat (ANA Bul. 147–AF Only)
NAS1253–1260	11/66	Bolt, .0312 Oversize Shank, 100° Head "Hi-Torque", Close Tolerance, 160,000 PSI, Short Thread (ANA Bul. 147–AF Only)
NAS1261–1265(2)	4/63	Bolt, Hex Head, Close Tolerance, Short Thread, Titanium Alloy (ANA Bul. 147–AF Only)
NAS1266–1270(2)	1/63	Bolt, Hex Head, Close Tolerance, Titanium Alloy (ANA Bul. 147–AF Only)
NAS1271–1280(2)	1/63	Bolt, Twelve Point, External Wrenching, Titanium Alloy (ANA Bul. 147–AF Only)
NAS1281	6/66	Hook Snap, Spring (ANA Bul. 147–AF Only)
NAS1282	6/66	Hook Snap, Bolt (ANA Bul. 147–AF Only)
NAS1283(4)	12/68	Fasteners, Male Threaded, Self-Locking
NAS1284	10/64	Buckle, Web Strap–Light Duty (ANA Bul. 147–AF Only)
NAS1285	1/63	Safety Hook (ANA Bul. 147–AF Only)
NAS1286	1/63	Connecting Link (ANA Bul. 147–AF Only)
NAS1287(2)	12/68	Nut, Extended Washer, Hexagon–Self-Locking, Low Height, C'Bore
NAS1288	6/66	Nut, Hex and Washer, Self-Aligning, Self-Locking–Shear Pin (ANA Bul. 147–AF Only)
NAS1289(8)	10/61	Threaded Shear Fastener, Flush and Protruding Type for Applications Up to 900° F Maximum–5% CR Ultra High Strength Alloy Steel (H-11)–Procurement Specification (ANA Bul. 147–AF Only)
NAS1290(8)	10/61	Threaded Shear Fastener, Flush and Protruding Type for Applications up to 1400°F. Corrosion and Heat Resistant Rene 41 Nickel Base Alloy–Procurement Specification (ANA Bul. 147–AF Only)
NAS1291(3)	12/68	Nut, Self-Locking, Hexagon–Low Height, Light Weight
NAS1292–1296	10/61	Pin–Threaded Shear 100° CSK Head (ANA Bul. 147–AF Only)
NAS1297(2)	7/68	Bolt, Shoulder–Hexagon Head
NAS1298(2)	7/59	Screw–Brazier Head, Shoulder (ANA Bul. 147–AF Only)
NAS1299(2)	7/59	Screw–100° Flat Head, Shoulder (ANA Bul. 147–AF Only)
NAS1300	7/59	Thumbscrew (ANA Bul. 147)
NAS1301	7/59	Screw, Pan Head, Assembled Washers (ANA Bul. 147–AF Only)

Number	Current Issue	Title
NAS1302	8/59	Identification–Equipment and Component Wire Termination
NAS1303–1320(3)	7/68	Bolt, Shear–Hexagon Head
NAS1321	8/66	Rivet, Slug, Sleeveless (ANA Bul. 147–AF Only)
NAS1322–1326	10/61	Pin–Threaded Shear Protruding Head (ANA Bul. 147–AF Only)
NAS1329(3)**	5/69	Nut, Blind Rivet–Flathead, Internal Thread, Non-Locking (Free-Running) and Self-Locking (Prevailing Torque)
NAS1330(3)**	5/69	Nut, Blind Rivet–Countersunk Head, Internal THD, Non-Locking (Free Running) or Self-Locking (Prevailing Torque)
NAS1331(3)	5/60	Cart, Utility Equipment (ANA Bul. 147–AF Only)
NAS1332(8)	11/66	Pin, Quick Release, Positive Locking Single & Double Acting (ANA Bul. 147–AF Only)
NAS1333–1346(2)	11/66	Pin, Quick Release, Positive Locking, Single Acting (ANA Bul. 147–AF Only)
NAS1347(2)	7/59	Identification of Fasteners
NAS1348	9/59	Fasteners–Recommended Tensile Stress Areas for External Threaded Alloy Steel
NAS1349	10/64	Caster, Light Duty, Rigid and Swivel (ANA Bul. 147–AF Only)
NAS1350(8)	7/59	Equipment Identification Nameplate & Marking
NAS1351(3)	11/62	Screw, Cap, Socket Head–Undrilled and Drilled, Plain and Self-Locking, Alloy Steel and Corrosion–Resisting Steel, UNF-3A Thread (ANA Bul. 147–AF Only)
NAS1352(3)	11/62	Screw, Cap, Socket Head–Undrilled and Drilled, Plain and Self-Locking, Alloy Steel and Corrosion-Resisting Steel, UNC-3A and UNC-2A Thread (ANA Bul. 147–AF Only)
NAS1353–1366(2)	11/66	Pin, Quick Release, Positive Locking, Double Acting (ANA Bul. 147–AF Only)
NAS1367(2)	11/67	Fittings–Coated Lubricant & Sealant
NAS1368(2)	9/61	Grommet–Plastic Flip Type (ANA Bul. 147)
NAS1369(12)	5/60	Duct, Air, Flexible & Semi-Rigid (ANA Bul. 147–AF Only)
NAS1370–1379(3)	8/61	Duct–Air, Flexible, Helix Wire Supported (ANA Bul. 147)
NAS1380	12/61	Stud, Welding, Internally Threaded (ANA Bul. 147–AF Only)
NAS1381(2)	1/63	Stud, Welding Externally Threaded (ANA Bul. 147–AF Only)
NAS1382	11/60	Fitting, Tie Down Strap End (Light Duty) (ANA Bul. 147–AF Only)
NAS1383	12/61	Strap Assembly, Two Part Adjustable (ANA Bul. 147–AF Only)
NAS1384	12/61	Strap Assembly, One Part (ANA Bul. 147–AF Only)
NAS1385	3/67	Strap Assembly, Loop Type
NAS1386*	12/65	Eyelet, Metallic, Electrical, Funnel, See NAS1453
NAS1387	8/64	Splice, Electric, Permanent, Crimp Style, Uninsulated (Class 2) (ANA Bul 147–AF Only)
NAS1388	8/64	Splice, Electric, Permanent, Crimp Style, Pre-Insulated, (Class 2) Wire Sizes 12 to 26 (ANA Bul. 147–AF Only)
NAS1389	8/64	Splice, Electric, Permanent, Crimp Style, Pre-Insulated (Class 2) Wire Sizes 8 to 1/0 (ANA Bul. 147–AF Only)
NAS1390	6/61	Splice, Electric, Permanent, Crimp Style, Moisture Resistant (Class 2) (ANA Bul. 147–AF Only)
NAS1391(9)	1/61	Wire, Electrical, Insulated Copper Solid Conductor Hook-Up and General Purpose (For 105° C Service)
NAS1392(5)X	11/67	Hi-Torque, Design Details for (Recesses, Drivers and Gages), See MS33750
NAS1393(2)	5/68	Screw, Machine–Flat 82 Deg Head, Full Thread, Torq-Set
NAS1394	2/68	Insert–Threaded Metal, Light Weight, Self-Locking, Non Self-Locking
NAS1395	2/68	Insert–Threaded Metal, Heavy Duty, Self-Locking, Non Self-Locking
NAS1396	5/61	Chain–Welded Twist Link, Short Link Pattern (ANA Bul. 147–AF Only)
NAS1397	6/61	Clamp–Loop, Nylon (ANA Bul. 147–AF Only)
NAS1398(2)	10/62	Rivet–Blind, Protruding Head, Locked Spindle (ANA Bul. 147)
NAS1399(2)	10/62	Rivet–Blind, 100° Flush Head, Locked Spindle (ANA Bul. 147)
NAS1400(15)	10/62	Rivet–Blind, Self-Plugging, Mechanically Locked Spindle (ANA Bul. 147)
NAS1401	6/62	Washer, Radius (ANA Bul. 147–AF Only)
NAS1402–1406(2)	1/62	Screw, Machine, Aircraft, Pan Head, Philips Recess, 160,000 PSI Tensile (ANA Bul. 147–AF Only)
NAS1407(2)	8/62	Pin–Spring, Coiled (ANA Bul. 147–AF Only)
NAS1408(2)	12/68	Nut, Self-Locking, Hexagon–Regular Height, Coarse Thread, Non-Airborne use
NAS1409(2)	12/68	Nut, Self-Locking, Hexagon–Regular Height, Fine Thread, Non-Airborne use
NAS1410(2)	11/67	Nut, Universal Port Fitting, Sup. NAS1098
NAS1411	3/66	Band, Identification and Marking (ANA Bul. 147–AF Only)
NAS1412(9)	11/64	Gasket, Sealing, Custom Design
NAS1413(13)	12/63	Pins and Collars, Swage Locking (ANA Bul. 147)
NAS1414–1422(2)	10/63	Pin, Swage Locking, Steel 100° Shear Head, Stump Type, Close Tol. (ANA Bul. 147)
NAS1423(2)	4/65	Nut, Plain, Hexagon, Drilled Jam, Thin (ANA Bul. 147–AF Only)
NAS1424–1432(2)	10/63	Pin, Swage Locking, Steel Protruding Head, Shear, Stump Type, Close Tol. (ANA Bul. 147)
NAS1429X	2/61	End–Control Rod (Steel Only), See NAS1483
NAS1431X	11/60	Rod, Continuous Thread, See NAS1454
NAS1433	5/64	Radiographic Standards for Aluminum Alloy Castings (ANA Bul. 147)
NAS1434	5/64	Radiographic Standards for Magnesium Alloy Castings (ANA Bul. 147–AF Only)
NAS1435	8/65	Terminal–Strap, Cable, Eye End and Fork End (ANA Bul. 147–AF Only)
NAS1436–1442(2)	10/63	Pin, Swage Locking, Steel 100° Shear Head, Pull Type, Close Tol. (ANA Bul. 147)
NAS1437X	2/61	Resistor, Fixed, Wirewound, Precision High Temperature, See NAS1484
NAS1439X	2/61	Resistor, Fixed, Wirewound, Precision High Temperature, See NAS1485
NAS1441X	3/61	Locknut–Bearing Retaining, See NAS1493
NAS1443	8/62	Lockwasher–Bearing Retaining (ANA Bul. 147)
NAS1444(19)	11/64	Resistance Seam and Roll Spot Welding Equipment–3 Phase
NAS1446–1452(2)	10/63	Pin, Swage Locking, Steel Protruding Head, Shear, Pull Type, Close Tol. (ANA Bul. 147)
NAS1453(2)	12/65	Eyelet, Metallic, Electrical, Printed Circuit Board, Funnel Type (ANA Bul. 147–AF Only)
NAS1454	11/60	Rod, Continuous Thread (ANA Bul. 147–AF Only)
NAS1455	4/61	Chain, Weldless Flat Type Link (ANA Bul. 147–AF Only)

Number	Current Issue	Title
NAS1456–1462(2)	10/63	Pin, Swage Locking, Steel 100° Head (AN509), Tension, Pull Type, Close Tol. (ANA Bul. 147)
NAS1463	7/60	Clip–Spring Tension (ANA Bul. 147–AF Only)
NAS1464(3)	1/69	Clip, Spring Tension
NAS1465–1472(2)	10/63	Pin, Swage Locking, Steel Protruding Head, Tension, Pull Type, Close Tol. (ANA Bul. 147)
NAS1473(3)	12/68	Nut, Self-Locking, Plate–Two Lug, Cap, Floating
NAS1474(3)	12/68	Nut, Self-Locking, Plate–Two Lug, Cap, Floating, Reduced Rivet Spacing
NAS1475–1482(2)	10/63	Pin, Swage Locking, Steel 100° Head (MS20426), Tension, Pull Type, Close Tol. (ANA Bul. 147)
NAS1483	7/64	End–Control Rod (Steel Only)
NAS1484	3/62	Resistor, Fixed, Wirewound, Precision High Temperature (ANA Bul. 147–AF Only)
NAS1485	3/62	Resistor, Fixed, Wirewound, Precision High Temperature (ANA Bul. 147–AF Only)
NAS1486–1492(2)	10/63	Pin, Swage Locking, Steel 100° Head (AN509), Tension, Stump Type, Close Tol. (ANA Bul. 147)
NAS1493	2/64	Locknut–Bearing Retaining (ANA Bul. 147)
NAS1494	4/61	Terminal, Lug–Single Hole Tongue, Single Barrel, Pressure Grip Type (ANA Bul. 147–AF Only)
NAS1495(4)	2/68	Recommended Qualification Test Procedure For Compliance With NAS Specifications and Standard Drawings
NAS1496–1502(2)	10/63	Pin, Swage Locking, Steel Protruding Head, Tension, Stump Type, Close Tol. (ANA Bul. 147)
NAS1503–1510(3)	6/64	Bolt, 100° Close Tolerance Head and Shank, 160,000 PSI Short Thread (ANA Bul. 147–AF Only)
NAS1511(8)	9/61	Heat Treatment and Processing Requirements for 5% CR Ultra High Strength Alloy Steel (H-11) Fasteners Only)
NAS1512–1513(2)	12/68	Nut Assembly, Self-Locking, Gang Channel–Low Height, C'Bore, Miniature
NAS1514(6)	6/63	Radiographic Standard for Classification of Fusion Weld Discontinuities
NAS1515	8/61	Washers–Plastic and Synthetic Rubber (ANA Bul. 147–AF Only)
NAS1516–1522(2)	10/63	Pin, Swage Locking, Aluminum Alloy 100° Head (AN509), Tension Pull Type, Close Tol. (ANA Bul. 147)
NAS1524(17)	7/68	Standardization Savings, Identification & Calculation
NAS1525–1532(2)	10/63	Pin, Swage Locking, Aluminum Alloy Protruding Head, Tension, Pull Type, Close Tol. (ANA Bul. 147)
NAS1535–1542(2)	10/63	Pin, Swage Locking, Aluminum Alloy 100° Head (MS20426), Tension, Pull Type, Close Tol. (ANA Bul. 147)
NAS1543(3)	8/61	Valve, Shutoff, Electric Motor Operated, Fuel and Oil System, General Specification for
NAS1544	8/61	Plate, Wall, Electrical, Weatherproof (ANA Bul. 147–AF Only)
NAS1546–1552(2)	10/63	Pin, Swage Locking, Aluminum Alloy 100° Head (AN509), Tension, Stump Type, Close Tol. (ANA Bul. 147)
NAS1553(2)	2/68	Knob, Hand, Ground Support Equipment

Number	Current Issue	Title
NAS1556–1562(2)	10/63	Pin, Swage Locking, Aluminum Alloy Protruding Head, Tension, Stump Type, Close Tol. (ANA Bul. 147)
NAS1563	10/63	Gages, Pins & Collar Inspection (ANA Bul. 147)
NAS1564(2)	11/67	Reducer, Flared Tube
NAS1565	6/61	Reducer, Flareless Tube (ANA Bul. 147–AF Only)
NAS1566	6/61	Reducer Assembly, Flareless Tube (ANA Bul. 147–AF Only)
NAS1567	4/63	Junction Box Assembly–Cast AL ALY, Weatherproof (ANA Bul. 147–AF Only)
NAS1568X	4/63	Junction Box–Sheet Metal, Non-Weatherproof
NAS1569*	4/63	Cover–Junction Box, See NAS1567
NAS1570	4/63	Gasket–Junction Box (ANA Bul. 147–AF Only)
NAS1571(3)	4/63	Switch Fitting Box, Single Gang, Weatherproof (ANA Bul. 147–AF Only)
NAS1572(3)	4/63	Switch Fitting Box, Two Gang, Weatherproof, Single Device (ANA Bul. 147–AF Only)
NAS1573(3)	4/63	Switch Fitting Box, Three Gang, Weatherproof, Single Device (ANA Bul. 147–AF Only)
NAS1574(2)	12/61	Switch Fitting Box, One, Two, or Three Gang, Non-Weatherproof (ANA Bul. 147–AF Only)
NAS1575(2)	4/63	Cover, Switch Fitting Box, Weatherproof (ANA Bul. 147–AF Only)
NAS1576	4/63	Cover, Switch Fitting Box, Non-Weatherproof (ANA Bul. 147–AF Only)
NAS1577	4/63	Gasket–Switch Fitting Box (ANA Bul. 147–AF Only)
NAS1578(3)	5/64	Bolt, Shear, Pan Head, 1200° F (ANA Bul. 147–AF Only)
NAS1579(2)	5/64	Screw, Pan Head, Full Threaded, 1200° F (ANA Bul. 147–AF Only)
NAS1580(3)	5/64	Bolt, Tension, Flush Head, 1200° F (ANA Bul. 147–AF Only)
NAS1581(3)	6/64	Bolt, Shear, Reduced 100° Flush Head, 1200° F (ANA Bul. 147–AF Only)
NAS1582(2)	5/64	Screw, Flush Head, Full Threaded, 1200° F (ANA Bul. 147–AF Only)
NAS1583	2/63	Pin, 100° Csk. Head, Hi-Shear Rivet, Close Tolerance, 1200° F (ANA Bul. 147–AF Only)
NAS1584	2/63	Pin, Flat Head, Hi-Shear Rivet, Close Tolerance, 1200° F (ANA Bul. 147–AF Only)
NAS1585	2/63	Collar–Hi-Shear Rivet, 1200° F (ANA Bul. 147–AF Only)
NAS1586(3)	2/63	Bolt–Tension, 12 Pt Ext Wrenching, 1200° F (ANA Bul. 147–AF Only)
NAS1587	2/63	Washer–Plain and Csk, 1200° F (ANA Bul. 147–AF Only)
NAS1588(2)	8/66	Bolt, Shear, Hex Head, 1200°F (ANA Bul. 147–AF Only)
NAS1589(13)	2/63	Fasteners–Threaded, Corrosion & Heat Resistant, 1200° F, Procurement Specification for (ANA Bul. 147–AF Only)
NAS1590(8)	2/63	Pin, Hi-Shear Rivet, 1200° F, Procurement Specification for (ANA Bul. 147–AF Only)
NAS1593(3)	12/62	Packing, Preformed–Mil-R-25897 Rubber, 75 Shore, "O" Ring (ANA Bul. 147–AF Only)
NAS1594(3)	12/62	Packing, Preformed–Mil-R-25897 Rubber, 90 Shore, "O" Ring (ANA Bul. 147–AF Only)
NAS1595	12/62	Packing, Preformed–Straight Thread Tube Fitting Boss, MIL-R-25897 Rubber, 75 Shore, "O" Ring (ANA Bul. 147–AF Only)
NAS1596	12/62	Packing, Preformed–Straight Thread Tube Fitting Boss, Mil-R-25897 Rubber, 90 Shore, "O" Ring (ANA Bul. 147–AF Only)
NAS1597(5)	4/63	Steel, Corrosion and Heat Resistant, Bar, Rod, and Forging

Number	Current Issue	Title
NAS1598	12/62	Washer–Sealing (ANA Bul. 147)
NAS1599(40)	11/66	Connectors, General Purpose, Electrical, Miniature Circular, Environment Resisting, 200° C Maximum Temperature (ANA Bul. 147–AF Only)
NAS1600(16)	12/65	Contacts, Crimp Type, For Electrical Connectors (ANA Bul. 147–AF Only)
NAS1603–1610(3)	8/66	Bolt, .0312 Oversize Shank, 100° Head, Close Tolerance, 160,000 PSI (ANA Bul. 147–AF Only)
NAS1611	5/61	Packing, Preformed O-Ring Phosphate Ester Resistant (-65° F to +160° F) (ANA Bul. 147–AF Only)
NAS1612	5/61	Packing, Preformed, Straight Thread Tube Fitting Boss, Phosphate Ester Resistant (-65° F to +160° F) (ANA Bul. 147–AF Only)
NAS1613(10)	5/61	Packing, O-Ring, Phosphate Ester Resistant (ANA Bul. 147–AF Only)
NAS1614(6)	5/64	Coating–Protective, Temporary, Water Reducible, Chemical Solution Removable (ANA Bul. 147–AF Only)
NAS1615(9)	5/64	Primer Coating Water Reducible, Corrosion Inhibiting (ANA Bul. 147–AF Only)
NAS1616(12)	9/60	Topcoat–Water Reducible (ANA Bul. 147–AF Only)
NAS1617(6)	9/60	Coating–High Temperature Resistant, Water Reducible (ANA Bul. 147–AF Only)
NAS1618(8)	5/64	Shop Soil Remover–Water Base (ANA Bul. 147–AF Only)
NAS1619(2)	5/64	Cleaner, Metal, Soak Type Water Based (ANA Bul. 147–AF Only)
NAS1620–1628(3)	5/68	Screw, Machine–Flat 100 Deg Head, Short Thread, Torq-Set
NAS1629(12)	5/61	O-Ring Elongation Modulus Test
NAS1630–1634(2)	5/68	Screw, Machine–Pan Head, Short Thread, Torq-Set
NAS1635(2)	4/63	Screw, Pan Head, Cross Recessed, Full Thread (ANA Bul. 147–AF Only)
NAS1636	6/63	Washer, Key, Dual Tab (ANA Bul. 147–AF Only)
NAS1637(2)	9/63	Handle, Latching Internal Release Components and Assembly (ANA Bul. 147–AF Only)
NAS1638(2)	1/64	Cleanliness Requirements of Part Used in Hydraulic Systems
NAS1639(6)	8/65	Compressive Strength Test (Edgewise For Sandwich Constructions)
NAS1640	2/66	Washer, Lock, Spring, Non-Magnetic (ANA Bul. 147–AF Only)
NAS1641(4)	9/65	Connector, Receptacle, Electrical, Flange Mount, Thread Coupling (ANA Bul. 147–AF Only)
NAS1642(4)	9/65	Connector, Receptacle, Electrical, Single-Hole Mount, Thread Coupling (ANA Bul. 147–AF Only)
NAS1643(4)	9/65	Connector, Plug, Electrical, Thread Coupling (ANA Bul. 147–AF Only)
NAS1644(2)	5/64	Insert Arrangements, Electrical Connector, Shell Size 10
NAS1645(3)	5/64	Insert Arrangements, Electrical Connector, Shell Size 12
NAS1646(4)	5/64	Insert Arrangements, Electrical Connector, Shell Size 14
NAS1647(2)	5/64	Insert Arrangements, Electrical Connector, Shell Size 16
NAS1648(3)	9/65	Insert Arrangements–Electrical Connector, Shell Size 18
NAS1649(4)	5/64	Insert Arrangements, Electrical Connector, Shell Size 22
NAS1650(4)	1/67	Connector, Receptacle, Electrical, Narrow Flange Mount, Bayonet Coupling (ANA Bul. 147–AF Only)
NAS1651(4)	1/67	Connector, Receptacle, Electrical, Wide Flange Mount, Bayonet Coupling (ANA Bul. 147–AF Only)
NAS1652(4)	1/67	Connector, Receptacle, Electrical, Single-Hole Mount, Bayonet Coupling (ANA Bul. 147–AF Only)
NAS1653(4)	1/67	Connector, Plug, Electrical, Bayonet Coupling (ANA Bul. 147–AF Only)
NAS1654(2)	9/65	Insert Arrangement–Electrical Connector, Shell Size 10
NAS1655(2)	5/64	Insert Arrangement, Electrical Connector, Shell Size 12
NAS1656(4)	5/64	Insert Arrangement–Electrical Connector, Shell Size 14
NAS1657(3)	5/64	Insert Arrangement, Electrical Connector, Shell Size 16
NAS1658(2)	5/64	Insert Arrangement, Electrical Connector, Shell Size 18
NAS1659(3)	9/65	Insert Arrangement, Electrical Connector, Shell Size 20
NAS1660(4)	9/65	Insert Arrangement–Electrical Connector, Shell Size 22
NAS1661(2)	5/64	Insert Arrangement, Electrical Connector, Shell Size 24
NAS1662	9/65	Contact–Pin, Electrical Connector, Crimp Type (ANA Bul. 147–AF Only)
NAS1663	9/65	Contact–Socket, Electrical Connector, Crimp Type (ANA Bul. 147–AF Only)
NAS1664	9/65	Tool Contact, Insertion-Extraction, Electrical Connector (ANA Bul. 147–AF Only)
NAS1665	1/67	Backshell, Electrical Connector, Straight, Less Strain Relief (ANA Bul. 147–AF Only)
NAS1666	1/67	Backshell, Electrical Connector, Straight, Strain Relief (ANA Bul. 147–AF Only)
NAS1667	1/67	Backshell, Electrical Connector, 90° Strain Relief (ANA Bul. 147–AF Only)
NAS1668	5/64	Plug, Grommet Sealing, Electrical Connector (ANA Bul. 147–AF Only)
NAS1669(2)	6/64	Fastener–Blind, Internally Threaded, External Sleeve, General Purpose, Protruding Head (ANA Bul. 147)
NAS1670(2)	6/64	Fastener–Blind Internally Threaded, External Sleeve, General Purpose, Flush Head (ANA Bul. 147)
NAS1671(2)	8/65	Fastener–Blind, Internally Threaded, External Sleeve, High Temperature Protruding Head (ANA Bul. 147)
NAS1672(2)	6/64	Fastener–Blind, Internally Threaded, External Sleeve, High Temperature, Flush Head (ANA Bul. 147)
NAS1673(2)	6/64	Fastener–Blind, Internally Threaded, External Sleeve, Light Weight, Protruding Head (ANA Bul. 147)
NAS1674(2)	6/64	Fastener–Blind, Internally Threaded, External Sleeve, Light Weight, Millable Head (ANA Bul. 147)
NAS1675(29)	6/64	Fastener–Blind, Internally Threaded, External Sleeve (ANA Bul. 147–AF Only)
NAS1676	4/64	Washer, Lock, Split, Helical, Hi-Collar (ANA Bul. 147–AF Only)
NAS1677(2)	12/68	Strap, Retaining–Cushioned, Rectangular, Two Piece
NAS1678(4)	3/66	Connector, Receptacle, Electrical, Hermetic, Solder Mount, Bayonet Coupling (ANA Bul. 147–AF Only)
NAS1679(4)	3/66	Connector, Receptacle, Electrical, Hermetic, Single-Hole Mount, Bayonet Coupling (ANA Bul. 147–AF Only)
NAS1680(4)	3/66	Connector, Receptacle, Electrical, Hermetic, Narrow Flange Mount, Bayonet Coupling (ANA Bul. 147–AF Only)
NAS1681(4)	3/66	Connector, Receptacle, Electrical, Hermetic, Solder Mount, Thread Coupling (ANA Bul. 147–AF Only)
NAS1682(4)	3/66	Connector, Receptacle, Electrical, Hermetic, Single-Hole Mount, Thread Coupling (ANA Bul. 147–AF Only)
NAS1683(4)	3/66	Connector, Receptacle, Electrical, Hermetic, Flange Mount, Thread Coupling (ANA Bul. 147–AF Only)
NAS1685(3)	3/67	Insert Arrangement, Electrical Connector, Shell Size 8
NAS1688(3)	11/64	Hose Assembly, Tetrafluoroethylene, High Temperature, Medium Pressure, Flared (ANA Bul. 147–AF Only)
NAS1689(3)	11/64	Hose Assembly, Tetrafluoroethylene, High Temperature, Medium Pressure, Flareless (ANA Bul. 147–AF Only)
NAS1690(3)	11/64	Hose Assembly, Polytetrafluoroethylene, High Temperature, High Pressure, Flared (ANA Bul. 147–AF Only)
NAS1691(3)	11/64	Hose Assembly, Polytetrafluoroethylene, High Temperature, High Pressure, Flareless (ANA Bul. 147–AF Only)
NAS1692(4)	11/66	Connector, Receptacle, Electrical, Flange Mount, Thread Coupling, Conductive Finish (ANA Bul. 147–AF Only)

Number	Current Issue	Title
NAS1693(4)	11/66	Connector, Receptacle, Electrical, Single-Hole Mount, Thread Coupling, Conductive Finish (ANA Bul. 147–AF Only)
NAS1694(4)	11/66	Connector, Plug, Electrical, Thread Coupling, Conductive Finish (ANA Bul. 147–AF Only)
NAS1695	11/66	Backshell, Electrical Connector, Straight, Less Strain Relief, Conductive Finish (ANA Bul. 147–AF Only)
NAS1696	11/66	Backshell, Electrical Connector, Straight, With Strain Relief, Conductive Finish (ANA Bul. 147–AF Only)
NAS1697	11/66	Backshell, Electrical Connector, 90° Strain Relief, Conductive Finish (ANA Bul. 147–AF Only)
NAS1698	5/67	Backshell, Electrical Connector, Straight, Crimp Type, Conductive Finish
NAS1699(4)	3/67	Connector, Receptcle, Electrical, Narrow Flange Mount, Bayonet Coupling, Conductive Finish
NAS1700(4)	3/67	Connector, Receptcle, Electrical, Wide Flange Mount, Bayonet Coupling, Conductive Finish
NAS1701(4)	3/67	Connector, Receptacle, Electrical, Single-Hole Mount, Bayonet Coupling, Conductive Finish
NAS1702(4)	3/67	Connector, Plug, Electrical, Bayonet Coupling, Conductive Finish
NAS1703–1710(3)	8/66	Bolt, .0156 Oversize Shank, 100° Head, Close Tolerance, 160 000 PSI (ANA Bul. 147–AF Only)
NAS1711	12/68	Strap, Retaining–Cushioned, Rectangular, Open Bottom
NAS1712(2)	12/68	Clamp, Loop–Cushioned, Rectangular
NAS1713(2)	12/68	Clamp, Loop–Cushioned, Center Mounting
NAS1714(2)	12/68	Clamp, Loop–Cushioned, Elongated
NAS1715(2)	7/68	Clamp, Loop–Cushioned
NAS1716(2)	7/67	Clamp Assembly, Cushioned, Saddle
NAS1717(15)	10/67	Reliability Demonstration of Liquid Propellant Rocket Engines
NAS1718(12)	10/67	Reliability Demonstration of Solid Propellant Rocket Motors
NAS1731(3)	4/68	Nut, Blind, Serrated Head, Self-Locking, 450°F
NAS1732(3)	4/68	Nut, Blind, Serrated Head, Self-Locking, Closed End, 450°F
NAS1733(2)	4/68	Nut, Blind, Serrated Head, Self-Locking, Installation Data For
NAS1734(3)	4/68	Nut, Blind, Elliptical Head, Self-Locking, 450°F
NAS1735(3)	4/68	Nut, Blind, Elliptical Head, Self-Locking, Closed End, 450°F
NAS1736(3)	4/68	Nut, Blind, Elliptical Head, Self-Locking, Installation Data For
NAS1738(2)	3/68	Rivet–Blind, Protruding Head, Mechanically Locked Spindle, Bulbed
NAS1739(2)	3/68	Rivet–Blind, 100° Flush Head, Mechanically Locked Spindle, Bulbed
NAS1740(14)	3/68	Rivet–Blind, Self-Plugging, Mechanically Locked Spindle, Bulbed
NAS1744(3)†	1/69	Splice, Conductor, Heat Shrinkable, Insulated, Immersion Resistant
NAS1745(3)†	1/69	Splice, Conductor, Heat Shrinkable, Insulated
NAS1746(3)†	1/69	Splice, Conductor, Heat Shrinkable, Insulated–Infrared Heat
NAS1747(10)†	1/69	Splice, Conductor, Heat Shrinkable, Insulated, Specification for
NAS1748(48)†	1/69	Interconnecting Devices, Electrical Wire Terminating
NAS1749†	1/69	Contact, Electrical, Pin
NAS1806–1816(2)	9/57	Rivet–Hi-Shear, Flat Head, Interference Fit, Titanium Alloy (ANA Bul. 147–AF Only)
NAS1906–1916(2)	9/57	Rivet–Hi-Shear, 100° Head Interference Fit, Titanium Alloy (ANA Bul. 147–AF Only)
NAS2005–2012(2)	6/66	Bolt–Lock, Tension, Protruding Head, Pull Type, Titanium Alloy (ANA Bul. 147–AF Only)
NAS2105–2112(2)	6/66	Bolt–Lock, Tension, 100° (AN509) Head, Pull Type, Titanium Alloy (ANA Bul. 147–AF Only)
NAS2206–2210(2)	12/59	Bolt–Lock, Tension, Protruding Head Stump Type, Titanium Alloy (ANA Bul. 147–AF Only)
NAS2306–2310(2)	12/59	Bolt–Lock, Tension, 100° (AN509) Head, Stump Type, Titanium Alloy (ANA Bul. 147–AF Only)
NAS2406–2412(2)	12/65	Bolt–Lock, Shear, Protruding Head, Pull Type, Titanium Alloy (ANA Bul. 147–AF Only)
NAS2506–2512(2)	12/65	Bolt–Lock, Shear, 100° Head, Pull Type, Titanium Alloy (ANA Bul. 147–AF Only)
NAS2606–2612(2)	12/59	Bolt–Lock, Shear, Protruding Head, Stump Type, Titanium Alloy (ANA Bul. 147–AF Only)
NAS2706–2712(2)	12/59	Bolt–Lock, Shear, 100° Head Stump Type, Titanium Alloy (ANA Bul. 147–AF Only)
NAS2803–2810(3)	5/65	Bolt, 100° Head, "Torq-Set", Close Tolerance, 180,000 PSI (ANA Bul. 147–AF Only)
NAS2903–2920(7)**	1/69	Bolt, Shear–Hexagon Head, .0156 Oversize Shank
NAS3003–3020(7)**	1/69	Bolt, Shear–Hexagon Head, .0312 Oversize Shank
NAS3103–3110(2)	6/62	Bolt–U (ANA Bul. 147–AF Only)
NAS3203–3210(2)	6/62	Bolt–Hook (ANA Bul. 147–AF Only)
NAS3303–3305(2)	3/67	Bolt–U, Strap Type
NAS3310(2)*	2/68	Gasket, Sealing, Aluminum Alloy and Corrosion-Resisting Steel, ASA Pipe Flanges
NAS3311x	3/67	Metal Stock List, Carbon & Alloy Steel Sheet & Strip
NAS3312x	3/67	Metal Stock List, Carbon & Alloy Steel Plate
NAS3313(5)x	3/67	Metal Stock List, Bar & Rod, Round & Hexagon–Alloy Steel
NAS3314(3)x	3/67	Metal Stock List, Bar, Square & Flat–Alloy Steel
NAS3315x	3/67	Metal Stock List, Tubing–Alloy Steel
NAS3316x	3/67	Metal Stock List, Wire–Carbon Steel
NAS3317(2)x	3/67	Metal Stock List, Sheet & Strip–Cres
NAS3318x	3/67	Metal Stock List, Plate–Cres
NAS3319(5)x	3/67	Metal Stock List, Bar & Rod, Round & Hexagon–Cres
NAS3320(11)x	3/67	Metal Stock List, Tubing–Cres
NAS3321x	3/67	Metal Stock List, Wire–Cres
NAS3322(12)	2/67	Metal Stock List, Aluminum Sheet and Plate
NAS3323(3)	2/67	Metal Stock List, Aluminum Alloy–Bar (Squares and Rectangles)
NAS3324(2)	2/67	Metal Stock List, Aluminum Alloy–Rod and Bar (Round and Hexagon)
NAS3325(5)	2/67	Metal Stock List, Aluminum Tubing
NAS3326	2/67	Metal Stock List, Copper and Copper Alloys, Sheet, Plate, and Strip
NAS3327	2/67	Metal Stock List, Copper and Copper Alloys, Strip
NAS3328(3)	2/67	Metal Stock List, Copper and Copper Alloys, Bar and Rod
NAS3329	2/67	Metal Stock List, Copper and Copper Alloys, Rod
NAS3330	2/67	Metal Stock List, Copper and Copper Alloys, Wire
NAS3331	2/67	Metal Stock List, Copper and Copper Alloys, Tubing
NAS3332(2)	2/67	Metal Stock List, Magnesium Sheet and Plate

Number	Current Issue	Title
NAS3333	2/67	Metal Stock List, Magnesium Bar
NAS3334	2/67	Metal Stock List, Magnesium Bar and Rod
NAS3335	2/67	Metal Stock List, Magnesium Tube
NAS3350(19)	4/63	Nuts, Self-Locking, 450° F, High Quality (ANA Bul. 147–AF Only)
NAS3351	4/63	Stud and Mandrel Test Fixture, Self-Locking Nuts (ANA Bul. 147–AF Only)
NAS3352	4/63	Fixture–Stress Embrittlement Test–Self-Locking Nuts (ANA Bul. 147–AF Only)
NAS3353	4/63	Fixture–Bearing Surface Squareness Test–Self-Locking Nuts (ANA Bul. 147–AF Only)
NAS3354	4/63	Fixtures–Vibration Test–Self-Locking Nuts (ANA Bul. 147–AF Only)
NAS3355	4/63	Fixture–Tensile Yield Test–Self-Locking Nuts (ANA Bul. 147–AF Only)
NAS3356(3)	4/63	Lubricant Coatings, Suitability of Self-Locking Nuts (ANA Bul. 147–AF Only)
NAS3400(2)	6/67	Volatile Corrosion Inhibitor
NAS3401	5/64	Wrap, With Stiffener (ANA Bul. 147–AF Only)
NAS3402	5/64	Bag, Transparent, Single Item (ANA Bul. 147–AF Only)
NAS3403	5/64	Bag, Transparent, Multiple Item (ANA Bul. 147–AF Only)
NAS3404	5/64	Bag, Transparent, Individual Box (ANA Bul. 147–AF Only)
NAS3405	5/64	Bag, Transparent, Single Item, With Accessories (ANA Bul. 147–AF Only)
NAS3406	5/64	Bag, Neutral, Multiple Item (ANA Bul. 147–AF Only)
NAS3407	5/64	Bag, Water Vaporproof (ANA Bul. 147–AF Only)
NAS3408	5/65	Box, Unit, Cushioned (ANA Bul. 147–AF Only)
NAS3409	5/64	Box, Plastic, Cushioned (ANA Bul. 147–AF Only)
NAS3410	5/64	Box, Bulk Quantities (ANA Bul. 147–AF Only)
NAS3411	10/65	Box, Slotted Partitions (ANA Bul. 147–AF Only)
NAS3412	5/64	Spool, Boxed (ANA Bul. 147–AF Only)
NAS3413	5/64	Tube, Mailing (ANA Bul. 147–AF Only)
NAS3414(2)	11/68	Package, Multiple, Axial Lead Components, For Automatic Equipment
NAS3415(2)	10/65	Bag, Transparent, With Stiffener (ANA Bul. 147–AF Only)
NAS3416	10/65	Box, With Fiberboard Stiffener (ANA Bul. 147–AF Only)
NAS3417(3)	10/65	Packaging of Electrical Cable-Cord-Wire (ANA Bul. 147–AF Only)
NAS3418	10/65	Bag, Neutral, Single Item (ANA Bul. 147–AF Only)
NAS3419	10/65	Box, With Die-Cut Pads/Fillers (ANA Bul. 147–AF Only)
NAS3420(2)	10/65	Containers, Water-Vaporproof, W/Desicant (ANA Bul. 147–AF Only)
NAS3421(2)	10/65	Wrap, Greaseproof, W/Preservative (ANA Bul. 147–AF Only)
NAS3422	3/66	Wrap, Greaseproof W/Preservative, Individual Box (ANA Bul. 147–AF Only)
NAS3423(2)	10/65	Contoured Items, Packaging of (ANA Bul. 147–AF Only)
NAS3424(2)	3/66	Hose, Hose Assemblies, Packaging of (ANA Bul. 147–AF Only)
NAS3425**	6/69	Bag, Anti-Static, Single Item
NAS3426(2)	3/66	Electrical Harness-Cable Assemblies, Supplier Packaging of (ANA Bul. 147–AF Only)
NAS3427(2)	8/66	Multiple Lead Devices, Individual Packaging of (ANA Bul. 147–AF Only)
NAS3428	3/67	Kit, Supplier Packaging of
NAS3429(3)	4/66	Axial/Radial Lead Components, Individual Packaging of (ANA Bul. 147–AF Only)
NAS3430	3/66	Axial/Radial Lead Components, Multiple Packaging of (ANA Bul. 147–AF Only)
NAS3431	3/66	Bag, Individual Item Capped, With or Without Accessories (ANA Bul. 147–AF Only)
NAS3432(2)**	6/69	Silver Tarnish Inhibitor
NAS3433	3/67	Close Tolerance Cylindrical Hardware Items, Packaging of
NAS3434(2)	9/67	Precision Machined Items, Packaging of
NAS3435	8/66	Box, With Multiple Individual Transparent Bags (ANA Bul. 147–AF Only)
NAS3436	8/66	Box, With Transparent Bag, Cylindrical Core (ANA Bul. 147–AF Only)
NAS3437	3/67	Wrap, Single Face Corrugated Paper
NAS3438	6/67	Vial, Plastic
NAS3439	3/67	Box, Multiple Quantities, Tightly Packed
NAS3440(2)	3/67	Package, Multiple Cavity
NAS3441	3/67	Bag, Cushioned, Single Item
NAS3442	3/67	Box, Interleave, Multiple Items
NAS3443	6/67	Box, Foam Cushioning Convoluted
NAS3444(2)	6/67	Box, W/Desiccant, Bag Water-Vaporproof, Box
NAS3445(2)	6/67	Skin Packaging
NAS3446	6/67	Tube, Fiber/Paperboard, Spiral Wound
NAS3447(3)	11/68	Precision Cleaned Items, Contamination Barrier For
NAS3448(2)	9/67	Optical Components, Packaging of
NAS3449 †	6/69	Hoses/Tubes, Preformed or Straight, Multiple Packaging of
NAS3450(2) †	6/69	Uncased Devices (Silicon Monolithic)
NAS3451 †	6/69	Castings, Packaging of
NAS3601(68)**	4/69	Recommended Standard Data Format of Transport Airplane Characteristics of Airport Planning
NAS4000(5)	12/68	Recess, Tri-Wing, Dimensions of Recess and Gages
NAS4001(6) †	4/69	Bit, Screwdriver, Tri-Wing, Specification For
NAS4002(20)	12/68	Fasteners, Alloy Steel, Externally Threaded
NAS4003(20)	12/68	Fasteners, A286 Cres, Externally Threaded
NAS4004(20)	12/68	Fasteners, 6AL-4V Titanium Alloy, Externally Threaded
NAS4005(6) †	4/69	Bit, Screwdriver, Tri-Wing Recess
NAS4104–4116(6)	12/68	Bolt, 100° Head, Tri-Wing Recess, Close Tolerance, Alloy Steel, Long Thread, Self-Locking and Nonlocking
NAS4204–4216(6)	12/68	Bolt, 100° Head, Tri-Wing Recess, Close Tolerance, A286 Cres, Long Thread, Self-Locking and Nonlocking
NAS4304–4316(6)	12/68	Bolt, 100° Head, Tri-Wing Recess, Close Tolerance, 6AL-4V Titanium Alloy, Long Thread, Self-Locking and Nonlocking
NAS4400–4416(6)	12/68	Bolt, 100° Head, Tri-Wing Recess, Close Tolerance, Alloy Steel, Short Thread, Self-Locking and Nonlocking
NAS4500–4516(6)	12/68	Bolt, 100° Head, Tri-Wing Recess, Close Tolerance, A286 Cres, Short Thread, Self-Locking and Nonlocking
NAS4600–4616(6)	12/68	Bolt, 100° Head, Tri-Wing Recess, Close Tolerance, 6AL-4V Titanium Alloy, Short Thread, Self-Locking and Nonlocking
NAS4703–4716(6)	12/68	Bolt, 100° Reduced Head, Tri-Wing Recess, Close Tolerance, Alloy Steel, Short Thread, Nonlocking
NAS4803–4816(6)	12/68	Bolt, 100° Reduced Head, Tri-Wing Recess, Close Tolerance, A286 Cres, Short Thread, Nonlocking
NAS4903–4916(6)	12/68	Bolt, 100° Reduced Head, Tri-Wing Recess, Close Tolerance, 6AL-4V Titanium Alloy, Short Thread, Nonlocking
NAS5000–5006(4)	12/68	Bolt, Pan Head, Close Tolerance, Short Thread, Tri-Wing Recess, Alloy Steel

Number	Current Issue	Title
NAS5100–5106(4)	12/68	Bolt, Pan Head, Close Tolerance, Short Thread, Tri-Wing Recess, Corrosion Resistant Steel, A-286
NAS5200–5206(4)	12/68	Bolt, Pan Head, Close Tolerance, Short Thread, Tri-Wing Recess, Titanium Alloy 6AL-4V
NAS5300–5306(2)	12/68	Screw, Machine–Flat Fillister Head, Full Thread, Tri-Wing Recess, Alloy Steel
NAS5400–5406(2)	12/68	Screw, Machine–Flat Fillister Head, Full Thread, Tri-Wing Recess, Corrosion Resistant Steel, A-286
NAS5500–5506(2)	12/68	Screw, Machine–Flat Fillister Head, Full Thread, Tri-Wing Recess, Titanium Alloy, 6AL-4V
NAS5600–5606(2)	12/68	Screw, 100° Head, Tri-Wing Recess, Alloy Steel, Full Thread, Self-Locking and Nonlocking
NAS5700–5706(2)	12/68	Screw, 100° Head, Tri-Wing Recess, A286 Cres, Full Thread, Self-Locking and Nonlocking
NAS5800–5806(2)	12/68	Screw, 100° Head, Tri-Wing Recess, 6AL-4V Titanium Alloy, Full Thread, Self-Locking and Nonlocking
NAS5900–5903(2)	12/68	Screw, Hex Head, Tri-Wing Recess, Alloy Steel, Full Thread, Nonlocking

Number	Current Issue	Title
NAS6000–6003(2)	12/68	Screw, Hex Head, Tri-Wing Recess, A286 Cres, Full Thread, Nonlocking
NAS6100–6103(2)	12/68	Screw, Hex Head, Tri-Wing Recess, 6AL-4V Titanium Alloy, Full Thread, Nonlocking
NAS6203–6220(5)	12/68	Bolt, Hex Head, Close Tolerance, Alloy Steel, Short Thread, Self-Locking & Non-Locking
NAS6303–6320(5)	12/68	Bolt, Hex Head, Close Tolerance, A-286, Short Thread, Self-Locking & Non-Locking
NAS6403–6420(5)	12/68	Bolt, Hex Head, Close Tolerance, 6-4 Titanium Alloy, Short Thread, Self-Locking & Non-Locking
NAS6500–6506(2)	12/68	Screw, 100° Oval Head, Tri-Wing Recess, A286 Cres, Full Thread, Self-Locking and Nonlocking
NAS6604–6620(6)†	2/69	Bolt, Hex Head, Close Tolerance, Alloy Steel, Long Thread, Self-Locking and Nonlocking
NAS6704–6720(6)†	2/69	Bolt, Hex Head, Close Tolerance, A286 Cres, Long Threads, Self-Locking and Nonlocking
NAS6804–6820(6)†	2/69	Bolt, Hex Head, Close Tolerance, 6AL-4V Titanium Alloy, Long Thread, Self-Locking and Nonlocking

NATIONAL AEROSPACE STANDARDS
ALPHABETICAL INDEX

NOTICE

National Aerospace Standards are proposed as guides to the aerospace industry in serving the public interest and the national defense program. The standards therein are designed to eliminate misunderstandings between manufacturers and purchasers, to facilitate interchangeability and improvement of products, and to assist purchaser in expeditiously selecting and obtaining the proper product for a particular need.

Symbols Used to Denote Status of National Aerospace Standards
**New revisions and amendments issued since date of last index
†New specifications or standards issued since date of last index

C

D

E

Q

R

AIR FORCE—NAVY AERONAUTICAL, U. S. AIR FORCE AND MILITARY (MS) STANDARDS NUMERICAL INDEX

Symbols Used to Denote Status of Standards
* Inactive for new design
** Applicable notice

Number	Current Issue	Title
AN3–		
20(4)	3/67	Bolt–Machine, Aircraft
AN21–		
36(3)	6/66	Bolt–Clevis
AN42–		
49(2)	2/67	Bolt, Eye
65**	8/60	Bolt–Hexagon Head, Coarse Thread, Use MS35291 and MS35309
AN73–		
81(2)	6/66	Bolt, Machine, Aircraft, Drilled Head
AN100**	2/62	Thimble–Wire Cable
AN115*	12/59	Shackle–Wire Rope, See MS20115
AN116**	2/62	Shackle, Anchor, Forged Carbon Steel Screw Pin
AN173–		
186(4)	6/66	Bolt–Machine, Close Tolerance, Aircraft
AN208*	6/45	Bearing–Heavy Duty Double Row Ball (Felt Seal), See AN205
AN218*	3/61	Bearing–Ball, Bellcrank, Anti-Friction, Airframe, Use MS20218
AN219*	5/57	Pulley–Control, Secondary, Aircraft, Use MS20219
AN226*		Fastener–Cowl, Post Type, See AND10076
AN227(3)	5/53	Fasteners–Snap and Curtain
AN229	8/68	Fasteners–Interlocking Slide
AN230*	8/57	Grommets–Metallic, Plain and Spur, With Washers, Use MS20230
AN231	10/48	Grommet–Drainage, Plastic
AN233*	5/65	Hook and Buckle
AN240	4/43	Eyelet–Lacing
AN250*	7/49	Hinge–Butt
AN251*	6/49	Hinge–Continuous, See AN257
AN254	6/66	Screw–Thumb, Necked
AN255	6/66	Screw–Externally Relieved Body
AN256	4/52	Nut–Self Locking, Right Angle Plate
AN280*	1/60	Key–Woodruff
AN286**	8/60	Fittings–Lubricator, Pressure-Grip, Type PT, Use MS15006
AN287**	8/60	Fittings–Lubricator, Pressure-Grip, Type D
AN295**	8/60	Cup–Oil
AN302		Pin–Escutcheon
AN310(2)	2/69	Nut, Plain, Castellated, Airframe, Sup. AN355
AN315	2/69	Nut, Plain, Hexagon, Airframe
AN316	2/69	Nut, Jam, Hexagon
AN320	2/69	Nut, Plain, Castellated, Shear
AN340*	3/67	Nut, Plain, Hexagon, Non-Structural, Coarse Thread, Use MS20341
AN345	6/66	Nut, Plain, Hexagon, Non-Structural, Fine Thread
AN361*	4/67	Nut–Self Locking, Plate, Countersunk 100°, 550°F, Use MS21049 or MS21050
AN362*	4/67	Nut, Plate, Self-Locking, Non-Countersunk, 550°F, Use MS21047 or MS21048
AN367*	4/49	Nut–Plate, Countersunk 82°
AN373*	4/67	Nut–Self Locking, Plate, Countersunk 100°, 250°F, Use MS21049
AN385	2/54	Pin–Tapered Plain
AN386	2/54	Pin–Threaded Taper
AN415	1/54	Pin–Lock
AN416	1/54	Pin–Retaining, Safety
AN425*	12/43	Rivet–78° Countersunk Head
460*	12/68	Rivet–100°; Countersunk Head, Blind, Type B, Class I, Use NAS1399
461*	12/68	Rivet–Brazier Head, Blind, Type B, Class I, Use NAS1398
AN490	10/66	Rod End–Threaded, Aircraft
AN500*	5/67	Screw, Machine, Fillister Head, Coarse Thread, Use MS35265, MS35273 or MS35275
AN501(2)*	11/65	Screw, Machine, Fillister Head, Fine Thread, Use MS35266, MS35274 and MS35276
AN502	5/53	Screw–Machine, Drilled Fillister Head, Fine Thread
AN503	5/53	Screw–Machine, Drilled Fillister Head, Coarse Thread
AN504(3)	2/59	Screw–Tapping, Thread Cutting, Round Head, Machine Screw Thread
AN505(2)	2/59	Screw–Machine, Flat Head, 82 Degree, Coarse Thread
AN508		Screw–Round Head Machine (For Electrical Use)
AN510(2)*	3/64	Screw, Machine, Flat Head, 82 Degree, Fine Thread, Use MS35191, MS35199, MS51960 and MS35203 MS51957
AN525*	9/63	Screw–Washer Head, Use MS27039
AN530(3)	2/59	Screw–Tapping, Thread Forming or Thread Cutting, Round Head, Spaced Thread
AN531(3)	2/59	Screw–Tapping, Thread Forming or Thread Cutting, 82-Degree Flat Head, Spaced Thread
AN545(2)	2/59	Screw–Wood, Round Head
AN550(2)	2/59	Screw–Wood, Flat Head, 82 Degree
AN555	6/66	Bolt–Shoulder, Anchor Light Mounting
AN565(2)	6/66	Setscrew–Hexagon and Fluted Socket, Headless
AN665	3/65	Terminal–Threaded Clevis Type Tie Rod
AN701	6/66	Tie Rod–Internal 1000 Lb (No. 6-40)
AN703	6/66	Tie Rod–Internal 2,100 Lb (No. 10-32)
AN704	6/66	Tie Rod–Internal 3,400 Lb (1/4-28)
AN705	6/66	Tie Rod–Internal 6,100 Lb (5/16-24)

Number	Current Issue	Title
AN706	6/66	Tie Rod–Internal 8,000 Lb (3/8-24)
AN707	6/66	Tie Rod–Internal 11,500 Lb (7/16-20)
AN708	6/66	Tie Rod–Internal 15,500 Lb (1/2-20)
AN735	6/66	Clamp, Loop Type Bonding
AN737	1/56	Clamp–Hose, Aircraft, Sup. AN748
AN742(3)	12/68	Clamp, Plain, Support, Loop-Type, Aircraft
AN743(2)	6/66	Bracket, Support Clamp
AN753	7/52	Splice–Electrical Disconnect
AN763(2)**	10/62	Gasket–Swivel Fitting, Flanged
AN774	2/67	Bolt–Flared Tube Fitting End Universal
AN775	2/67	Bolt, Fluid Passage–Universal Fitting
AN776	2/67	Elbow–Universal 90°
AN777	2/67	Elbow–Universal 75°
AN778	2/67	Elbow–Universal 45°
AN779	2/67	Tee–Universal Fitting
AN780	7/56	Nipple–Union
AN783	2/67	Tee, Flared Tube Internal Thread on Side
AN784	2/67	Tee, Flared Tube, Internal Thread on Run
AN785	8/56	Coupling–Union, Brazing
AN786	6/54	Coupling–Airspeed Tube Union (Brazing)
AN790	8/56	Elbow–Union
AN791	8/56	Elbow–Union 45°
AN792**	3/60	Fitting Assembly–Hose Swivel Detachable, Use MS28740
AN795	8/56	Tee–Union
AN800	6/54	Cone–Union
AN801(2)	8/56	Connection–Straight Flanged Hose
AN802(2)	8/56	Connection–45° Flanged Hose
AN803(2)	7/56	Connection–90° Flanged Hose
AN804	2/67	Tee–Flared Tube with Bulkhead on Run
AN805	8/56	Nut–Union
AN806	2/67	Plug–Flared Tube
AN807	2/67	Adapter–Hose to Universal
AN809*	9/64	Core–High Pressure Valve
811*	8/45	Fittings–Solderless
AN812*	9/48	Body–High Pressure Air Valve, See AN6287
AN814(2)	2/67	Plug and Bleeder–Screw Thread
AN815	2/67	Union–Flared Tube
AN816(2)	2/67	Nipple–Flared Tube and Pipe Thread
AN817(2)	2/67	Nut, Sleeve, Coupling
AN818(2)	7/68	Nut, Coupling
AN821	2/67	Elbow–Flared Tube, 90°
AN824	2/67	Tee–Flared Tube
AN827	2/67	Cross–Flared Tube
AN832	2/67	Union–Flared Tube, 3/8 Bulkhead and Universal
AN833	2/67	Elbow–Flared Tube and Bulkhead Universal 90°
AN834	2/67	Tee–Bulkhead and Universal, Flared Tube
AN837	2/67	Elbow–Flared Tube and Universal, 45°
AN838	2/67	Elbow, Hose to Universal 90°
AN839	2/67	Elbow–Hose to Universal 45°
AN840*	6/54	Adapter–Hose to Pipe Thread
AN842*	6/54	Elbow–Pipe Thread and Hose 90°
AN844*	6/54	Elbow–Pipe Thread and Hose 45°
AN846	2/67	Elbow–Hose Bulkhead 45°
AN848	2/67	Elbow–Hose Bulkhead 90°
AN849	2/67	Connector–Hose, Bulkhead
AN867**	3/60	Flange–Pipe Thread Welding
AN871	6/54	Flange–Straight Thread Boss Welding
AN872	10/53	Flange–Circular Formation Light Mounting
AN876	6/54	Collar–Tubing End Hose Connection
AN886	8/55	Gasket–Anchor-Light Cover
AN887	4/51	Gasket–Circular Formation Light Cover
AN889	8/55	Gasket–Wing Position Light Cover
AN891**	8/60	Cross–Primer and Gauge
AN893(2)	2/67	Bushing–Screw Thread Reducer
AN894(2)	2/67	Bushing–Screw Thread Expander
AN896**	8/55	Gasket–Cabin Dome-Light Cover
AN898	8/55	Gasket–Exterior Light Cover
AN899	8/55	Gasket–Fuselage Light-Cover Screw
AN901	6/68	Gasket-Metal Tube Connection Seal
AN903**	6/60	Gasket, Downward Recognition Light Cover
AN904	4/43	Gasket–Upward Recognition Light Cover
AN909	6/54	Plug–Square Head, Drilled for Locking
AN910	2/67	Coupling–Pipe Thread
AN911	2/67	Nipple–Pipe Thread
AN912	2/67	Bushing–Reducer
AN914	2/67	Elbow–Internal and External Pipe Thread 90°
AN915	2/67	Elbow–Internal and External Pipe Thread 45°
AN916	2/67	Elbow–Internal Pipe Thread 90°
AN917	2/67	Tee–Internal Pipe Thread
AN918	2/67	Cross–Internal Pipe Thread
AN919	2/67	Reducer–External Thread Flared Tube
AN924	5/67	Nut, Plain Hexagon, Tube, Bulkhead and Universal Fitting, Sup. MS24400
AN929(2)	2/67	Cap Assembly, Pressure Seal Flared Tube Fitting, Sup. AN820
AN932	2/67	Plug–Countersink Hex Head Pipe
AN933	2/67	Plug–Pipe, Hex Head
AN935*	9/67	Washer–Lock, Spring, Use MS35338
AN936(2)	8/59	Washer–Tooth Lock
AN937	2/67	Cross–Internal Screw Thread, Sup. AN925
AN938	2/67	Tee–Internal Screw Thread, Sup. AN926
AN939	2/67	Elbow–Internal Screw Thread, 90°, Sup. AN927
AN941	2/67	Elbow–Internal Screw Thread, 45°, Sup. AN928
AN942	8/51	Bearing–Plain, Heavy Duty, Self-Aligning, Airframe
AN943		Bearing–Rod End, Internal Thread, Self-Aligning, Plain, Airframe
AN947		Bearing–Rod End, External Thread, Self-Aligning, Plain, Airframe
AN949		Bearing–Rod End, Hollow Shank, Self-Aligning, Plain, Airframe
AN960(2)	4/69	Washer, Flat
AN961(2)	8/67	Washer–Flat, Electrical, Sup. MS45901 (Ord)
AN970	4/69	Washer, Flat, For Wood, Sup. MS63040 (Ord)
AN975	6/66	Washer–Taper Pin
AN995**	6/66	Wire, Lock, Use MS20995
AN996	6/66	Ring–Lock
AN2508**	5/58	Light–Runway Marker
AN2512	9/51	Base Assembly–Runway Marker Light
AN2519		Power Plant–Electric Direct Current, 3 Kilowatt, 110 Volt
AN2520		Power Plant–Electric Direct Current, 500 Watt, 110 Volt, Portable
AN2528	12/49	Cabinet and Relay Assembly–Distribution
AN2531		Indicator Assembly–Wind Cone (Unlighted and Externally Lighted)
AN2534	6/54	Cable Assembly–Power, Electrical, Runway Marker Lamp
AN2541	6/66	Lens, Aircraft Obstruction Marker Light
AN2547(2)	4/68	Light, Marker, Aircraft Obstruction
AN2551	6/57	Cable Assembly, Power, Electrical, 28 Volt DC
AN2552	6/57	Receptacle–External Power, 28 Volt DC
AN2555	9/44	Nozzle–Aircraft Fueling
AN2578	2/57	Cover–Drum-Type Rotating Beacon, 24-Inch
AN2579	10/54	Frequency Changer–60 to 400 Cps Rotary Type for Ground Use
AN3022	2/52	Switch, Toggle, Single Pole, Three Hole Mounting, Sup. AN3014, AN3015, AN3016, AN3017, AN3018, AN3019, AN3020
AN3030	1/54	Light Assembly–Circular Formation
AN3032*	1/67	Light Assembly, Wing Position Recessed
AN3033	12/56	Light Assembly–Wing Position, Wing Tip
AN3036*	1/67	Light Assembly, Berth

Number	Current Issue	Title
AN3037	7/56	Light Assembly–Cabin Dome
AN3038	1/49	Light Assembly–Ultraviolet Fluorescent Cockpit
AN3039*	7/64	Light Assembly, Cabin Dome, Extra Strength Cover, Use MS28916
AN3040	11/50	Cover–Circular Formation Light
AN3042	5/53	Cover–Position Light
AN3043	7/59	Cover–Anchor Light
AN3047	7/52	Light–Bracket, Work Table
AN3048	10/53	Ferrule–Conduit, Flexible, One Step, Synthetic Covered
AN3049*	1/67	Light Assembly–Open Cabin Dome
AN3050	10/53	Ferrule–Conduit, Flexible
AN3051	10/53	Ferrule–Conduit, Flexible, One Step
AN3052	11/53	Ferrule–Conduit, Flexible, Two Step
AN3053	5/52	Ferrule–Rigid Conduit
AN3054	9/56	Nut–Coupling, Electrical Conduit
AN3059	4/53	Bushing–Electrical Conduit
AN3060	4/53	Elbow–Electrical Conduit, 45°
AN3061	4/53	Elbow–Electrical Conduit, 45° Junction Box
AN3062	4/53	Elbow–Electrical Conduit, 90°
AN3063	4/53	Elbow–Electrical Conduit, 90° Junction Box
AN3066	9/56	Nut–Plain, Hexagon, Conduit Coupling
AN3067	7/54	Seal Assembly–3/8 Inch Conduit Boss
AN3068	4/53	Coupling–Electrical Conduit
AN3069	5/52	Adapter Assembly–3/16 Inch Conduit
AN3081		Probe–High Frequency Electro-Magnetic
AN3087*	1/67	Hood–Berth Light
AN3088	5/55	Light Assembly–Anchor
AN3091	2/49	Light Assembly–Exterior, Edge Mounting
AN3092	7/53	Light Assembly–Exterior, Surface Mounting
AN3093	10/52	Cover–Exterior Light Assembly
AN3095	1/55	Light Assembly–Electrically Retractable Landing
AN3104	8/52	Elbow–Shielded Spark Plug
AN3105	12/47	Terminal–Ground
AN3111		Ring–Bonding, Electrical Connector
AN3114	6/57	Receptacle–External Power, 115/200 Volts, 3 Phase
AN3115		Receptacle–Airspeed Tube Electrical
AN3116	10/52	Plug Assembly–Instrument Electrical
AN3119	1/51	Gasket–Common, Tail Position Light
AN3120	7/56	Lamp–RP-11 Bulb, Single Contact Candelabra Bayonet Base, Signaling
AN3125**	8/60	Lamp–Fluorescent, Double Contact Bayonet Candelabra Indexing, RP-12 Bulb
AN3146		Light Assembly–Passing
AN3147		Light Assembly–Fixed Landing (Par-46 Bulb, Flat)
AN3148	6/45	Light Assembly–Fixed Landing (Par-64 Bulb, Flat)
AN4149*	5/67	Pump–3 GPM at 3000 PSI Power Driven Hydraulic
AN3150	6/50	Battery–Storage Shielded 24-Volt, 36-Ampere-Hour
AN3151	9/50	Battery–Storage, Shielded, 24-Volt, 24-Ampere-Hour
AN3154	4/46	Battery–24 Volt, 11 Ampere Hour, Shielded Storage
AN3155	2/48	Rheostat–Aircraft Power
AN3156	8/53	Clamp–Battery Hold-Down
AN3177	12/56	Light Assembly–Top and Bottom Fuselage (Two Lamps)
AN3201	9/45	Ammeter–Direct Current
AN3203	7/48	Voltmeter–Direct Current, Panel
AN3206	9/50	Regulator–Voltage, Direct Current Generator, 30 Volt
AN3212		Switch–Single Engine Ignition
AN3213*	5/58	Switch–Two Engine Ignition
AN3214*	5/58	Switch–Four Engine Ignition
AN3236	4/50	Switch Assembly–12 Position Selector, Triple Unit
AN3249**	9/60	Jack–Tip Voltmeter Test
AN3250	4/47	Coupling–Conduit Y

Number	Current Issue	Title
AN3303*	4/59	Relay, 3 Amp, 4PDT, Sealed, Connector Mounting, Class 0
AN3304*	10/59	Relay–3 Amp, 4PDT, Sealed, Panel Mounting, Class 0, Use MS25267
AN3306*	7/56	Relay–10 Amp, 2PDT, Sealed, Connector Mounting, Class 0
AN3307*	2/60	Relay, 10 Amp, 2PDT, Sealed, Panel Mounting, Class 0
AN3308*	7/56	Relay–10 Amp, 3PDT, Sealed, Connector Mounting, Class 0
AN3310*	7/56	Relay–10 Amp, 4PDT, Sealed, Connector Mounting, Class 0
AN3312*	7/56	Relay–10 Amp, 3PDT, Sealed, Class 0, Use MS25024
AN3314*	5/56	Relay–10 Amp, 1PDT, Unsealed, Class 0
AN3315*	5/56	Relay–10 Amp, 2PDT, Unsealed, Class 0
AN3316*	9/57	Relay–10 Amp, 3PDT, Unsealed, Class 0
AN3319*	6/55	Relay–20 Amp, 1PST, Normally Closed, Unsealed, Class 0
AN3320*	10/57	Relay–25 Amp, 1PST, Unsealed, Class 0
AN3324*	6/55	Relay–25 Amp, 1 PDT, Unsealed, Class 0
AN3335*	12/57	Relay–5 Amp, 3PST, Unsealed, Class 0
AN3337*	12/57	Relay–25 Amp, 1PST, Unsealed, Class 0
AN3354*	11/57	Relay–50 Amp, 2PDT, Intermittent, Unsealed, Class 0, See MS25030
AN3362*	8/57	Relay–100 Amp, 2PST, Unsealed, Class 0, Use MS24170
AN3364*	11/57	Relay–100 Amp, 2PDT, Intermittent, Unsealed, Class 0, See MS25031
AN3372*	11/57	Relay–200 Amp, 2PST, Class 0, Unsealed
AN3374*	11/57	Relay–200 Amp, 2PDT, Intermittent, Class 0, Unsealed, See MS25032
AN3430	4/67	Cable Assembly, Power, Electrical, 115/200 Volts, 3 Phase Use MS24342
AN3532*	4/57	Inverter–Class A, 250 Volt-Ampere, 3/1 Phase, Use MS25093
AN3621	9/51	Generator–Direct Current, 100 Ampere 30 Volt, High Speed Range
AN3623	10/51	Generator–Direct Current, 300 Ampere 30 Volt, High Speed Range
AN3632	1/53	Generator–Direct Current, 200 Ampere 30 Volt, Wide Speed Range
AN3633	12/52	Generator–Direct Current, 300 Ampere 30 Volt, Wide Speed Range
AN3634	4/55	Generator–Direct Current, 400 Amperes, 30 Volt, Wide Speed Range
AN3650	6/53	Generator–Direct Current, 400 Amperes, 30 Volts, Wide Speed Range, 11 Inch Diameter Mounting Flange
AN4008**		Filler–Cooling System
AN4022**	3/60	Nipple Assembly–Priming System Jet
AN4023**	4/60	Elbow Assembly–Priming System Jet
AN4024*	4/53	Tee Assembly–Priming System Jet
AN4027	12/43	Gasket–Spark Plug
AN4028	11/45	Marker–Ignition Cable
AN4030**	4/60	Distributor–Engine Primer
AN4032**	4/60	Clamp–Primer Line
AN4042*	8/44	Plug–Spark Plug Hole (18mm–1.5mm Pitch)
AN4044*	6/65	Gasket–Type XI Engine Accessory Drive, Use MS9135
AN4045*	6/65	Gasket–Type X or XV Engine Accessory Drive, Use MS9134

Number	Current Issue	Title
AN4047*	6/65	Gasket–Type XII, or Type XIV -A and -B Engine Accessory Drive, Use MS9136
AN4052	11/46	Bushing–Generator and Power Take-Off Pad and Drive Adapter (For Converting to Type I or IA)
AN4058*	6/65	Gasket–Type XIV-C Engine Accessory Drive, Use MS9138
AN4059*	6/65	Gasket–Type XIII Engine Accessory Drive, Use MS9137
AN4060	12/63	Protector–Ignition Lead Terminal
AN4061	7/54	Plug–Crankcase Dehydrator
AN4062	9/56	Plug–Engine Cylinder Dehydrator, 18-MM Thread
AN4064	1/52	Plug–Engine Cylinder Dehydrator, (14-MM Thread)
AN4068*	9/49	Envelope–Engine Protective (For V-770 Engine)
AN4071*	9/49	Envelope–Engine Protective (For R-985 and R-1340 Engines)
AN4072*	9/49	Envelope–Engine Protective (For R-1820, R-1830 and R-2000 Engines)
AN4073*	9/49	Envelope–Engine Protective (For R-2600 Engine)
AN4076**	12/61	Fitting–Engine Cylinder Head Temperature Measuring Element
AN4081*	6/53	Tube–Oil Cooler
AN4103	7/52	Valve–Oil Cooler Temperature Regulating with Surge Protection
AN4111*	5/53	Connection–Engine Oil 90° (4 Bolt Flange), See MS28750
AN4112*	9/49	Envelope–Engine Protective (For R-2800 Engine)
AN4116(3)	10/67	Starter–Aircraft, Direct Cranking, 28 V DC
AN4119*	6/65	Gasket–Type XIX Engine Accessory Drive, Use MS9140
AN4120*	4/50	Plug–Crankcase Dehydrator Short, See AN4061
AN4121*	9/49	Envelope–5KW Aircraft Accessory Power Plant Protective
AN4122**	4/60	Adapter–Fuel or Oil Tank Filler Cap, Flush, 3 Inch I. D.
AN4123**	4/60	Adapter–Fuel or Oil Tank Filler Cap Raised 3 Inch ID
AN4137**	4/60	Yoke–Fuel Selector Valve
AN4138	3/53	Adapter–Types XII-J, XII-K or XII-W Drive Flange
AN4139	3/53	Adapter–Type XVI Drive Flange
AN4141(2)	11/50	Adapter–Type XII-J and Type XII-K Drive Spline
AN4144*	6/65	Gasket–XVI Engine Accessory Drive, Use MS9139
AN4146		Pump–1/2 GPM at 3000 PSI Power Driven Hydraulic
AN4149		Pump–3 GPM at 3000 PSI Power Driven Hydraulic
AN4154*	9/49	Envelope–Engine Protective (For J-33 Engine)
AN4155*	9/49	Envelope–Engine Protective (For 0-435 Engine)
AN4159*	9/49	Envelope–Engine Protective (For J-34 Engines)
AN4165**	4/60	Valve–Fuel Selector, Manually Operated 3 Port, 4 Position, Angle Flow
AN4166**	4/60	Valve–Fuel Selector, Manually Operated 3 Port, 5 Position
AN4167**	4/60	Valve–Fuel Selector, Manually Operated 3 Port, 6 Position
AN4168**	4/60	Valve–Fuel Selector, Manually Operated 4 Port, 4 Position
AN4169**	4/60	Valve–Fuel Selector, Manually Operated 5 Port, 5 Position
AN4170		Gasket–Flange Hose Connection
AN4171*	5/53	Gasket–Engine Oil Connection 3 Bolt
AN4172*	5/53	Gasket–Engine Oil Connection 4 Bolt
AN4173**	4/60	Valve–Fuel Selector, Manually Operated 3 Port, 4 Position, Tee Flow
AN4181	7/59	Vibrator–Aircraft Engine Starting, Radio-Interference Suppressed
AN4182		Adapter–Type XVII Drive Spline
AN5025	7/52	Cover–Propeller Governor Drive
AN5026		Cap–Inboard Dual Propeller Shaft Thread
AN5528**	8/60	Indicator–Fuel Flowmeter 1-7/8 Inch Dial
AN5531**	8/60	Generator–Electric Tachometer
AN5532**	1/66	Indicator–Engine Synchroscope
AN5533	9/54	Shaft–Tachometer Flexible Drive
AN5534	2/54	Resistor–Thermocouple Lead Spool
AN5537	5/48	Connector Assembly–Thermocouple Lead
AN5538	8/46	Terminal–Thermocouple Lead Soldering
AN5539	4/53	Terminal–Thermocouple, Brass
AN5540	12/47	Thermocouple–Aircraft Engine Gasket Type
AN5542	4/53	Terminal–Thermocouple
AN5548	1/50	Terminal–Lug, Thermocouple, Chromel and Alumel
AN5733**	8/60	Compass–Pilot's Card Type Magnetic
AN5744	6/50	Inclinometer–Aircraft, Ball Style
AN5767(3)	9/53	Tube–Extension, Gyro Stabilized Drift Meter
AN5768	9/53	Rod Assembly–Extension Drift Meter Prism Control
AN5785	11/53	Transmitter–Electrical, Wheels and Flaps Position
AN5786	4/52	Transmitter–Position, 28 Volt DC
AN5787	4/52	Transmitter–Position, Gearless
AN5792**	12/66	Indicator–Temperature, Electrical-Resistance, -50° to +300° C
AN5800	7/53	Holder Assembly–Correction Card
AN5805**	6/60	Tube–Single Throat Venturi
AN5806**	6/60	Tube–Double Throat Power Venturi for Turn and Bank Indicator
AN5807**	6/60	Tube–Double Throat Power Venturi for Directional Gyro and Gyro Horizon Indicators
AN5811	3/56	Tube, Pitot, Electrically Heated, "L" Shaped, Inverted
AN5812	3/56	Tube, Pitot, Electrically Heated, "L" Shaped
AN5813	10/52	Tube, Pitot, Electrically Heated
AN5814	11/55	Tube, Pitot Static, "L" Shaped, Electrically Heated
AN5815**	6/60	Tube Pilot-Static, Unheated Biplane Type
AN5816	9/53	Tube–Pitot Static, Straight, Electrically Heated
AN5820	3/55	Indicator, Bank and Turn
AN5823	1/67	Card–Compass Correction, Sup. AN5803
AN5829*	12/55	Valve–Instrument Vacuum Control, Use MS28056
AN5830*	9/55	Valve–Low Pressure Check, Internal Pipe Threads, Use MS28050
AN5831*	9/55	Valve–Static Pressure Selector, Use MS28051
AN5832*	9/55	Valves–Pressure Gage Lines Bleeder and Filler Check, Use MS28052
AN5835**	6/60	Computer–Dead Reckoning, Use AN5834
AN5841**	8/60	Computer–Air Position Indicator
AN5842**	8/60	Pump–Air Position Indicator
AN5843**	8/60	Controller–Air Position Indicator
AN5844**	8/60	Amplifier–Air Position Indicator
AN5846**	8/60	Shaft–Flexible (Air Position Indicator)
AN5847**	8/60	Drive–Right Angle, Air Position Indicator
AN5851**	6/60	Sextant–Bubble Type (With Altitude Averaging Device)
AN5853		Collimator–Sextant
AN5861*	9/53	Indicator–Airspeed, Pitot Static, 40-400 Knots, Use MS28046
AN6004	10/45	Regulator–Diluter Demand Oxygen
AN6009	7/45	Coupling–Automatic Oxygen
AN6010	10/45	Regulator–Automatic Continuous Flow Oxygen
AN6011	7/59	Gage–Panel Mounting High Pressure Oxygen
AN6012	9/55	Valve, High Pressure Oxygen Line
AN6014	11/49	Valve–High Pressure Oxygen Check, Style A
AN6015	8/52	Valve–High Pressure Oxygen Check (Style B)
AN6016	11/49	Valve–High Pressure Oxygen Check (Style C)
AN6017	11/49	Valve–High Pressure Oxygen Check, Style D
AN6018	11/49	Valve–High Pressure Oxygen Check, Style E
AN6021	2/69	Gage, Panel Mounting Low Pressure Oxygen
AN6024(4)	10/66	Valve, Filler Low Pressure, Oxygen
AN6026	2/69	Gage–Regulator Mounting Low Pressure Oxygen
AN6027	11/66	Adapter, Filler Valve, Low Pressure Oxygen
AN6029*	2/65	Indicator–Oxygen Flow

Number	Current Issue	Title
AN6043*	5/53	Connection Assembly–Demand Mask to Regulator Tube Internal, See MS22009
AN6044	2/48	Adapter–Portable Oxygen Recharger Orifice
AN6045		Extinguisher Assembly–Fire, Carbon Dioxide, Portable
AN6046		Bracket–Mounting, Extinguisher, Fire
AN6100**	4/60	Pump Assembly–2 G. P. H. Fluid Metering, Type A
AN6101**	4/60	Filter Assembly–Fluid Metering Pump
AN6102**	4/60	Pump Assembly–5 G. P. H. Fluid Metering, Type B
AN6110	8/45	Pump–4-1/2 C.F.M. Engine Driven Air
AN6111(2)	8/45	Pump–7-1/2 C.F.M. Engine Driven Air
AN6112	8/45	Pump–15 C.F.M. Engine Driven Air
AN6121(3)		Separator–Air Pump Oil
AN6201		Pump–Hydraulic Hand
AN6204	6/56	Valve, Hydraulic Bleeder
AN6207	10/45	Valve–1500 PSI Hydraulic Check
AN6210*	7/50	Valve–In Line Type 3-1/2 GPM Hydraulic Directional Control, See AN6293 or AN6294
AN6211*	7/50	Valve–In Line Type 6 GPM Hydraulic Directional Control, See AN6293 or AN6294
AN6213*	7/50	Valve–Radial Type 3-1/2 G.P.M. Hydraulic Directional Control, See AN6293 or AN6294
AN6214*	7/50	Valve–Radial Type 6 GPM Hydraulic Directional Control, See AN6293 or AN6294
AN6215*	7/50	Valve–Radial Type 16 GPM Hydraulic Directional Control, See AN6293 or AN6294
AN6225*	12/52	Packing–"V" Ring Hydraulic
AN6226*	2/53	Packing–"U" Cup Hydraulic
AN6227	6/67	Packing, "O" Ring Hydraulic
AN6228*	12/52	Adapter–"V" Ring Female Hydraulic Packing
AN6229*	12/52	Adapter–"V" Ring Male Hydraulic Packing
AN6230	6/67	Gasket, "O" Ring Hydraulic
AN6232**	8/60	Felt–Strip Packing Gland, Use MS28932
AN6234*	2/52	Filter–Hydraulic Replaceable Element Line Type, See MS28720
AN6235	11/52	Filter Element–Hydraulic Replaceable Micronic Line Type
AN6236	3/51	Filter Element–Hydraulic Replaceable Micronic Reservoir Type
AN6237	9/48	Filter Element–Hydraulic Replaceable Micronic Vent Type
AN6238	11/53	Gasket–Hydraulic Replaceable Reservoir Type Filter Element
AN6240	9/49	Filter–Hydraulic Replaceable Element Vent Type
AN6244		Ring–Hydraulic Gasket Back-Up
AN6245	10/53	Valve–Hydraulic Thermal Expansion Relief
AN6247	10/67	Valve Controllable Hydraulic Check (Rotating Action)
AN6248	1/49	Pump–Hydraulic Hand, Type 3000
AN6249	9/59	Valve, 3000 PSI Hydraulic Check
AN6250	12/48	Pump–1 GPM at 1500 PSI Power Driven Hydraulic
AN6270	2/54	Hose Assembly–Detachable Swivel Fitting, Low Pressure, Sup. AN856
AN6277	11/47	Valve–Hydraulic Shuttle, 3000 PSI (Internal Thread Tube Fitting Outlet)
AN6278	11/55	Valve, Hydraulic Shuttle, 3000 PSI (Universal and Internal Thread Fitting Outlet)
AN6280	10/49	Valve–3000 PSI Hydraulic Check Internal Ports
AN6281	12/52	Fuse–Hydraulic Quantity Measuring Class 3000 Type I
AN6282	4/53	Fuse–Hydraulic Quantity Measuring Class 3000 Type II
AN6285	6/54	Adapter–Air Pressure Gage
AN6289	2/67	Nut–Flared Tube Universal Fitting
AN6293(2)	12/55	Valve, 4-Way Rotary Selector, 3,000 PSI, Non-Interflow
AN6294(2)	12/55	Valve, 4-Way Rotary Selector, 3,000 PSI, Interflow
AN6295	4/50	Regulator–10 GPM Hydraulic Pressure
AN6296	5/67	Valve, Controllable Check, Plunger-Type 3,000 PSI
AN6506	1/69	Belt Assembly–Troop and Turret Gunner's Safety
AN6508	10/52	Ring–Parachute Harness Quick Fit "V"
AN6532(5)	6/67	Connector, Hose, Quick Disconnect, Female, Anti "G" Suit
AN6538*	7/65	Valve, Regulating, Jet Type Aircraft, Anti "G" Suit, Use MS24350
AN6539	3/49	Tester–Anti "G" Suit Regulating Valve
AN6568		Pin–Parachute Rip Cord Locking
AN6571	5/53	Grip–Parachute Rip Cord, Trapezoidal
AN6572		Fastener–Parachute Pack Tab (Oval)
AN7008**	8/60	Spool–Aircraft Camera Film (5-1/4 x 2-1/8) (20 Ft)
AN7009**	8/60	Spool–Aircraft Camera Film (5-1/2 x 2-5/8) (56 Ft)
AN7508	4/52	Belt–Gunner's Safety
AN7511**	10/55	Indicator–Humidity, Large, Use MS20003
AN7513**	10/55	Indicator–Humidity, Small, Use MS20003
AN7514**	10/55	Card–Humidity-Indicator Color Comparison, Use MS20003
AN7516*	10/64	Fitting–Cargo Airplane Tie Down, Use MIL-A-8905
AN8004	6/44	Cap–Flexible Push-Pull Control Casing
AN8005		Stop–Flexible Push-Pull Control Casing
AN8008**	4/61	Insulator–Vibration
AN8009	10/52	Anchor–Sea
AN8010	7/49	Buoy–Seaplane Mooring (Circular Atmospheric Pressure Type)
AN8011**		Case–Flight Report Holder and Map, Use MS26556
AN8013	7/43	Washer–Vibration Insulator Stop
AN8015	12/47	Kit–Airplane Mooring
AN8016	12/51	Casing–Flexible, Steel, Push-Pull Control
AN8018	9/54	Horn, Flyer's Relief Tube
AN8019	8/54	Tee–Flyer's Relief Tube
AN8020	4/46	Bracket–Flyer's Relief Tube Horizontal
AN8028	7/54	Ax, Pick Head, Rubber Coated Grip
AN8030**	3/61	Reel, Tow Target, 12,000 Foot Cable Capacity, Airborne
AN8031		Pad–Tank Strap
AN8505(2)		Wrench–Thin Double End Open End
AN8506(2)**	5/60	Wrench–Thin Crow Foot
AN8507**	5/60	Bar–Telescopic, (1" OD, 43-3/4" Extended Length)
AN8509**	5/60	Socket–12 Point 1/2 Inch Square Drive 7/8 Inch Broached Opening Extra Deep Spark Plug
AN8510	2/47	Knockout–Threaded Taper Pin
AN8511**	5/60	Adapter–Telescoping Bar, (7/8 to 1 Inch OD)
AN8512**	5/60	Wrench–Spark Plug Elbow Square Drive (3/4 Inch Opening)
AN8513		Wrench–Spark Plug Elbow Square Drive (13/16 Inch Opening)
AN8514	9/48	Wrench–3/8 Inch Square Drive Adjustable Spanner (.75 to 2 Inches Dia.)
AN8515	9/48	Wrench–1/2 Inch Square Drive Adjustable Spanner (1.75 to 4 Inches Dia.)
AN8516		Wrench–1/2 Inch Square Drive Adjustable Spanner (3.5 to 6 Inches Dia.)
AN100036	1/58	Cover–Type XI Engine Accessory Drive
AN100038*	3/59	Cover–Type X Engine Accessory Drive, See AN100043
AN100039	1/58	Cover–Type XIII Engine Accessory Drive
AN100041	1/58	Cover–Type XII, XIV-A, XIV-B, XIV-E, XVII-A and XVII-B Engine Accessory Drive
AN100043	7/59	Cover–Type X & XV Engine Accessory Drive
AN100044	1/58	Cover–Type XVI, XVII-C, D, E and F Engine Accessory Drive
AN101001–101100	3/69	Bolt–Hex Head, .190-32
AN101101–101200	3/69	Bolt–Hex Head, .250-28
AN101201–101300	3/69	Bolt–Hex Head, .3125-24
AN101301–101400	3/69	Bolt–Hex Head, .375-24

Number	Current Issue	Title
AN101401–101500*	3/69	Bolt–Hex Head, .4375-20
AN101501–101600*	3/69	Bolt–Hex Head, .500-20
AN101601–101700*	3/69	Bolt–Hex Head, .5625-18
AN101701–101800*	3/69	Bolt–Hex Head, .625-18
AN101801–101900*	3/69	Bolt–Hex Head, .750-16
AN101901–102000*	3/69	Bolt–Hex Head, Drilled Shank, .190-32
AN102001–102100*	3/69	Bolt–Hex Head, Drilled Shank, .250-28
AN102101–102200*	3/69	Bolt–Hex Head, Drilled Shank, .3125-24
AN102201–102300*	3/69	Bolt–Hex Head, Drilled Shank, .375-24
AN102301–102400*	3/69	Bolt–Hex Head, Drilled Shank, .4375-20
AN102401–102500*	3/69	Bolt–Hex Head, Drilled Shank, .500-20
AN102501–102600*	3/69	Bolt–Hex Head, Drilled Shank, .5625-18
AN102601–102700*	3/69	Bolt–Hex Head, Drilled Shank, .625-18
AN102701–102800*	3/69	Bolt–Hex Head, Drilled Shank, .750-16
AN102801–102900*	3/69	Bolt–Drilled Hex Head, 1 Hole, .190-32
AN102901–103000*	3/69	Bolt–Drilled Hex Head, 1 Hole, .250-28
AN103001–103100*	3/69	Bolt–Drilled Hex Head, 1 Hole, .3125-24
AN103101–103200*	3/69	Bolt–Drilled Hex Head, 1 Hole, .375-24
AN103201–103300*	3/69	Bolt–Drilled Hex Head, 1 Hole, .4375-20
AN103301–103400*	3/69	Bolt–Drilled Hex Head, 1 Hole, .500-20
AN103401–103500*	3/69	Bolt–Drilled Hex Head, 1 Hole, .5625-18
AN103501–103600*	3/69	Bolt–Drilled Hex Head, 1 Hole, .625-18
AN103601–103700*	3/69	Bolt–Drilled Hex Head, 1 Hole, .750-16
AN103701–103800*	3/69	Bolt–Drilled Hex Head, 6 Holes, .190-32
AN103801–103900*	3/69	Bolt–Drilled Hex Head, 6 Holes, .250-28
AN103901–104000*	3/69	Bolt–Drilled Hex Head, 6 Holes, .3125-24
AN104001–104100*	3/69	Bolt–Drilled Hex Head, 6 Holes, .375-24
AN104101–104200*	3/69	Bolt–Drilled Hex Head, 6 Holes, .4375-20
AN104201–104300*	3/69	Bolt–Drilled Hex Head, 6 Holes, .500-20
AN104301–104400*	3/69	Bolt–Drilled Hex Head, 6 Holes, .5625-18
AN104401–104500*	3/69	Bolt–Drilled Hex Head, 6 Holes, .625-18
AN104501–104600*	3/69	Bolt–Drilled Hex Head, 6 Holes, .750-16
AN104601–104700*	12/66	Bolt–Cres Hex Head, .190-32
AN104701–104800*	12/66	Bolt–Cres Hex Head, .250-28
AN104801–104900*	12/66	Bolt–Cres Hex Head, .3125-24
AN104901–105000*	12/66	Bolt–Cres Hex Head, .375-24
AN105001–105100*	12/66	Bolt–Cres Hex Head, .4375-20
AN105101–105200*	12/66	Bolt–Cres Hex Head, .500-20
AN105201–105300*	12/66	Bolt–Cres Hex Head, .5625-18
AN105301–105400*	12/66	Bolt–Cres Hex Head, .625-18
AN105401–105500*	12/66	Bolt–Cres Hex Head, .750-16
AN105501–105600*	12/66	Bolt–Cres Hex Head, Drilled Shank, .190-32
AN105601–105700*	12/66	Bolt–Cres Hex Head, Drilled Shank, .250-28
AN105701–105800*	12/66	Bolt–Cres Hex Head, Drilled Shank, .3125-24
AN105801–105900*	12/66	Bolt–Cres Hex Head, Drilled Shank, .375-24
AN105901–106000*	12/66	Bolt–Cres Hex Head, Drilled Shank, .4375-20
AN106001–106100*	12/66	Bolt–Cres Hex Head, Drilled Shank, .500-20
AN106101–106200*	12/66	Bolt–Cres Hex Head, Drilled Shank, .5625-18
AN106201–106300*	12/66	Bolt–Cres Hex Head, Drilled Shank, .625-18
AN106301–106400*	12/66	Bolt–Cres Hex Head, Drilled Shank, .750-16
AN106401–106500*	12/66	Bolt–Cres Drilled Hex Head, 1 Hole, .190-32
AN106501–106600*	12/66	Bolt–Cres Drilled Hex Head, 1 Hole, .250-28
AN106601–106700*	12/66	Bolt–Cres Drilled Hex Head, 1 Hole, .3125-24
AN106701–106800*	12/66	Bolt–Cres Drilled Hex Head, 1 Hole, .375-24
AN106801–106900*	12/66	Bolt–Cres Drilled Hex Head, 1 Hole, .4375-20
AN106901–107000*	12/66	Bolt–Cres Drilled Hex Head, 1 Hole, .500-20
AN107001–107100*	12/66	Bolt–Cres Drilled Hex Head, 1 Hole, .5625-18
AN107101–107200*	12/66	Bolt–Cres Drilled Hex Head, 1 Hole, .625-18
AN107201–107300*	12/66	Bolt–Cres Drilled Hex Head, 1 Hole, .750-16
AN107301–107400*	12/66	Bolt–Cres Drilled Hex Head, 6 Holes, .190-32
AN107401–107500*	12/66	Bolt–Cres Drilled Hex Head, 6 Holes, .250-28
AN107501–107600*	12/66	Bolt–Cres Drilled Hex Head, 6 Holes, .3125-24
AN107601–107700*	12/66	Bolt–Cres Drilled Hex Head, 6 Holes, .375-24
AN107701–107800*	12/66	Bolt–Cres Drilled Hex Head, 6 Holes, .4375-20
AN107801–107900*	12/66	Bolt–Cres Drilled Hex Head, 6 Holes, .500-20
AN107901–108000*	12/66	Bolt–Cres Drilled Hex Head, 6 Holes, .5625-18

Number	Current Issue	Title
AN108001–108100*	12/66	Bolt–Cres Drilled Hex Head, 6 Holes, .625-18
AN108101–108200*	12/66	Bolt–Cres Drilled Hex Head, 6 Holes, .750-16
AN115401–115450	10/54	Screw–Flat Fillister Head, .190-32
AN115451–115500	7/63	Screw, Machine–Steel, Flat Fillister Head, .250–28
AN115501–115550	10/54	Screw–Flat Fillister Head, .3125-24
AN115551–115600	10/54	Screw–Flat Fillister Head, .375-24
AN115601–115650	10/54	Screw–Flat Fillister Head, Drilled Shank, .190-32
AN115651–115700	8/55	Screw–Flat Fillister Head, Drilled Shank, .250-28
AN115701–115750	10/54	Screw–Flat Fillister Head, Drilled Shank, .3125-24
AN115751–115800	10/54	Screw–Flat Fillister Head, Drilled Shank, .375-24
AN115801–115850	10/54	Screw–Flat Fillister, Drilled Head, .190-32
AN115851–115900	10/54	Screw–Flat Fillister, Drilled Head, .250-20
AN115901–115950	10/54	Screw–Flat Fillister, Drilled Head, .250-28
AN115951–116000	10/54	Screw–Flat Fillister, Drilled Head, .3125-18
AN116001–116050	10/54	Screw–Flat Fillister, Drilled Head, .3125-24
AN116051–116100	10/54	Screw–Flat Fillister, Drilled Head, .375-16
AN116101–116150	10/54	Screw–Flat Fillister, Drilled Head, .375-24
AN116901–116912	12/54	Screw–Oval Fillister Head, .112-40
AN116913–116924	12/54	Screw–Oval Fillister Drilled Head, .112-40
AN116925–116960	12/54	Screw–Oval Fillister Head, .138-32
AN116961–117000	3/58	Screw–Oval Fillister Drilled Head, .138-32
AN117001–117040	12/54	Screw–Oval Fillister Head, .164-32
AN117041–117080	12/54	Screw–Oval Fillister Drilled Head, .164-32
AN121501–121525*	3/55	Nut–Plain
AN121551–121575*	3/55	Nut–Castle
AN121601–121650	8/62	Pin, Straight, Headed–.125 Nom.
AN121651–121700	8/62	Pin, Straight, Headed–.187 Nom.
AN121701–121750	8/62	Pin, Straight, Headed–.250 Nom.
AN121751–121800	8/62	Pin, Straight, Headed–.312 Nom.
AN121801–121850	8/62	Pin, Straight, Headed–.375 Nom.
AN121851–121925	8/62	Pin, Straight, Headed–.500 Nom.
AN122576–122775		Pin–Dowel, Steel
AN122776–122802*	7/59	Key–Chamfered Woodruff
AN123020–123119	5/68	Gasket–Aluminum–Asbestos, Annular
AN123151–123300(2)	5/63	Rivet, Solid, Universal Head, AMS7229
AN123301–123450(2)	5/63	Rivet, Solid, Universal Head, AMS7232
AN123451–123600(2)	5/63	Rivet, Solid, 100° Flush Head, AMS7229
AN123601–123750(2)	5/63	Rivet, Solid, 100° Flush Head, AMS7232
AN123851–123950	2/60	Packing, Preformed–AMS7274, "O" Ring
AN123951–124050	12/64	Packing, Preformed–AMS 7270, 'O' Ring
AN124951–125100	5/54	Rivet–100° Countersunk Head, Corrosion Resistant Steel
AN125401–125550	5/54	Rivet–Universal Head, Corrosion Resistant Steel
AN125551–125700	5/54	Rivet–Universal Head, Mild Steel
AN125951–126274	6/58	Stud–Straight, .250-20 x .250-28
AN126275–126586	6/58	Stud–Straight, .3125-18 x .3125-24
AN126587–126652	6/58	Stud–Straight, .375-16 x .375-24
AN126653–126712	6/58	Stud–Straight, .4375-14 x .4375-20
AN126713–126766	6/58	Stud–Straight, .500-13 x .500-20
AN126767–126820	6/58	Stud–Straight, .5625-12 x .5625-18
AN126821–126880	6/58	Stud–Straight, .625-11 x .625-18
AN126881–127192	6/58	Stud–Straight, Necked, .375-16 x .375-24
AN127193–127492	6/58	Stud–Straight, Necked, .4375-14 x .4375-20
AN127493–127786	6/58	Stud–Straight, Necked, .500-13 x .500-20
AN127787–128086	6/58	Stud–Straight, Necked, .5625-12 x .5625-18
AN128087–128362	6/58	Stud–Straight, Necked, .625-11 x .625-18
AN128363–128686	6/58	Stud–Straight, Drilled, .250-20 x .250-28
AN128687–128998	6/58	Stud–Straight, Drilled, .3125-18 x .3125-24
AN128999–129064	6/58	Stud–Straight, Drilled, .375-16 x .375-24
AN129065–129124	6/58	Stud–Straight, Drilled, .4375-14 x .4375-20
AN129125–129178	6/58	Stud–Straight, Drilled, .500-13 x .500-20
AN129179–129232	6/58	Stud–Straight, Drilled, .5625-12 x .5625-18
AN129233–129292	6/58	Stud–Straight, Drilled, .625-11 x .625-18
AN129293–129604	6/58	Stud–Straight, Drilled, Necked, .375-16 x .375-24
AN129605–129904	6/58	Stud–Straight, Drilled, Necked, .4375-14 x .4375-20
AN129905–130198	6/58	Stud–Straight, Drilled, Necked, .500-13 x .500-20
AN130199–130498	6/58	Stud–Straight, Drilled, Necked, .5625-12 x .5625-18

Number	Current Issue	Title
AN130499–130774	6/58	Stud–Straight, Drilled, Necked, .625-11 x .625-18
AN148551–148650*	8/58	Bolt–Socket Head, 6-Hole Drilled, .190-32, See MS9088
AN148651–148750*	8/58	Bolt–Socket Head, 6 Holes Drilled, .250-28, See MS9089
AN148751–148850*	8/58	Bolt–Socket Head, 6 Holes Drilled, .3125-24, See MS9090
AN148851–148950*	8/58	Bolt–Socket Head, 6 Holes Drilled, .375-24 See MS9091
AN148951–149050*	8/58	Bolt–Socket Head, 6 Holes Drilled, .4375-20, See MS9092
AN149051–149150*	8/58	Bolt–Socket Head, 6 Holes Drilled, .500-20, See MS9093
AN149151–149250*	8/58	Bolt–Socket Head, 6 Holes Drilled, .5625-18, See MS9094
AN149251–149350	8/58	Bolt–Socket Head, 6 Holes Drilled, .625-18
AN150301–150400		Pin–Lock, Brass
AN150401–150425		Nut–Hex, Check
AN150426–150450		Nut–Hex, Shear, Slotted
AN150501–150800	6/57	Stud–Stepped, 1.5 Dia Engagement, .250-20 x .190-32
AN150801–151100	3/55	Stud–Stepped, 1.5 Dia Engagement, .3125-18 x .250-28
AN151101–151400	3/55	Stud–Stepped, 1.5 Dia Engagement, .375-16 x .3125-24
AN151401–151700	3/55	Stud–Stepped, 1.5 Dia Engagement, .4375-14 x .375-24
AN151701–152000	3/55	Stud–Stepped, 1.5 Dia Engagement, .500-13 x .4375-20
AN152001–152300	3/55	Stud–Stepped, 1.5 Dia Engagement, .5625-12 x .500-20
AN152301–152600	3/55	Stud–Stepped, 1.5 Dia Engagement, .625-11 x .5625-18
AN152601–152900	3/55	Stud–Stepped, Drilled, 1.5 Dia Engagement, .250-20 x .190-32
AN152901–153200	3/55	Stud–Stepped, Drilled, 1.5 Dia Engagement, .3125-18 x .250-28
AN153201–153500	3/55	Stud–Stepped, Drilled, 1.5 Dia Engagement, .375-16 x .3125-24
AN153501–153800	3/55	Stud–Stepped, Drilled, 1.5 Dia Engagement, .4375-14 x .375-24
AN153801–154100	3/55	Stud–Stepped, Drilled, 1.5 Dia Engagement, .500-13 x .4375-20
AN154101–154400	3/55	Stud–Stepped, Drilled, 1.5 Dia Engagement, .5625-12 x .500-20
AN154401–154700	3/55	Stud–Stepped, Drilled, 1.5 Dia Engagement, .625-11 x .5625-18
AN154701–155000	3/55	Stud–Stepped, 2 Dia Engagement, .250-20 x .190-32
AN155001–155300	3/55	Stud–Stepped, 2 Dia Engagement, .3125-18 x .250-28
AN155301–155600	3/55	Stud–Stepped, 2 Dia Engagement, .375-16 x .3125-24
AN155601–155900	3/55	Stud–Stepped, 2 Dia Engagement, .4375-14 x .375-24
AN155901–156200	3/55	Stud–Stepped, 2 Dia Engagement, .500-13 x .4375-20
AN156201–156500	3/55	Stud–Stepped, 2 Dia Engagement, .5625-12 x .500-20
AN156501–156800	3/55	Stud–Stepped, 2 Dia Engagement, .625-11 x .5625-18
AN156801–157100	3/55	Stud–Stepped, Drilled, 2 Dia Engagement, .250-20 x .190-32
AN157101–157400	3/55	Stud–Stepped, Drilled, 2 Dia Engagement, .3125-18 x .250-28
AN157401–157700	3/55	Stud–Stepped, Drilled, 2 Dia Engagement, .375-16 x .3125-24
AN157701–158000	3/55	Stud–Stepped, Drilled, 2 Dia Engagement, .4375-14 x .375-24
AN158001–158300	3/55	Stud–Stepped, Drilled, 2 Dia Engagement, .500-13 x .4375-20
AN158301–158600	3/55	Stud–Stepped, Drilled, 2 Dia Engagement, .5625-12 x .500-20
AN158601–158900	3/55	Stud–Stepped, Drilled, 2 Dia Engagement, .625-11 x .5625-18
AN158901–159200	3/55	Stud–Stepped, Necked, 1.5 Dia Engagement, .4375-14 x .375-24
AN159201–159500	3/55	Stud–Stepped, Necked, 1.5 Dia Engagement, .500-13 x .4375-20
AN159501–159800	3/55	Stud–Stepped, Necked, 1.5 Dia Engagement, .5625-12 x .500-20
AN159801–160100	3/55	Stud–Stepped, Necked, 1.5 Dia Engagement, .625-11 x .5625-18
AN160101–160400	3/55	Stud–Stepped, Drilled, Necked, 1.5 Dia Engagement, .4375-14 x .375-24
AN160401–160700	3/55	Stud–Stepped, Drilled, Necked, 1.5 Dia Engagement, .500-13 x .4375-20
AN160701–161000	3/55	Stud–Stepped, Drilled, Necked, 1.5 Dia Engagement, .5625-12 x .500-20
AN161001–161300	3/55	Stud–Stepped, Drilled, Necked, 1.5 Dia Engagement, .625-11 x .5625-18
AN161301–161600	3/55	Stud–Stepped, Necked, 2 Dia Engagement, .4375-14 x .375-24
AN161601–161900	3/55	Stud–Stepped, Necked, 2 Dia Engagement, .500-13 x .4375-20
AN161901–162200	3/55	Stud–Stepped, Necked, 2 Dia Engagement, .5625-12 x .500-20
AN162201–162500	3/55	Stud–Stepped, Necked, 2 Dia Engagement, .625-11 x .5625-18

Number	Current Issue	Title
AN162501–162800	3/55	Stud–Stepped, Drilled, Necked, 2 Dia Engagement, .4375-14 x .375-24
AN162801–163100	3/55	Stud–Stepped, Drilled, Necked, 2 Dia Engagement, .500-13 x .4375-20
AN163101–163400	3/55	Stud–Stepped, Drilled, Necked, 2 Dia Engagement, .5625-12 x .500-20
AN163401–163700	3/55	Stud–Stepped, Drilled, Necked, 2 Dia Engagement, .625-11 x .5625-18
AN163701–164000	3/55	Stud–Stepped, Necked, Wrench Pad, 1.5 Dia Engagement, .3125-18 x .250-28
AN164001–164300	3/55	Stud–Stepped, Necked, Wrench Pad, 1.5 Dia Engagement, .375-16 x .3125-24
AN164301–164600	3/55	Stud–Stepped, Necked, Wrench Pad, 1.5 Dia Engagement, .4375-14 x .375-24
AN164601–164900	3/55	Stud–Stepped, Necked, Wrench Pad, 1.5 Dia Engagement, .500-13 x .4375-20
AN164901–165200	3/55	Stud–Stepped, Necked, Wrench Pad, 1.5 Dia Engagement, .5625-12 x .500-20
AN165201–165500	3/55	Stud–Stepped, Necked, Wrench Pad, 1.5 Dia Engagement, .625-11 x .5625-18
AN165501–165800	3/55	Stud–Stepped, Drilled, Necked, Wrench Pad, 1.5 Dia Engagement, .3125-18 x .250-28
AN165801–166100	3/55	Stud–Stepped, Drilled, Necked, Wrench Pad, 1.5 Dia Engagement, .375-16 x .3125-24
AN166101–166400	3/55	Stud–Stepped, Drilled, Necked, Wrench Pad, 1.5 Dia Engagement, .4375-14 x .375-24
AN166401–166700	3/55	Stud–Stepped, Drilled, Necked, Wrench Pad, 1.5 Dia Engagement, .500-13 x .4375-20
AN166701–167000	3/55	Stud–Stepped, Drilled, Necked, Wrench Pad, 1.5 Dia Engagement, .5625-12 x .500-20
AN167001–167300	3/55	Stud–Stepped, Drilled, Necked, Wrench Pad, 1.5 Dia Engagement, .625-11 x .5625-18
AN167301–167600	3/55	Stud–Stepped, Necked, Wrench Pad, 2 Dia Engagement, .3125-18 x .250-28
AN167601–167900	3/55	Stud–Stepped, Necked, Wrench Pad, 2 Dia Engagement, .375-16 x .3125-24
AN167901–168200	3/55	Stud–Stepped, Necked, Wrench Pad, 2 Dia Engagement, .4375-14 x .375-24
AN168201–168500	3/55	Stud–Stepped, Necked, Wrench Pad, 2 Dia Engagement, .500-13 x .4375-20
AN168501–168800	3/55	Stud–Stepped, Necked, Wrench Pad, 2 Dia Engagement, .5625-12 x .500-20
AN168801–169100	3/55	Stud–Stepped, Necked, Wrench Pad, 2 Dia Engagement, .625-11 x .5625-18

Number	Current Issue	Title
AN169101–169400	3/55	Stud–Stepped, Drilled, Necked, Wrench Pad, 2 Dia Engagement, .3125-18 x .250-28
AN169401–169700	3/55	Stud–Stepped, Drilled, Necked, Wrench Pad, 2 Dia Engagement, .375-16 x .3125-24
AN169701–170000	3/55	Stud–Stepped, Drilled, Necked, Wrench Pad, 2 Dia Engagement, .4375-14 x .375-24
AN170001–170300	3/55	Stud–Stepped, Drilled, Necked, Wrench Pad, 2 Dia Engagement, .500-13 x .4375-20
AN170301–170600	3/55	Stud–Stepped, Drilled, Necked, Wrench Pad, 2 Dia Engagement, .5625-12 x .500-20
AN170601–170900	3/55	Stud–Stepped, Drilled, Necked, Wrench Pad, 2 Dia Engagement, .625-11 x .5625-18
AND10000**	3/60	Drive–Type I Fluid Power Pumps, Use AND20000
AND10001**	3/60	Mounting Pad and Drive–Type II Vacuum and Hydraulic Power Pumps, Use AND20001
AND10002**	3/60	Mounting Pad and Drive–Aircraft Engine Generator and Power Take-Off, Use AND20002
AND10003**	3/60	Mounting Pad and Drive–Aircraft Engine Fuel Pump, Use AND20003
AND10004**	3/60	Drive–Aircraft Engine Starter, Use AND20004
AND10005**	3/60	Mounting Pad and Drive–Aircraft Tachometer, Use AND20005
AND10008**	3/60	Mounting Pad and Drive–Aircraft Engine Power–Take-Off (Flange Type), Use AND20008
AND10009(2)**	3/60	Mounting Pad and Drive–Small Aircraft Engine Starter, See AND20009
AND10010**	3/60	Mounting Pad and Drive–Aircraft Engine Propeller Governor or Hydro-Control Valve, Use AND20010
AND10051	6/54	Bosses–Conduit Connection, Standard Dimensions for
AND10052	6/54	Bosses–Standard Dimensions for External Pipe Thread
AND10053*	1/55	Bosses–Pipe–Thread, Internal, Standard Dimensions for
AND10059(4)	1/64	Fittings–Styles of Pipe Threaded, Flared Tube and Hose
AND10064(4)	2/56	Fittings–Installation of Flared Tube, Straight Threaded Connectors
AND10065	3/46	Connection Assembly–Flexible, Fluid, Standard Design of
AND10067	6/54	Valve Installation–Hydraulic Bleeder (Standard Dimensions for)
AND10068*	3/57	Nuts and Plate Nuts–Self Locking, Functional Limitations of, Use MS33588
AND10069**	8/60	Casing Assembly–Installation of Control Flexible
AND10070	7/44	Adapter–Oxygen Filler
AND10071(2)	12/66	Boss and Installations–Air Connection (AN6287)
AND10074		Boss Spacing–Hydraulic
AND10079**	8/60	Terminals–Eye and Fork Cable
AND10080		Fittings–Installation of Universal Elbow
AND10081(2)	7/65	Terminal Shank–Swaging, Dimensions for
AND10089	7/56	Fitting End–Standard Dimensions for Cone Connection
AND10102	7/65	Tubing–Standard Dimensions for Round Alloy Steel
AND10104	5/67	Tubing, Steel, Corrosion-Resistant, Round, Standard Dimensions for
AND10106	4/48	Tubing–Standard Sizes for Aluminum Alloy (52SO) Round
AND10107	10/42	Tubing–Standard Sizes for Aluminum Alloy (24ST) Round
AND10108	2/45	Tubing–Standard Sizes for Aluminum Alloy (61S) Round
AND10119	10/42	Aluminum Alloy–Standard Conditions and Sizes for Sheet and Strip (24S)
AND10125		Aluminum Wire–Standard Alloys, Tempers & Sizes of Round and Hexagon (For Welding Rod & General Use)

Number	Current Issue	Title
AND10126	2/43	Aluminum Wire—Standard Alloys, Tempers and Sizes of (For Rivets, Bolts, Screws, and General Use)
AND10130	7/43	Aluminum Rod & Bar—Standard Alloys, Tempers and Sizes of Round and Hexagon
AND10131		Aluminum Bar—Standard Alloys, Temper and Sizes of Square
AND10132		Aluminum Bar—Standard Alloy and Temper (24ST), and Sizes of Rectangular
AND10133	3/45	Angle—Equal Leg Extruded
AND10134(2)	3/45	Angle—Unequal Leg Extruded
AND10135	3/45	Angle—Bulb Section Extruded
AND10136(2)	3/45	Tee—Extruded
AND10137(4)	3/45	Channel—Extruded
AND10138(2)	3/45	Zee—Equal Leg Extruded
AND10139(2)	3/45	Zee—Unequal Leg Extruded
AND10140	3/45	I-Section—Extruded
AND10141	3/45	Tee—Bulb Section Extruded
AND10152(13)**	6/53	Shaft Ends—Propeller
AND10201	6/48	Screens and Gaskets—Design Requirements for Carburetor
AND10210**	3/60	Flange—Single 12 Cylinder 3 Bolt Magneto
AND10211**	3/60	Mounting Flange—Double Type Magneto
AND10214**	3/60	Coupling Assemblies—Magneto Drive
AND10215**	3/60	Mounting Flanges—Single Type Magneto (9 and 14 Cyl)
AND10216**	3/60	Throttle-Handle Design for Engine
AND10220		Vapor Pressure Chart
AND10221(5)		Vapor Prrssure Correction Chart
AND10230(2)	11/54	Reference Chart—Aircraft Engine Accessory Drives
AND10237	5/57	Spline—Serrated
AND10260	3/53	Flange—Type X Accessory Mounting
AND10261	3/53	Flange—Type XI Accessory Mounting
AND10262(2)	3/53	Flange—Type XII Accessory Mounting
AND10263	3/53	Flange—Type XIII Accessory Mounting
AND10264(2)	11/54	Flange—Type XIV Accessory Mounting
AND10265	3/53	Flange—Type XV Accessory Mounting
AND10266	3/53	Flange—Type XVI Accessory Mounting
AND10267(2)	3/53	Flange—Type XVII Accessory Mounting
AND10268(2)	3/53	Flange—Type XVIII Accessory Power Input
AND10269	3/53	Flange—Type XIX Accessory Mounting
AND10302	8/49	Installation Clearance—Type XI Drive
AND10303(2)**	4/60	Installation Clearance—Type XIV Drive
AND10305	8/49	Installation Clearance—Type XII and Type XVII Drive (Generator—Power Take-Off)
AND10306**	4/60	Propeller Governor Installation Clearance—Aircraft Engine (For Engines of 999 Take Off H.P. or Less)
AND10307**	4/60	Propeller Governor Installation Clearance—Aircraft Engine (For Engines of 1000 Take Off HP and Over)
AND10310(2)	3/50	Installation Clearance—Type XV Drive
AND10312**	8/60	Element Installation Clearance—Engine Cylinder Head Temperature Measuring
AND10313		Relief Valve Installation Clearance—Hydraulic
AND10320**	4/60	Installation Clearance—Type XIII Drive
AND10325	4/49	Self Tapping Screws—Coarse Thread (Hole Sizes for)
AND10326	4/43	Self Tapping Screws—Machine Screw Thread (Hole Sizes for)
AND10328**	4/60	Installation Clearance—Type XI-B Drive Magneto
AND10331	8/49	Installation Clearance—Type XII-J Drive (Fluid Power Pump)
AND10332**	4/60	Starter—Generator Installation Clearance—Type XII Drive
AND10335**	4/60	Installation Clearance—Type XIX Drive
AND10336	7/49	Installation Clearance—Type XVIII Drive
AND10338	8/49	Installation Clearance—Type XII-K (Fluid Power Pump)
AND10341**	6/60	Transmitter Installation Clearance—Fuel and Oil Pressure, Use MS33587
AND10342	7/50	Regulator Installation Clearance—DC Generator Voltage
AND10343	8/49	Installation Clearance—Type XVI Drive
AND10355(2)		Tolerances—Carbon Steel Sheet and Strip

Number	Current Issue	Title
AND10360(3)	4/51	Tolerances—Corrosion and Heat Resistant Steel Bars, Rods and Wire
AND10362(2)	2/52	Tolerances—Carbon Steel Bar and Rod
AND10367(2)		Tolerances—Alloy Steel Sheet and Strip
AND10380	10/49	Fitting Installations—Standard AN Conduit
AND10387	5/53	Drill Sizes and Drilled Hole Tolerances—Twist
AND10391		Conduit Assembly—Flexible Metallic
AND10396		Numbers—Aircraft Engine Cylinder Location Designation
AND10399		Number—Aircraft Engine Location Designation
AND10405	9/65	Instrument Board—Installation of Vibration Insulated
AND10406	8/49	Thermocouple Leads—Iron-Constantan, Installation of
AND10407		Insulators—Installation of Vibration
AND10410	9/52	Pitot-Static and Pitot Tube—Wiring Diagram for
AND10411		Wiring Diagram—Percent Type Tachometer Indicator and Two Pole Tachometer Generator
AND10441(2)		Battery Installation
AND10445		Sump Jar—Installation of Battery
AND10446**	8/60	Compass Wiring Diagram—Remote Indicating Magnetic
AND10448**	8/60	Indicator Wiring Diagram—Air Position
AND10451**	8/60	Compass Wiring Diagram—Gyro Fluxgate
AND10453**	6/60	Switch—Toggle, Installation of, Use MS33630
AND10458**	4/60	Mounting Pad—Oil Connection, 3 and 4 Stud, Use MS33786
AND10460(2)		Indentors and Nests—Copper Electric Terminal Power Operated Press
AND10476	5/67	Washer—Limitations on Usage of Lock
AND10481(2)		Standard Dimensioning System for Extruded Sections
AND10527(3)	3/53	Propeller Hubs—Standard Dimensions for Ground, Adjustable
AND10529	7/52	Propeller Hubs—Standard Dimensions for Involute Tooth Spline Controllable Pitch
AND10530	2/51	Propeller Hubs—Standard Dimensions for Straight Tooth Spline Controllable Pitch
AND10562	3/69	Rim Contours—8.00-4, 6.00-6, 6.50-10, 7.50-10, 8.50-10, 11.00-12, 15.00-16 and 19.00-23, Type III, Low Pressure, Main Landing and Beaching Gear Wheel
AND10563	3/69	Rim Contour—17.00-16, Type III, Low Pressure, Main Landing Gear Wheel
AND10566	3/69	Rim Contours—15.50-20 and 17.00-20, Type III, Low Pressure, Main Landing Gear Wheel
AND10567	3/69	Rim Contours—5.00-4 and 9.00-6, Type III, Low Pressure, Tail Wheel
AND10571	3/69	Rim Contour—10.00-7, Type III, Low Pressure, Tail Wheel
AND10573	10/64	Rim Contours—26x6.6, 30x7.7, 32x8.8, 34x9.9, 36x11, 38x11, 40x12, 44x13, Type VII, Extra High Pressure, Main Landing Gear Wheels
AND10575	9/64	Rim Contour—9.00-6, Type III, Low Pressure, Beaching Gear Tail Wheel
AND10576	9/64	Rim Contour—24x7.7, Type VII, Extra High Pressure, Main Landing Gear Wheel
AND10578	9/64	Rim Contour—5.00-5, Type III, Low Pressure, Helicopter Landing Gear Wheel
AND10581	9/64	Rim Contour—20x4.4, Type VII, Extra High Pressure, Landing Gear Wheel
AND10582**	8/60	Rim Contours—22 x 5.5 and 24 x 5.5, Type VII, Extra High Pressure, Landing Gear Wheel
AND10583	9/64	Rim Contour—25.00-28, Type VIII, Low Pressure, Main Landing Gear Wheel
AND10598		Chamfer or Radius—Valve Holes in Airplane Wheel Rims
AND10599**	8/60	Knurl Detail—Airplane Wheel Rim Flanges
AND20000	4/59	Drive—Type X Engine Accessory
AND20001	3/53	Drive—Type XI Engine Accessory
AND20002(3)	3/53	Drive—Type XII Engine Accessory
AND20003	3/53	Drive—Type XIII Engine Accessory